Applied Tree Biology

Applied Tree Biology

Andrew D. Hirons
University Centre Myerscough
Preston, UK

Peter A. Thomas
Keele University
Newcastle-under-Lyme, UK

This edition first published 2018
© 2018 John Wiley & Sons Ltd

The right of Andrew D. Hirons and Peter A. Thomas to be identified as the authors of this work has been asserted in accordance with law.

Registered Office(s)
John Wiley & Sons, Inc., 111 River Street, Hoboken, NJ 07030, USA
John Wiley & Sons Ltd, The Atrium, Southern Gate, Chichester, West Sussex, PO19 8SQ, UK

Editorial Office
9600 Garsington Road, Oxford, OX4 2DQ, UK

For details of our global editorial offices, customer services, and more information about Wiley products visit us at www.wiley.com.

Wiley also publishes its books in a variety of electronic formats and by print-on-demand. Some content that appears in standard print versions of this book may not be available in other formats.

Library of Congress Cataloging-in-Publication Data

Names: Hirons, Andrew D., author. | Thomas, Peter A., 1957– author.
Title: Applied tree biology / by Dr. Andrew D. Hirons, Myerscough College, Preston, UK,
    Dr. Peter A. Thomas, Keele University, Newcastle-under-Lyme, UK.
Description: Hoboken, NJ : Wiley, 2018. | Includes bibliographical references and index. |
Identifiers: LCCN 2017039501 (print) | LCCN 2017040134 (ebook) | ISBN 9781118296363 (pdf) |
    ISBN 9781118296370 (epub) | ISBN 9781118296400 (pbk.)
Subjects: LCSH: Trees, Care of. | Trees.
Classification: LCC QK475 (ebook) | LCC QK475 .H534 2018 (print) | DDC 582.16–dc23
LC record available at https://lccn.loc.gov/2017039501

Cover Design: Wiley
Cover Image: Image courtesy of Andrew Hirons

Set in 10/12pt Warnock by SPi Global, Pondicherry, India

SKY3EA48FE4-AE72-4110-80DF-8860262AF781_031822

To our families, Ruth and Cyril; Judy, Matthew and Daniel

In a word, and to speak a bold and noble truth, trees and woods have twice saved the whole world; first by the ark, then by the cross; making full amends for the evil fruit of the tree in paradise, by that which was born on the tree in Golgotha.

From *Sylva*
by John Evelyn, 1664

Contents

List of Contributors

Richard C. Beeson Jr.
Mid-Florida Research and Education
Center
University of Florida
Apopka, FL, USA

A. Roland Ennos
School of Biological, Biomedical and
Environmental Sciences
The University of Hull
Hull, UK

David Lonsdale
Alton
Hampshire, UK

Glynn C. Percival
R.A. Bartlett Tree Research Laboratory
University of Reading
Reading, UK

Henrik Sjöman
Department of Landscape Architecture,
Planning and Management
Swedish University of Agricultural
Sciences
Alnarp, Sweden;
Gothenburg Botanical Garden
Gothenburg, Sweden

Duncan Slater
Department of Greenspace
University Centre Myerscough
Bilsborrow
Preston, UK

Foreword

Practical arboriculture is the 'art' and 'science' of tree management brought together by skilled arborists. When the two principles are applied correctly at the right time, the results that follow lead to healthier and less stressed trees in a beautiful treescape.

As arborists we learn the disciplines of tree work, such as how to prune trees correctly and where the final pruning cut should be in relation to the attachment point on the parent branch by using the 'target pruning' principles rather than the old 'flush cutting' techniques. Many of us just accept these principles at face value, taking them for granted without really understanding the science of plant physiology and knowing why we target prune.

One of my highly respected predecessors was William Dallimore, who worked in the Arboretum at the Royal Botanic Gardens (RBG), Kew, in the late nineteenth century. He was one of the first arborists employed at Kew by the Director, Sir William Turner Thisleton-Dyer, to refine forestry principles and adapt them to suit and improve the maintenance of the specimen trees in the arboretum collections at RBG, Kew. This he did with very successful results, but he based most of his work on what he observed in the gardens following the pruning operations he carried out, without understanding the science behind it. He noted in his journals how different species of trees responded to the various pruning techniques that he used with his array of hand tools, stating that leaving long stumps caused dieback and eventual decay. He would follow his team, finishing off the cuts properly if he was not happy with their efforts. He even wrote a book based on these observations, *The pruning of trees and shrubs; being a description of the methods practiced in the Royal Botanic Gardens, Kew*, which was first published in 1926. Unlike Dallimore in his era, today we base our pruning practices on the scientific research work of the 1960s and 1970s. The 'compartmentalisation of decay in trees' (CODIT) by Dr Alex Shigo, the North American plant pathologist, has changed arboricultural practices around the world for the better.

When arborists know how they should be pruning, such as the correct positioning of the saw blade when making that final cut, coupled with the science behind the principles, it makes more logical sense and is easier to carry out the operation, knowing that this is better for the longevity and health of the tree. Without this scientific knowledge it would be much harder to understand and practice.

The same goes for planting a tree. Most people think they can plant a tree, but there are many 'rights' and 'wrongs' and several general practical principles such as correct planting depth, addition of soil ameliorants and mycorrhizal products, suitable plant support in the form of staking, effective weed control and adequate aftercare. All these are used successfully today based on sound scientific research which has led to much higher success rates with tree establishment in urban tree planting. When these principles are more widely accepted and used in everyday arboriculture, our treescape will be a much better and healthier one.

There are many reference works that specialise in the practical and scientific principles of the various disciplines of arboriculture, but there are few that bring them all together in one work. This is such a book, and will help arborists at all levels to understand why we do what we do. Andrew Hirons, senior lecturer at University Centre Myerscough, and Peter Thomas, a reader in plant ecology at Keele University, are without doubt most competent to do this successfully with their broad knowledge of applied tree biology. I hope that every practising arborist and horticulturist uses this work to help them understand practical arboriculture.

Tony Kirkham
Head of the Arboretum, Gardens and
Horticultural Services
Royal Botanic Gardens
Kew, UK

Preface

This book comes about from a desire to create a text on tree biology that is accessible to anyone looking to understand how trees work or who manages landscapes that contain trees. It is written for those studying arboriculture and tree management, whether as part of a formal course or simply as a result of their own interest. The overall aim is to provide knowledge about trees that can be used to underpin management recommendations so that the health and vitality of trees in our gardens, parks, streets and courtyards can be promoted. We have tried to include just the information that is needed to meet these aims rather than give a comprehensive guide of all that is known about how trees work. Our text is supported by a series of 'Expert boxes' authored by a range of leading practitioners and academics, namely: Richard Beeson, Roland Ennos, David Lonsdale, Glynn Percival, Henrik Sjöman and Duncan Slater.

We would like to give our heartfelt thanks to Ruth Hirons, Tony Kirkham, David Lonsdale, Hugh Morris, Glynn Percival, Keith Sacre and Duncan Slater for helpful discussions and for their reading of early drafts.

We are grateful to Laura Power and Adrian Capstick, Myerscough College, and Andy Lawrence, Keele University, for assistance in redrawing some of the figures. We are also thankful to Myerscough College for the financial support that helped secure the use of a number of the figures used. The following are gratefully acknowledged for supplying and helping to develop figures: Lukas Ball, Richard Beeson, Alex Chau, Roland Ennos, Linda Hirons, David Lonsdale, Kevin Martin, Kevin McGinn, Keith Sacre, Francis Schwarze, Fritz Schweingruber, Henrik Sjöman and Duncan Slater. Your contributions have greatly enriched the text; thank you.

All photographs are our own unless otherwise acknowledged.

ADH would like to acknowledge the support provided by the RHS Coke Bursary Trust Fund that enabled the development of materials used in this book whilst on sabbatical from Myerscough College. ADH would also like to acknowledge the support provided by his past and present colleagues at Myerscough College; in particular, Mark Johnston for nourishing his interest in books about trees, David Elphinstone for his mentorship and Duncan Slater for the thought-provoking discussions too numerous to count. PAT thanks the help and encouragement given while a Bullard Fellow at Harvard Forest, Harvard University. Much of his contribution was written while resident at Harvard.

A Note on the Text

Italics are used to emphasize key words and concepts when first used. The abbreviations **sp.** and **spp.** are used for one or more species, respectively. The units of measurement used in this book are explained at various points but it might help to know that a micrometre (μm) is a thousandth of a millimetre (mm), and ppm are parts per million.

Where the works of others are quoted, the names of the authors are given together with dates of publication so that the article or book can be looked up in the references at the end of each chapter.

1

Introduction

Value of Trees Globally

The three trillion trees around the world (Crowther *et al.* 2015) are hugely important to us and to the well-being of our planet (Figure 1.1). Their value is usually described in terms of ecosystem services – what trees and forests can do to help us humans. A detailed list of ecosystem services provided by trees and forests would fill this book (the UK National Ecosystem Assessment 2011 provides a very good summary) so, by way of illustration, here are just three major services.

One of the major services is storing carbon. Forests hold around 45% of the carbon stored on land (i.e. not including the reserves held in oceans) which amounts to 2780 Gt of carbon (Giga has nine zeros; i.e. billions). This is about 3.3 times the amount already in the atmosphere (829 Gt). Carbon dioxide in the atmosphere has increased from 280 ppm in pre-industrial times to 404 ppm at the time of writing, an increase of 42%. If all the world's trees died and decomposed to release their carbon into the atmosphere, the atmospheric level of carbon dioxide would rise to 1700 ppm (>600% pre-industrial) with catastrophic effects on our world (UNEP 2008), so global carbon storage in trees and forests is a hugely important service.

Forests also help to determine weather patterns. This is partly by forests evaporating large amounts of water, producing clouds that release rain downwind. Furthermore, it has recently been discovered that a chemical released by trees, pinene (one of the monoterpenes), can help 'seed' clouds by acting as nuclei for water to condense around, and so help clouds to form and rain to fall (Kirby *et al.* 2016). It seems plausible that other volatile organic compounds (VOCs) emitted by trees have a similar effect. Trees and forests are also beneficial by acting as sponges, slowing the journey of rainfall to the ground and helping to improve soil structure, both of which encourage water to sink into the soil rather than run off the surface. This delays water discharge to streams and rivers, helping to reduce flooding and soil erosion.

Most of the world's biodiversity is held in forests. Tropical forests, which cover 7% of land surface, hold more than 60% of the world's species of terrestrial animals and plants (Bradshaw *et al.* 2009), and all the world's forests hold more than 80% of species (Balvanera *et al.* 2014).

Applied Tree Biology, First Edition. Andrew D. Hirons and Peter A. Thomas.
© 2018 John Wiley & Sons Ltd. Published 2018 by John Wiley & Sons Ltd.

Figure 1.1 Forests are globally important to mankind for storing carbon, helping to determine weather patterns and providing a habitat for a vast range of life. This scene is of the temperate forest in Robert H. Treman State Park, New York.

Value of Urban Trees

On a smaller scale, urban trees and woodlands also have an important role in our well-being, but for slightly different reasons. Fundamentally, urban trees make our towns and cities better places to live. Quite apart from making urban areas look more appealing, trees can provide a sense of place and time. They help provide outdoor recreation opportunities and make the urban environment more pleasant. Economic benefits of urban trees include higher property values; reduced energy costs of buildings; and reduced expenditure on air pollution removal and storm water infrastructure (Roy *et al.* 2012; Mullaney *et al.* 2015). There are also many environmental benefits, the most important of which are summarised in Expert Box 1.1.

With more than half of the world's population now living in cities, one of the most important contributions that trees and green spaces make is to our health. There is a growing body of information that shows that exposure to trees and green spaces improves wellness and our sociability (Wolf and Robbins 2015). Studies have also shown that the positive health impact of trees is independent of access to green space in general. For example, in Sacramento, California, higher tree cover within 250 m of home was associated with better general health, partially mediated by lower levels of obesity and better neighbourhood social cohesion (Ulmer *et al.* 2016). There is also a body of information that shows that psychological benefits of trees can affect the physiology of our bodies by reducing pulse rate and levels of cortisol, a major stress hormone (Ochiai *et al.* 2015). This works even when looking at pictures of trees. There is also a physiological response because chemicals released by some trees affect us directly. For example, Ikei *et al.* (2015) found that oil from the Hinoki cypress *Chamaecyparis obtusa*, widely used in fragrances in soap, toothpaste and cosmetics in Japan, positively affects brain activity and induces a feeling of 'comfortableness'. This is the basis for shinrin-yoku (forest-air breathing or forest bathing), a popular form of relaxation in Japan, walking through wooded areas or standing beneath a tree and slowly breathing (Figure 1.2). The same monoterpenes that cause cloud formation are known to reduce tension and mental stress, reducing aggression and depression and increasing feelings of well-being. Even a short lunchtime walk of 1.8 km through green areas can improve sleep patterns that night (Gladwell *et al.* 2016). Moreover, the physiological effects stay with us. A study by Li (2010) found that a 3-day forest visit had positive effects on the immune system up to 30 days later.

The loss of trees from urban environments has also been demonstrated to have negative outcomes for human health. Over 100 million ash *Fraxinus* spp. trees have been lost in the north-eastern USA since 2002 as a result of the emerald ash borer (EAB), an invasive beetle. This huge loss of trees has been linked to increased human mortality as a result of higher levels of cardiovascular and respiratory diseases (Donovan *et al.* 2013). Social costs, such as an increase in crime, have also been associated with the loss of trees

Figure 1.2 A sign encouraging people to breathe in the air in a forest in northern Honshu Island, Japan. This shinrin-yoku (forest-air breathing) is a popular form of relaxation in Japan.

caused by EAB (Kondo *et al.* 2017). Consequently, there is a growing body of evidence that the presence of trees in and around our urban environments provides major public health and societal benefits.

However, in some cases, the much-championed value of urban trees is perhaps not all that is claimed. Examples of this include oxygen production and carbon sequestration (the locking-up of carbon). It is true that trees produce an abundance of oxygen. For example, urban forests in the USA have been estimated to produce enough oxygen (61 Mt of it) annually to keep two-thirds of the US population breathing (Nowak *et al.* 2007). However, given the enormous reserves of oxygen in the atmosphere, this is a fairly minor benefit of urban trees. Another benefit of urban trees that is often over-played is their role in mitigating carbon emissions. Roland Ennos, Expert Box 1.1, points out that Greater London's 8.4 million trees are estimated to store 2.4 million tonnes of carbon (t C) and sequester about 77 200 t C each year (Rogers *et al.* 2015). This amounts to about 3% of the city's annual carbon emissions or, to put it another way, enough to cover the city's emissions for about 12 days. London's trees sequester only about 0.2% of annual carbon emissions. This is not to disparage carbon sequestration in urban trees, but just to put it into perspective; urban trees are very valuable to us but planting them will not be a solution for climate change or even offset the carbon emissions of our towns and cities to any great extent. In this regard, conservation of the world's forests is of much greater significance.

Although trees are overwhelmingly beneficial for our landscapes and for us, they can also create problems, particularly if they are inappropriately planted, the wrong species is selected for the site or the site is poorly designed with respect to tree development. Trees can get too big for their location; they can conflict with buildings, utilities and sightlines. At certain times of year, pollen from trees can contribute to discomfort amongst those with hay-fever; litter from flowers, fruit and leaves can create slip haz-ards or block drains. Tree roots sometimes cause damage to pavements, making them uneven, and they may exacerbate damage to pipes by exploiting them as a source of water and nutrition. Occasionally, in dry years, certain species growing on shrinkable clay soils can extract enough water to cause subsidence damage to built structures. Trees may also pose a risk to persons or property if they are structurally unstable or develop extensive decay. But should these potential problems prevent us keeping and planting urban trees? Emphatically not.

Even though many of the problems associated with trees in urban landscapes can be linked to poor planning, design and workmanship, the tree is invariably blamed. Despite the evidence for the benefits of trees, widespread loss of trees from our urban environ-ments is often reported. In the USA it has been estimated that four million urban trees are lost per year (Nowak and Greenfield 2012) and a similar trend can be seen across Europe. More insidiously, even where the total number of trees is not appreciably declining, the size of the tree is changing. In the UK, the number of large trees, such as London plane *Platanus* × *acerifolia*, is declining while the smaller hawthorns *Crataegus* spp., cherries *Prunus* spp., whitebeams and rowans *Sorbus* spp. and birch *Betula* spp. are increasingly common (Trees and Design Action Group 2008). This reduces the ben-efits derived from the urban forest. Larger trees intercept more rainfall (Xiao and McPherson 2002) and reduce temperatures more than small trees (Gratani and Varone 2006; Gómez-Muñoz *et al.* 2010), especially when they have denser crowns (Sanusi *et al.* 2017). It is therefore vital that we strive to provide opportunities and the right conditions for large trees across our landscapes.

Managing Trees

Forests have survived for millions of years without us 'managing' them, so why is it necessary to look after trees at all? The answer is threefold. First, in a healthy mature forest, the next generation of trees is established from thousands of seeds. Most of these are eaten, develop in unsuitable growing conditions or are out-competed by other species. The fact that only a fraction of these seeds ever develop into mature trees is insignificant to the bigger picture of a forest: such losses are just part of a forest's natural ecology. However, in parks and gardens the success of each individual tree is tightly coupled to the success of the planting scheme. We need to actively manage the selection, planting and establishment of the tree to ensure that each tree can make a long-term contribution to the landscape.

Secondly, whilst stable forest environments represent ideal conditions, many trees in gardens, parks and streets have to cope with human-induced problems or conflicts imposed on them by our built environments. Trees often occupy space that is shared with humans; this erodes the quality of the environment for the tree. In some cases, even our admiration of trees or desire to be amongst them is detrimental to the tree. Visitors to parks and gardens, drawn by the appeal of the landscape, can cause high levels of soil compaction; buried utilities lead to excavation of rooting environments; the need for safe roads and paths in winter leads to high levels of salt being applied close to trees; the list goes on. Trees and their environment need managing so that these conflicts (and others) are not detrimental to tree health.

Thirdly, normal patterns of tree development mean that trees can become too large for their position, or their condition can decline over time to such an extent that they endanger people or property. In these cases, trees need managing to control their size and safety.

If the many benefits from trees are to be realised, we must do what we can to ensure the health and longevity of trees across our landscapes and provide well-designed space for new trees. A fundamental requisite for these aims is a sound understanding of tree biology.

Conditions for trees within our towns and cities are highly variable. It is wrong to think of the urban environment as being always hostile to trees: there are many parks and gardens that provide excellent conditions for tree growth which may well exceed the quality of the tree's natural habitat. However, many sites provide very challenging conditions for trees. Soils may be infertile and compacted; sealed surfaces can restrict water infiltration and limit soil aeration; and the rooting environment may need to be shared with utilities. Above ground, branches are removed to reduce conflict with buildings, traffic, cables (Figure 1.3) and sightlines, particularly given the rise in the number of CCTV cameras. Natural processes are also disrupted, leaves are swept off to some remote location far away from the roots that they were intended to nourish. Most of these constraints, however, can be ameliorated with a little informed foresight.

If we expect trees to add value to our landscapes, then it is vital that we seek to emulate the forest environment wherever possible in the design and construction of planting sites. Applying the concept of *forest mimicry*, mimicking the way that trees work in their natural environment, and being aware of the tree's biology is crucial. An appreciation of the conditions that trees naturally thrive in and an understanding of the tree's biology make the difference between successful management that promotes tree health and interventions that simply accelerate tree decline. In this way it is possible to develop

(a)   (b)

Figure 1.3 (a) An ash tree *Fraxinus* sp. conflicting with overhead wires in northern Japan. This tree now requires intensive management if it is to persist on this site. (b) A mature oak *Quercus* sp. in Atlanta, USA that has had to endure decades of pruning because it was planted in an unsuitable location. *Source:* (b) Courtesy of Lukas Ball.

sustainable landscapes with trees that provide communities with a link to their past, as well as a vision of their future.

An excellent example of how understanding tree biology can positively influence our management of trees relates to tree pruning practice. For much of the twentieth century, pruning guidelines recommended taking the branch back so that the final cut was flush with the tree's stem. By studying decay behind pruning wounds and looking at the process of natural branch shedding, Alex Shigo was able to promote the idea of *natural target pruning* (Shigo 1989). This transformed our approach to tree pruning and has been of immeasurable benefit to trees as they are now able to seal pruning wounds and restrict the development of decay more effectively (see Chapters 3 and 9). Applying tree biology to practice can also lead to improved rooting environments; more effective management of water and nutrient resources; improved tree establishment rates; and more accurate assessments of tree condition. Further, understanding how trees grow in different environments greatly assists our ability to anticipate their likely limitations and tolerances when we place them in human landscapes.

Such variable growing conditions across our parks, gardens and hard landscapes means that a highly prescriptive book on tree management would be left wanting. Instead, our approach is to give the reader an understanding of tree biology and ecology so that this can be used to better inform management decisions, whether in a small garden, a large public park, a street or a courtyard.

The chapters are divided into key themes: tree growth and development (Chapter 2); leaves and crowns (Chapter 3); roots (Chapter 4); the next generation of trees (Chapter 5); tree water relations (Chapter 6); tree carbon relations (Chapter 7); tree nutrition (Chapter 8); interactions with other organisms (Chapter 9) and, environmental challenges for trees (Chapter 10).

Expert Box 1.1 The Environmental Benefits of Urban Trees
Roland Ennos

Trees are, of course, marvellous organisms – you would not be reading this book if you did not realise that – and there is no doubt that urban trees beautify surroundings that would otherwise look bare and soulless. However, many claims are also made about the environmental benefits of urban trees: that they reduce traffic noise; absorb pollution; take up and store carbon; provide shade and cooling; and help prevent flooding. These claims all seem plausible, but it is only recently that experimental investigations have tested these claims and started to quantify the environmental benefits of trees.

Noise Reduction

You might expect trees to be as good at shielding noise as they are at visual screening. However, experiments have shown that trees are actually poor at reducing noise levels (Fang and Ling 2003). The structures of trees, even their trunks, are simply too narrow to affect sound waves – particularly the long waves that carry the deep hum of traffic noise – and sound simply goes right through them. However, trees do help reduce the nuisance of traffic noise in other ways. They shield us from seeing passing vehicles, so helping reduce our awareness of them, and they make their own, more restful rustling noise in the wind, further distracting us from the noise of traffic. There is even evidence that people drive more slowly in tree-lined streets, lowering the noise their vehicles make.

Absorption of Pollution

Trees reduce pollution, particularly the particulate pollutants produced by the engines of motor vehicles, by intercepting them with their leaves. Rainwater then washes the particles off on to the ground and down the drain. Modelling and experimental studies have suggested that this process could reduce the concentration of particulates by 5–20% in a typical city (McDonald *et al.* 2007). However, as trees are not adapted to absorb pollution, this reduction is not very great, and depends very much on wind-speeds and the fine details of the airflow around the city. In some cases, trees can reduce wind-speeds so much that they keep particulates trapped in urban streets, and *increase* pollution levels. Some species of tree – especially willows, poplars and oaks – emit volatile organic compounds (VOCs) from their leaves which react with nitrogen oxides from vehicle emissions to produce harmful ozone (Donovan *et al.* 2011). Trees' anti-pollution credentials are therefore fairly weak.

Carbon Storage and Sequestration

There is no doubt that trees are important stores of carbon, and growing trees actively take up the greenhouse gas carbon dioxide. They use it to make sugars in the process of photosynthesis and lay down and store the carbon in the form of wood. A hectare of urban forest typically stores around 10–30 tonnes of carbon above-ground, while the roots store a further 2–5 tonnes. A growing stand of trees will take up and sequester a further 0.5–0.9 tonnes per hectare per year (McPherson *et al.* 2013). Over a whole town, this builds up to a large quantity. Over Greater London, for instance, trees store around 2.4 million tonnes of carbon and take up a further 72 thousand tonnes every year (Rogers *et al.* 2015). However, these benefits need to be put into perspective. Some of the sequestered carbon is lost because of the death and removal of old and diseased trees, while a further percentage is counterbalanced by the carbon emitted from the power tools used for management operations. The amounts sequestered are also far smaller than the amounts of carbon dioxide emitted by vehicles and by the heating systems of buildings in the city, which exceed 10 million tonnes annually. It is really only the huge areas of forest in the countryside that make a real contribution to removing carbon from the atmosphere.

Shade and Cooling

If some of their other environmental benefits seem disappointing, trees really do have major shading and cooling benefits. The leaves of trees are adapted to intercept and absorb light, so they do it extremely well. Moreover, to allow carbon dioxide to enter for photosynthesis, the leaves have to keep their stomata open during the day, and that allows large quantities of water to evaporate from them. This cools down the leaves and the air surrounding them, just as the evaporation of sweat from our bodies keeps us cool in hot weather.

At a local level, the cooling benefits of trees are largely a result of the shade that they provide. Radiation from the sun is reduced by up to 90% under the canopy of a tree, and this shading cools the roads and pavements beneath (Figure EB1.1). Tarmac can reach temperatures of 50–60 °C in the sun on a hot summer's day but in the shade of a tree it can be kept below 30 °C (Ennos *et al.* 2014). Both effects greatly improve the comfort of people, because how hot we feel depends far more on the radiation balance with our surroundings than on air temperature. So, although a single tree has a negligible effect on the air temperature around it, a person in tree-shade will take up far less short-wave radiation from the sun and emit much more long-wave radiation to the surroundings; all this means that we can actually feel 10–15 °C cooler.

The shading effect is also important in helping make buildings more habitable and cheaper to run in hot summer weather. Tree shading, particularly by trees situated on the western and eastern sides of buildings, can reduce wall temperatures by up to 30 °C in sunny weather. Studies in the USA and Canada

Figure EB1.1 Infrared image showing people resting in the cool shade provided by park trees. The red in the background shows an area of tarmac, while the yellow shows grass in the sun. *Source:* Courtesy of Roland Ennos.

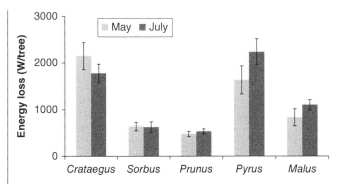

Figure EB1.2 The relative cooling performance of five small street trees in Manchester, UK. *Source:* Ennos *et al.* (2014). Licensed Open Government Licence v3.0, http://www.nationalarchives. gov.uk/doc/open-government-licence/version/3/.

have shown that optimally positioned trees can reduce air conditioning costs by around 30% (Akbari *et al.* 1997). Conversely, the effect that trees can have on the wind – reducing wind speeds and turbulence – can help reduce *heating* costs, particularly at high latitudes such as Canada and Scandinavia. In northern Europe, air conditioning is rare, and here trees provide natural air conditioning; they keep our buildings cooler in summer, helping to improve thermal comfort and prevent the deaths seen in heatwaves such as those that occurred in the summer of 2003. Unfortunately, the effectiveness of trees in this role has not been quantified.

On a larger scale – the entire city – the cooling effect of trees is actually brought about by the evaporation of water from their leaves, which cools the air around them. This is important because in periods of hot, calm, sunny weather, cities heat up far more than the surrounding countryside, an effect known as the urban heat island. City streets absorb a little more of the sun's radiation, and the heat gets trapped within steep urban canyons. The result is that cities heat up more during the day, by up to 2 °C, and remain up to 7 °C warmer at night than the surrounding countryside. Experiments have shown that trees can reverse this effect to some extent by using one-third to half of the energy hitting the leaves to evaporate water, rather than heat up the air (Ennos *et al.* 2014).

However, the cooling effectiveness of trees is very variable; fast-growing species with a dense canopy, such as Callery pears *Pyrus calleryana*, can be four times as effective as sparsely leafed cherry *Prunus* 'Umineko' (Figure EB1.2). Trees also need to be actively growing and transpiring to provide cooling, and hence need well-watered and aerated soil. We showed, for instance, that *Pyrus calleryana* trees growing in compaction-resistant 'Amsterdam structural soil' can produce five times the cooling of trees of the same species in conventional topsoil within a well-used pavement (Figure EB1.3). The basic principle seems to be that the healthier and faster-growing a tree is, the more cooling it will provide.

It is relatively easy to determine how much cooling individual trees are providing – you just have to measure the amount of water they are losing. However, the overall effect of trees on the city temperature is harder to gauge because it is simply not possible to compare two cities that are identical apart from their vegetation. Many studies have sought to overcome this problem by comparing the temperature of city parks with their

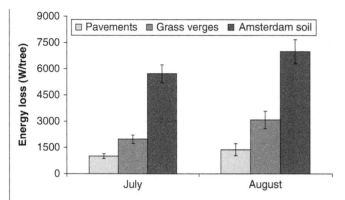

surrounding streets. Unfortunately, these measurements always give disappointing results: a park is typically only 1 °C cooler (Bowler *et al.* 2010). The problem is that winds blow cool air out of parks and warm air into them, so the cooling effect is spread over the city. To get better estimates of the cooling effects of trees, it would be more effective to incorporate their evaporative cooling into a large meteorological model of the city. Unfortunately, because of the complexity of cities, this is incredibly difficult and expensive to do.

Flood Prevention

Urban trees also have important benefits in helping reduce flood risk. During rainstorms, the leaves of trees intercept some of the rain before it reaches the ground and, of the rainfall that does get through or that drips down from the canopy, much of it infiltrates into the soil. This reduces the run-off – the rain that is diverted into the drains, and which could result in surface flooding – by around 50% (Wang *et al.* 2008). Trees perform better than areas of grass for two reasons: first, the canopy prevents 1–2 mm of rain from even reaching the soil; secondly, the tree roots dry out the soil and break it up, increasing the rate at which water can permeate into it by a factor of up to 60. City-wide, the effects of trees are usually estimated to be relatively small, around 5–6%, because of the relatively low cover of trees, which is typically in the region of 15–20% of the urban area (Gill *et al.* 2007). However, some experiments performed in Manchester suggested that these effects might be underestimates; even water that falls outside the canopy of a tree can enter its planting hole and be diverted away from the drains (Ennos *et al.* 2014). There is also great potential to increase the anti-flooding benefits of trees by growing them within sustainable urban drainage systems (SUDS) schemes. These schemes divert rainfall from buildings and roads into grassy swales and infiltration basins, and so are more effective at reducing flooding than simply adding areas of vegetation. Given the likely effects of climate change, using trees within such schemes could have two main benefits: they would help irrigate trees automatically, helping them survive the long summer droughts that are forecast; and they would help absorb and drain the water diverted into the area, so reducing waterlogging during the increasingly larger storms that are predicted to occur.

As we have seen therefore, the environmental benefits of trees are at last starting to be properly understood and quantified. We are even starting to be able to measure their benefits in monetary terms. The i-Tree software developed in the USA uses the results of tree surveys to estimate the cash worth of the carbon storage, pollution removal and flood prevention benefits of trees. i-Tree surveys certainly demonstrate the great benefits of trees. The London survey, for instance, came up with a figure of £132 m (Treeconomics 2015). However, this software leaves out many benefits and does not account for the great variability in the performance of trees, both within and between species.

Finally, of course, trees also cause environmental problems. Fast-growing species such as poplar and willow, which provide the most cooling and flood prevention, also emit the most volatile organic compounds; they are also more likely to cause building subsidence because they use more water, causing clay soils to dry out and shrink in the summer. Some limes (*Tilia* spp.) tend to be unpopular in cities because aphids suck sap from their leaves and secrete a sticky sap that drips on to cars. People can also object to trees reducing light levels and dropping their leaves in autumn. The key to deploying trees as environmental engineers is to grow the right tree in the right place, maximising the benefits they supply, while keeping everyone happy.

References

Akbari, H., Kurn, D.M., Bretz, S.E. and Hanford, J.W. (1997) Peak power and cooling energy savings of shade trees. *Energy and Buildings*, 25: 139–148.

Balvanera, P., Siddique, I., Dee, L., Paquette, A., Isbell, F., Gonzalez, A., *et al*. (2014) Linking biodiversity and ecosystem services: Current uncertainties and the necessary next steps. *BioScience*, 64, 49–57.

Bowler, D.E., Buyung-Ali, L., Knight, T.M. and Pullin, A.S. (2010) Urban greening to cool towns and cities: A systematic review of the empirical evidence. *Landscape and Urban Planning*, 97: 147–155.

Bradshaw, C.J.A., Sodhi, N.S. and Brook, B.W. (2009) Tropical turmoil: A biodiversity tragedy in progress. *Frontiers in Ecology and the Environment*, 7, 79–87.

Crowther, T.W., Glick, H.B., Covey, K.R., Bettigole, C., Maynard, D.S., Thomas, S.M., *et al*. (2015) Mapping tree density at a global scale. *Nature*, 525: 201–205.

Donovan, G.H., Butry, D.T., Michael, Y.L., Prestemon, J.P., Liebhold, A.M., Gatziolis, D., *et al*. (2013) The relationship between trees and human health: Evidence from the spread of the emerald ash borer. *American Journal of Preventive Medicine*, 44: 139–145.

Donovan, R., Owen, S., Hewitt, N., Mackenzie, R. and Brett, H. (2011) *The Development of an Urban Tree Air Quality Score (UTAQS): Using the West Midlands, UK Conurbation as a Case Study Area*. VDM Verlag Dr. Müller, Düsseldorf, Germany.

Ennos, A.R., Armson, D. and Rahman, M.A. (2014) How useful are urban trees? The lessons of the Manchester Research Project. In: Johnston, M. and Percival, G. (eds). *Trees, People and the Urban Environment II*. Institute of Chartered Foresters, Edinburgh, UK, pp. 62–70.

Fang, C.F. and Ling, D.L. (2003) Investigation of the noise reduction provided by tree belts *Landscape and Urban Planning*, 63: 187–195.

Gill, S., Handley, J.F., Ennos, A.R. and Pauleit, S. (2007) Adapting cities for climate change: The role of the green infrastructure. *Built Environment*, 33: 97–115.

Gladwell, V.F., Kuoppa, P., Tarvainen, M.P. and Rogerson, M. (2016) A lunchtime walk in nature enhances restoration of autonomic control during night-time sleep: Results from a preliminary study. *International Journal of Environmental Research and Public Health*, 13: article 280.

Gómez-Muñoz, V.M., Porta-Gándara, M.A. and Fernández, J.L. (2010) Effect of tree shades in urban planning in hot-arid climatic regions. *Landscape and Urban Planning*, 94: 149–157.

Gratani, L. and Varone, L. (2006) Carbon sequestration by *Quercus ilex* L. and *Quercus pubescens* Willd. and their contribution to decreasing air temperature in Rome. *Urban Ecosystems*, 9: 27–37.

Ikei, H., Song, C. and Miyazaki, Y. (2015) Physiological effect of olfactory stimulation by Hinoki cypress (*Chamaecyparis obtusa*) leaf oil. *Journal of Physiological Anthropology*, 34: article 44.

Kirby, J., Duplissy, J., Sengupta, K. *et al.* (2016) Ion-induced nucleation of pure biogenic particles. *Nature*, 533: 521–526.

Kondo, M.C., Han, S., Donovan, G.H. and MacDonald, J.M. (2017) The association between urban trees and crime: Evidence from the spread of the emerald ash borer in Cincinnati. *Landscape and Urban Planning*, 157: 193–199.

Li, Q. (2010) Effect of forest bathing trips on human immune function. *Environmental Health and Preventive Medicine*, 15: 9–17.

McDonald, A.G., Bealey, W.J., Fowler, D., Dragosits, U., Skiba, U., Smith, R.I., *et al.* (2007) Quantifying the effect of urban tree planting on concentrations and depositions of PM10 in two UK conurbations *Atmospheric Environment*, 41: 8455–8467.

McPherson, E.G., Xiao, Q.F. and Aguaron, E. (2013) A new approach to quantify and map carbon stored, sequestered and emissions avoided by urban forests. *Landscape and Urban Planning*, 120: 70–84.

Mullaney, J., Lucke, T. and Trueman, S.J. (2015) A review of benefits and challenges in growing street trees in paved urban environments. *Landscape and Urban Planning*, 134: 157–166.

Nowak, D.J. and Greenfield, E.J. (2012) Tree and impervious cover change in US cities. *Urban Forestry and Urban Greening*, 11: 21–30.

Nowak, D.J., Hoehn, R. and Crane, D.E. (2007) Oxygen production by urban trees in the United States. *Arboriculture and Urban Forestry*, 33: 220–226.

Ochiai, H., Ikei, H., Song, C., Kobayashi, M., Miura, T., Kagawa, T., *et al.* (2015) Physiological and psychological effects of a forest therapy program on middle-aged females. *International Journal of Environmental Research and Public Health*, 12: 15222–15232.

Rogers, K., Sacre, K., Goodenough, J. and Doick, K. (2015) *Valuing London's Urban Forest: Results of the London i-Tree Eco Project*. Treeconomics, London, UK.

Roy, S., Byrne, J. and Pickering, C. (2012) A systematic quantitative review of urban tree benefits, costs, and assessment methods across cities in different climatic zones. *Urban Forestry and Urban Greening*, 11: 351–363.

Sanusi, R., Johnstone, D., May, P. and Livesley, S.J. (2017) Microclimate benefits that different street tree species provide to sidewalk pedestrians relate to differences in Plant Area Index. *Landscape and Urban Planning*, 157: 502–511.

Shigo, A.L. (1989) *Tree Pruning: A Worldwide Photo Guide*. Shigo and Trees, Durham, New Hampshire, USA.

Trees and Design Action Group (2008) *No Trees, No Future: Trees in the Urban Realm*. Trees and Design Action Group, London, UK.

UK National Ecosystem Assessment (2011) *The UK National Ecosystem Assessment: Technical Report*. UNEP-WCMC, Cambridge, UK.

Ulmer, J.M., Wolf, K.L., Backman, D.R., Tretheway, R.L., Blain, C.J., O'Neil-Dunne, J.P. *et al.* (2016) Multiple health benefits of urban tree canopy: The mounting evidence for a green prescription. *Health and Place*, 42: 54–62.

UNEP (2008) *Kick the Habit: A UN Guide to Climate Neutrality*. United Nations Environment Programme, Arendal, Norway.

Wang, J., Endreny, T.A. and Nowak, D.J (2008) Mechanistic simulation of tree effects in an urban water balance model *Journal of the American Water Resources Association*, 44: 75–85.

Wolf, K.L. and Robbins, A.S. (2015) Metro nature, environmental health, and economic value. *Environmental Health Perspectives*, 123: 390.

Xiao, Q. and McPherson, E.G. (2002) Rainfall interception by Santa Monica's municipal urban forest. *Urban Ecosystems*, 6: 291–302.

2

The Woody Skeleton: Trunk and Branches

What is a Tree?

It seems sensible at the start of this chapter to identify what exactly is meant by the term 'tree'. Dictionary definitions often describe a tree as a perennial woody plant that has a self-supporting main stem and branches forming a distinct elevated crown; some even go on to suggest a minimum height of around 5 m. While there are aspects of these definitions – such as the perennial life cycle – which can be agreed upon, many of the other characteristics must be open to debate. Many 'trees' have multiple stems, yet we would not call them shrubs; strangler figs are not self-supporting for much of their life; palm trees have no distinct branching; and trees such as the Arctic willow *Salix arctica* never have a hope of reaching 5 m in their natural environment. The definition of a tree can therefore provide plenty of debate; however, in general, it is unrewarding: there are just too many exceptions. Arguably, the most useful definition comes not from a biologist but a judge, Lord Denning, who had to define a tree as part of a judgement under the UK Town and Country Planning Act. He stated, 'anything that one would ordinarily call a tree is a "tree" within this group of sections in the Act'. It is simple but surprisingly effective: if you would call it a tree, then it is a tree. There is a place for precise definitions but for most purposes this pragmatic view works just fine. Box 2.1 provides some useful definitions for parts of a tree.

How Does a Tree Grow?

Across most of the natural world, trees are the dominant form of vegetation. They are biomechanical and physiological triumphs that, in many cases, have endured for centuries or millennia despite storms, drought, freezing conditions and a host of biological assailants. Of the estimated 100 000 tree species around the globe, about 50 species (<0.005%) have reached gigantic proportions, measuring over 70 m in height (Tng *et al.* 2012). Currently, the title of 'tallest tree' goes to a coastal redwood *Sequoia sempervirens* measuring a staggering 115.7 m in height. It shares the west coast of North America with other conifer giants such as Douglas fir *Pseudotsuga menziesii*, giant sequoia *Sequoiadendron giganteum*, Sitka spruce *Picea sitchensis*, and noble fir *Abies procera*. The southern hemisphere plays host to the tallest flowering plants (angiosperms) which

Applied Tree Biology, First Edition. Andrew D. Hirons and Peter A. Thomas.
© 2018 John Wiley & Sons Ltd. Published 2018 by John Wiley & Sons Ltd.

Box 2.1 Parts of a Tree: Useful Definitions

Trees have a *shoot system* above ground and a *root system* below ground. The mass of branches and leaves make up the *crown* that is held on a trunk or trunks that arise from the root system. When the crowns of multiple trees join up, they are collectively referred to as the tree *canopy*.

Figure B2.1 illustrates the key regions found in a mature tree trunk. Starting from the outside, the *outer bark* is the waterproof skin that keeps the inside of the tree moist and helps prevent pests and pathogens getting into the tree. Inside this is the *inner bark* that is dominated by the *phloem*. This is living tissue that transports sugars and other useful compounds around the trees. Most often these are sugars produced by the leaves being taken to where they are needed around the tree, but can include many other compounds such as plant hormones. Inside the bark is the *wood* or *xylem* which contains small tubes through which water is transported up through the tree. The xylem may be composed of some form of *heartwood* that contains no living tissue, and the *sapwood* that is fundamental to water transport. The sapwood also contains living parenchyma cells that run axially, the *axial parenchyma*, and radially, the *rays* or *radial parenchyma*. These two systems of living cells connect to provide a complex interconnected three-dimensional network of living cells, known as the *symplast*, that has roles in storage, transport, support and defence. Between the phloem and xylem is the *vascular cambium*, a thin layer of tissue that produces new xylem on the inside and new phloem on the outside.

The xylem produced by the vascular cambium makes up what we commonly refer to as wood. Its cell walls are made up from a mix of cellulose, lignin and a cellulose-like compound called hemicellulose. Together, these compounds provide the strength that characterises wood.

Gymnosperm wood contains vertical tubes called *tracheids* that act to conduct water and also to physically support the trunk. By contrast, the wood of angiosperms (strictly, the dicotyledonous angiosperms because monocots – palms, and so on – have a different structure) divides up the functions, using *vessels* to conduct water and *fibres* to provide structural strength. The collective term for all the tubes that conduct water (i.e. vessels and tracheids) is *tracheary elements*.

Further information on these different elements of the tree is given in this chapter.

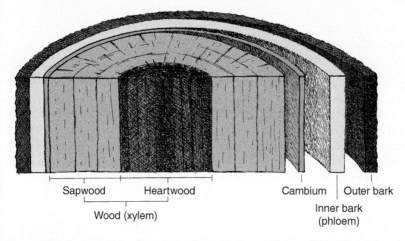

Figure B2.1 Cross-section of a mature tree trunk. *Source:* Thomas (2014). Reproduced with permission of Cambridge University Press.

reside in fertile areas of southern Australia (Victoria and Tasmania) where the tallest mountain ash *Eucalyptus regnans* stands at 99.6 m. The tropics too host a few giants with the emergent trees yellow meranti *Shorea faguetiana* and klinki pine *Araucaria hunsteinii* both reaching over 85 m (Figure 2.1).

(a)

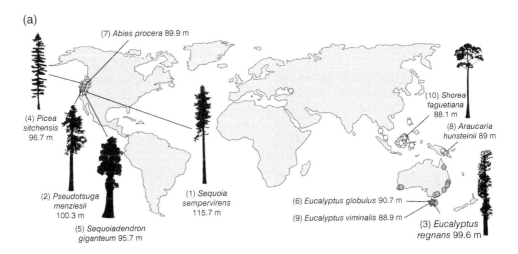

(b)

Figure 2.1 (a) Global distribution of tree species known to reach 70 m in height. Most of the tallest species are either conifers from the west coast of North America (represented by blue stars for the top five species and light blue dots for the remainder) or eucalypts in Tasmania (red stars for the three tallest species and light red dots for the remainder), although one dipterocarp species from Borneo (yellow star) and one conifer from New Guinea (blue star) rank among the top 10. Other angiosperm species that can exceed 70 m (pale yellow dots) are found in Southeast Asia, especially Borneo. One tall conifer (pale blue dot) occurs in Eurasia. *Source:* Tng *et al.* (2012). Reproduced with permission of John Wiley and Sons. (b) Mountain ash *Eucalyptus regnans*, one of the tallest species in the world, growing in Sherbrooke Forest, Victoria, Australia.

As a consequence of genetics or environment, most trees never reach such heights. In fact, in order to avoid icy Arctic winds, dwarf birch *Betula nana* only lifts its leaves about 20 cm above the ground. Yet, in spite of the vast contrast in size, the largest and the smallest species grow in the same way. *Primary growth* is responsible for stems and roots getting longer; *secondary growth* is responsible for the sideways (radial) expansion of the woody plant body. An exception to this is found in arborescent monocotyledons (e.g. palm trees) which only have primary growth. While primary growth is vital to all plants, it is secondary growth that allows trees to accumulate a rigid structure capable of supporting an expanding shoot system. In simple terms, secondary growth creates the woody skeleton that has enabled trees to become so large. It is therefore central to the identity of trees as well as vital for their ability to compete for light, resist mechanical failure and resist degradation from biological agents.

Regardless of the ultimate size a tree reaches, growth in height and diameter occur through the activity of *meristems*: plant tissues capable of cell division. Although meristems make up only a tiny fraction of the plant body, they are essential to its development because they create new cells that expand and produce the specialised tissues required for plant function. In trees, a series of apical meristems (usually contained in buds and root tips) and the sheath-like lateral meristems (the cambia that coat the tree under the bark) are responsible for the growth of the entire plant body.

Tree Design

One of the most remarkable things about trees is that the tree habit has evolved independently over a large number of plant families, making it one of the best examples of *convergent evolution* in nature. The quest to compete for light has resulted in many different plant species coming up with the same solution: grow taller and occupy more space than your neighbours. However, this solution does present a number of challenges. An increase in height requires additional biomass to support the above-ground parts. It is no good successfully occupying the space above neighbouring plants only to collapse in the first gust of wind. In order to remain competitive, the tree should not simply grow more rapidly than other plants (although there is a place for this) but it must survive and persist within that environment. Therefore, it must be resilient to biotic (e.g. pests and diseases) and abiotic (e.g. wind, drought and cold) stress, and it must be biomechanically stable over extended periods of time. To a large extent, these are competing outcomes that require trade-offs against optimum performance in one particular area. As a consequence, inherent in the design of all trees is a compromise between growth, biomechanical stability and resilience to the stresses presented by its environment. A useful way to visualise these competing outcomes is the tree trade-off triangle (Figure 2.2). Since carbon gained in the form of sugars produced through photosynthesis can only be used once, trees must determine where the 'carbon budget' should be spent and this creates a tension between competing outcomes of growth. For example, should the carbon expenditure be used to increase the strength of the wood, improve the resilience of wood to decay, expand the root system or maximise the leaf area? If growth is to be prioritised, this invariably means that something is lost in terms of biomechanical stability and stress resilience. Furthermore, the most stress resilient trees tend to have a reduced growth rate. No one design dominates in all environments

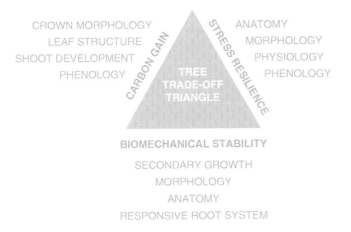

CROWN MORPHOLOGY
LEAF STRUCTURE
SHOOT DEVELOPMENT
PHENOLOGY

ANATOMY
MORPHOLOGY
PHYSIOLOGY
PHENOLOGY

CARBON GAIN

STRESS RESILIENCE

TREE
TRADE-OFF
TRIANGLE

BIOMECHANICAL STABILITY
SECONDARY GROWTH
MORPHOLOGY
ANATOMY
RESPONSIVE ROOT SYSTEM

Figure 2.2 Tree design is a compromise between carbon gain, stress resilience and biomechanical stability; these primary factors are informed by a number of other factors identified by the subheadings.

and trade-offs to specialise in one environment can make the same plant less competitive in another environment. It is, therefore, unlikely that tree design can ever be described as optimal; rather, trees represent efficient designs given their evolutionary origins (their phylogeny), the history of individual trees and environmental constraints (Crawley 1994). Ultimately, it is the outcome of these compromises that give trees such diverse shapes and sizes to their outer morphology (shape) and their internal anatomy.

How Shoots Grow

Building Blocks: Meristems and Buds

Shoots of woody plants grow rhythmically with periods of rapid shoot growth (flushing) alternating with periods of apparent inactivity during which new buds containing minia-ture shoots are formed. Environmental control of this periodic growth is indicated by fairly synchronous shoot development across a wide range of species. In temperate climates, the most important variable for the onset of growth is favourable temperatures. In more tropical climates, where temperatures are conducive for plant development all year round, it may be the arrival of heavy rain that stimulates growth.

During favourable growth periods, shoots seek to occupy space so that leaves may capture light for photosynthesis. This needs to be achieved whilst providing a system capable of gas exchange, water conduction and biomechanical support. Self-shading must also be minimised so that the photosynthetic efficiency of the whole plant is not diminished by the success of its own growth. Trees achieve this through repeated pat-terns of development that can be seen in the crown of any tree. The largest of these is the branch, which is made up of *modules* consisting of a shoot with its leaves and buds. These modules in turn are made up of a series of smaller subunits (termed a *metamer*) consisting of a leaf and bud at a node on a stem separated by intervening sections of branch called internodes. The final form of a tree is the result of the reiterative

(a) (b)

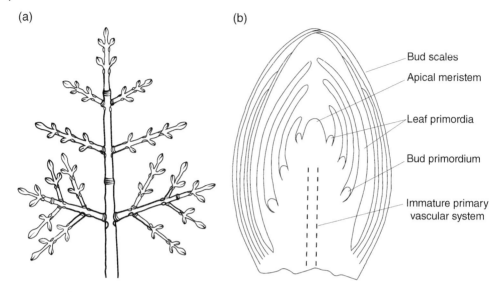

Figure 2.3 (a) A woody shoot showing 3 years' growth. Bud scale scars can be seen as rings at the base of the current and preceding years' extension growth. Apical meristems can be found at the apex of the main shoot and its side branches. *Source:* Ward (1904). (b) An apical meristem viewed in cross-section showing the meristem, young leaves (leaf primordia), surrounded by the protective bud scales. The new growth will be linked to the internal plumbing (the vascular tissue) shown developing at the base of the bud. *Source:* Beck (2010). Reproduced with permission of Cambridge University Press.

accumulation of metamers and modules that develop under genetic control in response to their environment and the space they are able to fill. Consequently, each tree has a number of possible but not random shapes. However, despite the myriad of shapes and sizes presented by trees, the fundamental processes of tree growth are common across otherwise unrelated species. As the Cambridge University botanist H.M. Ward described, 'What a complex matter in its summation but what a simple one in its graduated steps, the shaping of a tree is' (Ward 1909).

New shoot modules or branches arise from the *apical meristems* (regions of cells characterised by high levels of cell division) typically found at the end of existing shoots or leaf bases. Within the meristem, new leaves are produced as leaf primordia, seen as very small protuberances. Leaf primordia are produced in regular sequences in opposite pairs or spirals around the stem; these ultimately define nodes (sites of leaf attachment) and internodes (Figure 2.3).

Internodal stem sections are also developed in the bud by the apical meristem. Shortly after origin, distinct tissues begin to develop as shown in Figure 2.4. The central pith of the young stem is generally surrounded

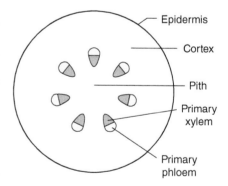

Figure 2.4 Cross-section of a young dicotyledonous stem showing the primary structure including the vascular tissue made up of 'vascular bundles' or rods, each containing xylem and phloem. *Source:* Beck (2010). Reproduced with permission of Cambridge University Press.

Figure 2.5 An expanding mixed bud of horse chestnut *Aesculus hippocastanum* showing the immature flower structure in the centre of the bud that is surrounded by the characteristic palmate leaves of the species. The bud scales can be seen just hanging on at the base of the developing shoot.

by a cortex of parenchyma (general purpose cells) that hold the vascular tissue (xylem and phloem). At first, these form a series of rod-shaped vascular bundles that later join together during secondary growth (see Secondary Growth later in this chapter). These tissues are enclosed by the cortex and some form of epidermis that provides a skin-like barrier between the external environment and the plant. Each tissue region may be composed of several simple tissues containing one cell type, or may be made up of more complex tissues containing multiple specialised cells. For example, phloem contains sieve tube elements and companion cells. All cells derived from the apical meristems are known as primary tissues; all those derived from the lateral meristems (such as the vascular cambium – discussed later) are known as secondary tissues.

Once fully developed, buds may be classified by their contents (vegetative, flower or mixed), location (terminal, lateral, axillary or adventitious) and level of activity (resting or dormant). Vegetative buds contain a miniature shoot metamer held in delayed development. Flower buds contain embryonic flowers but will also typically contain rudimentary leaves. Mixed buds will contain both embryonic leaves and flowers (Figure 2.5). Prior to expansion, the bud may be enclosed by bud scales (cataphylls) that give some degree of protection from desiccation, cold temperatures and herbivory during its dormant state. In species without bud scales (e.g. Eucalypts), leaf bases or stipules from older leaves may provide some degree of protection to the sensitive meristematic tissue. Other species (e.g. wayfaring tree *Viburnum lantana*, Caucasian wingnut *Pterocarya fraxinifolia* and many from (sub)tropical regions) have naked buds that are fully exposed to the environment during periods of rest or dormancy (Figure 2.6).

Not all the buds produced by the apical meristem will expand into shoots in the subsequent growth period. Some buds will remain dormant, some die or are aborted and flower buds obviously produce flowers. As dormant buds are essentially unexpanded vegetative buds developed in the leaf axils, they contain a bud trace to the pith. This distinguishes them from adventitious buds that originate irregularly from peripheral or relatively deep parenchyma in mature stem portions spatially separated from the meristems of the shoot apex or leaf axil.

(a)

(b)

Figure 2.6 Naked buds of (a) frangipani *Plumeria rubra* and (b) candlenut *Aleurites moluccana*.

New Shoots From Buds

As vegetative buds expand, two basic types of growth can be widely observed on trees, *determinate* (also known as fixed or flush-type) and *indeterminate* (also known as free or succeeding-type). Determinate growth found in a wide range of trees including at least some species from *Acer, Fagus, Fraxinus, Juglans, Picea, Pinus, Pseudotsuga* and *Quercus* involves the elongation of the new shoots and leaves from the resting buds. The preformed leaves are able to all flush during a relatively short period of time (days), as illustrated by the fixed species line in Figure 2.7, where leaves emerge and shoots elongate more or less simultaneously in a rapid burst of growth that may only last for a few weeks. The next bud then forms straight away and may sit for many months before it opens. In contrast, indeterminate growth, found in *Acer* (most), *Alnus, Betula, Eucalyptus, Larix, Lindera, Liquidambar, Malus* and *Populus*, is characterised by much of the shoot being newly formed (neoformed) as it develops. This may lead to

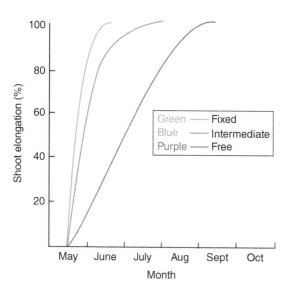

Figure 2.7 Three patterns of shoot growth (elongation) of temperate tree species. Fixed growth or determinate species show very rapid elongation and reach their maximum elongation with a few weeks. Free or indeterminate species display some shoot elongation throughout most of the growing season. Some species will also occupy intermediate positions. *Source:* Adapted from Kikuzawa and Lechowicz (2011). Reproduced with permission of Springer.

Figure 2.8 Mature leaves of various maples. (A) *Acer rubrum* (one leaf per node). Early leaves in upper row, youngest (node 1) on the left. (B) Heterophyllous leaves in (a) *Acer tataricum* [nodes 2, 5]; (b) *A. rufinerve* [1, 3]; (c) *A. campestre* var. *leiocarpum* [2, 5]; (d) *A. buergerianum* [2, 7]; (e) *A. monspessulanum* [1, 8] and (f) *A. sempervirens* [1, 4]. *Source:* Adapted from Critchfield (1971). Reproduced with permission of Harvard University.

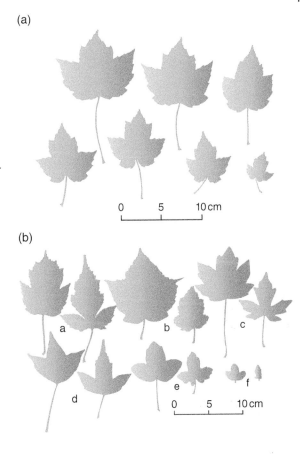

shoot development extending to most of the growth season; whilst in other species, most of the growth is achieved quite quickly despite a long period of shoot elongation (Figure 2.7). In indeterminate species, the lower leaves on a shoot are preformed in the bud and the outermost leaves are formed as the shoot grows (neoformed). The first leaves preformed in the bud are called *early leaves*, while leaves neoformed are called *late leaves*. Therefore, in temperate climates, tree shoot development occurs either as a function of the expansion of preformed shoot modules or as a combination of pre-formed and neoformed shoot modules. Unfavourable growth periods dictate that entirely neoformed shoots are absent from temperate climates because they develop a preformed shoot in a bud that is dormant during winter. Where morphological differences between the early leaves and late leaves occur, these species are termed *heterophyllous*. Examples of heterophyllous species include many *Acer* species (Figure 2.8), *Betula* and *Eucalyptus* species. Where differences in leaf morphology and anatomy occur between juvenile and adult shoots, they can also be termed heterophyllous.

Shoot growth that results simply from the expansion of the terminal bud on the main stem and branches can be termed *monopodial*, resulting in long, straight shoots. After the terminal shoot(s) elongate there is a short period of inactivity whilst new terminal buds form. Consequently, there can be several flushes of growth in any one growing season. Trees that show monopodial growth include species of *Pinus, Picea, Quercus* and *Carya*. In *sympodial* species, shoots do not expand from a terminal bud but originate from a lateral bud on the side of the shoot. This type of growth often develops when flowers occur at the end of a branch or when a shoot tip aborts. Characteristically, sympodial growth takes on a slightly zig-zag form. Trees that show sympodial growth include species of *Aesculus, Betula, Carpinus, Catalpa, Corylus, Diospyros, Gleditsia, Platanus, Robinia, Salix, Tilia* and *Ulmus*.

In some species (e.g. *Acer rubrum Malus domestica, Betula pendula, Fagus sylvatica, Carpinus betulus, Cercidiphyllum japonicum, Ginkgo biloba* and *Populus balsamifera*), shoots are divided into short and long shoots. Fully preformed shoots in the bud expand into *short shoots* that typically elongate less than 2 cm per year and do not bear lateral branches. These create characteristic spurs on branches. Shoots that are not fully formed in the resting bud develop into *long shoots* that typically elongate more than 2 cm per year and will normally bear lateral branches. Occasionally, due to changes in local environmental conditions, short shoots may revert to long shoots and *vice versa*. Having these two types of shoot allows long shoots to build the framework of the crown while short shoots fill the volume of the crown to efficiently position leaves for capturing light. As trees mature, the ratio of long shoots to short shoots changes, with older trees generally producing more short shoots. Although only a few millimetres in length, the bundles of needles found on pines *Pinus* spp. are borne on short shoots that are abscised along with the needles after a few years. Rosettes of needles found on cedars *Cedrus* spp. and larch *Larix* spp. are also held on short shoots that persist for one or more years.

Where conditions for growth continue to be favourable, some trees expand shoots from newly formed buds that have not become dormant. These are referred to as *sylleptic shoots* (Figure 2.9). Often these form earlier in the growth season so can be difficult to distinguish from normal extension growth. In temperate climates, fast growing species such as birch *Betula* spp., cherries *Prunus* spp. and poplars *Populus* spp. frequently

Figure 2.9 Black poplar *Populus nigra* often develops sylleptic shoots (arrows) under the favourable growth conditions such as those found at this field trial site in northern Italy.

develop sylleptic shoots. In the tropics, consistently favourable growing conditions mean that lateral buds develop into sylleptic shoots across a wide range of species. *Lammas shoots* that develop from recently formed terminal buds not usually expected to flush until the following growth season can also be termed sylleptic. *Proleptic shoots*, by contrast, arise from buds that have passed through a period of dormancy. Where pronounced seasonality exists, late season sylleptic or proleptic shoots are particularly vulnerable to winter injury as they have little time to sufficiently harden to low temperatures (see Chapter 10). Examples of genera that may exhibit late season shoots include: *Abies, Alnus, Carya, Fagus, Pinus, Pseudotsuga, Quercus, Ulmus.*

Apical Dominance and Apical Control

The shape of trees changes substantially through their lifetime, from structurally simple seedlings consisting of a few leaves on a single stem to complex crowns with multiple orders of branches extending in many directions. Although there are many external influences, such as wind and light, that help create the final shape of a tree, the processes of apical dominance and apical control have a profound effect on the pattern of crown development (Sterck 2005).

Apical dominance describes the physiological process that enables the apical bud at the apex of a shoot to govern whether lateral buds further down the shoot can start growing out into a branch. It therefore helps to determine the number of branches and ultimately, tree form (Cline 1991). This feature of shoot development is quite easily demonstrated by the fact that when an apical bud is removed from a shoot by pruning or storm damage, previously repressed lateral bud(s) begin to elongate: they are released from the dominance of the apical bud. The apical bud and young leaves export the plant hormone auxin to inhibit the development of lateral buds and exert dominance. After decapitation, the transport of auxin ceases and the lateral buds begin to elongate (Cline 2000). As these lateral buds develop into a shoot, its own apical bud begins to inhibit the outgrowth of its own lateral buds.

As a shoot grows and gets further away from lateral buds, the dominance of the apical bud begins to diminish. Generally, the apical dominance of lateral buds is broken one year after they are formed. It is for this reason that the first year of new shoot development is often limited to the extension of a single shoot and only in the second year can small lateral branches be seen to develop. In contrast to this, in species with sylleptic branches (see above) the apical and lateral buds are activated simultaneously.

As lateral shoots begin to grow, the leader shoot further controls their development by *apical control* (Brown *et al.* 1967; Wilson 2000). At maturity, many conifers and a few angiosperm trees have a dominant vertical leader shoot with relatively uniform lateral branches developing below. Since new lateral branches often grow in the same season as the leader, they can be said to have weak apical dominance but strong apical control; lateral branches grow more slowly and more horizontally than the leader, apparently under the control of the leader. The older lateral branches lower down the central stem have had more time to grow and so are longer than those higher up the stem, so the overall effect is a rather conical crown that can be referred to as *excurrent* (Figure 2.10a). In contrast, most angiosperm trees display high apical dominance but weak apical control. They have laterals that may be repressed for more than one year (strong apical

(a)   (b)

Figure 2.10 (a) The excurrent tree form of a young giant sequoia *Sequoiadendron giganteum* determined by weak apical dominance and strong apical control. (b) The decurrent form of a mature sycamore *Acer pseudoplatanus* determined by strong apical dominance but weak apical control.

dominance) but quite quickly will lose their central leader as lateral branches seek to expand the crown in whichever direction possible (weak apical control). The effect of this is a much more rounded crown that can be described as *decurrent* (Zimmerman and Brown 1971) (Figure 2.10b).

It should be noted that most young trees develop the conical excurrent habit as this helps them to grow upwards quickly to secure their place within the forest canopy or provide an advantage over competing vegetation. As the tree matures, the relative strength of apical dominance and control is more fully expressed, and crown architecture becomes more characteristic of the species.

Epicormic Shoots and Sprouting

The perennial nature of trees and their inability to move makes them vulnerable to natural disturbance by fire, storms, landslides and similar catastrophic events that may cause serious injury to the tree. Senescence and less intense disturbances, such as superficial wind damage and herbivory, may also warrant the replacement of lost leaf area or adjustment in the crown architecture. Consequently, many trees can sprout from dormant buds and/or adventitious buds formed in response to the disturbance. This trait gives the tree an 'insurance policy' capable of replacing lost leaves. Over time, sprouts may develop into substantial components of the tree crown and, in circumstances where they replace the main stem, even compete for apical dominance. In rare cases, significant portions of the tree crown can be made up from mature epicormic or adventitious sprouts that seek to reiterate the shape of the parent tree. Long-lived tree species, such as coastal redwood *Sequoia sempervirens*, often develop epicormic structures that lead to complex crowns which provide niche habitats for an array of different bryophytes, lichens, epiphytic ferns, as well as a host of animals (Sillet and Van Pelt 2007).

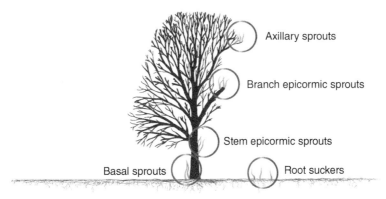

Axillary sprouts

Branch epicormic sprouts

Stem epicormic sprouts

Basal sprouts

Root suckers

Figure 2.11 Names given to types of sprouts in standing trees. Not all sprouting forms will be found in every species.

Sprouts may arise from stems, the basal region and, in some cases, the root system of trees (Figure 2.11). Almost all tree species possess the ability to sprout on young portions of stem i.e. a young tree or a young branch on an older tree, though many lose this ability as stems get older (or at least require increasing levels of stimulus to sprout). Other species maintain a prolific ability to sprout throughout most of their life, even on mature stems.

Epicormic Shoots (Sprouts)

All vegetative buds and meristems that are found on previously expanded stems can be referred to as epicormic buds or meristems. These structures are capable of producing shoots (more commonly referred to as sprouts) subsequent to the regular development of shoots (described above). Two types of epicormic bud can be identified based on their developmental history. *Preventitious* buds originate on new shoots in the apical shoot meristem, and maintain their viability by incremental annual growth that keeps pace with the radial expansion of the stem so that they stay just below or at the surface of the bark. In general, an *epicormic trace* of parenchyma cells can be seen connecting the bud to the pith. *Adventitious* buds originate from previously non-meristematic tissue; so are not connected to the pith and typically develop *in situ*, most frequently from callus produced in response to wounding.

Preventitious buds are usually normal axillary buds that fail to grow out. But since bud scales are just modified leaves, small axillary buds at their base may persist to also become preventitious buds. The population of epicormic buds on a tree is, therefore, closely integrated with the normal development of the shoots. Consequently, tree condition and the genetic control are likely to be the major sources of variation in the number of epicormic buds found in one tree. Adventitious buds tend not to constitute a significant portion of the epicormic bud bank in the absence of wounding.

Close inspection of a stem cross-section often reveals an archive of epicormic bud development in mature stems (Figure 2.12). Single continuous traces indicate a single bud that has neither sprouted or died (b in Figure 2.12). Single traces showing a sudden, pronounced widening are evidence of sprout formation (c in Figure 2.12), while a trace

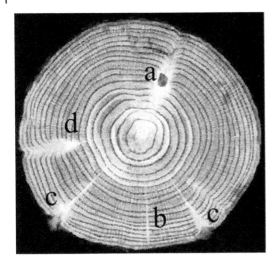

Figure 2.12 Cross-section of an oak log showing branch and epicormic structures: (a) a knot of a regular branch; (b) living epicormic bud with only minor trace expansion, signifying that it has not sprouted; (c) sprouted epicormic buds showing distinct bud traces and expansion of the bud trace at the time of sprouting, as well as development across the cambium; (d) primary epicormic sprout that sprouted 7 years after initiation and subsequently died. *Source:* Meier *et al.* (2012). Reproduced with permission of Oxford University Press.

that terminates within the wood shows that the bud has died (d in Figure 2.12). Evaluation of these traces can be useful in developing a profile of tree disturbance history in either natural or urban environments.

These shoots are important in the context of risk assessments as epicormic sprouts (particularly those of adventitious origin) are weakly attached and may be prone to failure as the sprout develops in size and sail area. This is especially true where multiple sprouts arise close to each other and compete for the same point of attachment. However, many species are able to form biomechanically stable crowns that include stems of epicormic origin. For example, despite being of epicormic origin, the reiterated stems on sweet chestnut *Castanea sativa* can form a new crown as the original crown declines (Figure 2.13). See also David Lonsdale's Expert Box on *Ancient and Veteran Trees* in Chapter 9.

(a)　　　　　　　　　　　　　　　　　　　(b)

Figure 2.13 As the original crown declines, stem epicormic shoots can often help to develop a new crown that can help the tree survive for many more years, as seen in this sweet chestnut *Castanea sativa* in the Royal Botanic Gardens, Kew. (a) Remnants of the original tree crown can be seen just above the new crown, formed by epicormic development. (b) A closer view of the epicormic branches formed on the lower portion of the trunk.

Four *epicormic strategies* have been identified based on the composition of epicormic structures (Meier *et al.* 2012). In *external clustering*, small axillary buds develop during normal shoot development into epicormic complexes consisting of numerous buds and shoots, often associated with burrs. Buds may be located on the bark surface with a meristem enclosed in bud scales (high bud) or may protrude only slightly above the bark surface with bud scales incorporated into the bark (flat bud) (Fink 1980). These can be maintained for at least 40 years in some temperate angiosperms, such as most oaks *Quercus* spp., but in conifers bud life-spans tend to be shorter. *Isolated bud* species are characterised by the production of larger epicormic buds (predominantly high buds) that tend not to form larger clusters and are less persistent, typically lasting <15 years. In some species of this group, such as white willow *Salix alba* and Norway spruce *Picea abies*, some original meristems buried in the bark are capable of producing buds for decades (>50 years), thereby compensating for shorter bud longevity. *Detached meristem* species have only been described in conifers and are characterised by minimally developed preventitious meristems in leaf axils. These require some sort of stimulus if they are to develop further into epicormic buds and shoots. In the absence of stimulus, these are sloughed with the bark after a few years. *Epicormic strand* species, typical in the Myrtaceae family, contain extensive meristematic strands within the bark that are capable of producing a continuous series of relatively short-lived epicormic buds. These tend to get larger with time and persist throughout the life of the tree, but still require stimulus to develop epicormic buds and sprouts. Additional epicormic strategies may well be identified as the field of epicormic research continues to develop.

Basal Sprouts

Most angiosperm trees have the ability to produce basal sprouts from buds located at the stem-root transition zone, often referred to as the *root collar* (Figure 2.14). Buds within this zone originally develop from the stem tissue above the cotyledons when the

(a) (b)

Figure 2.14 Basal sprouting on narrow leaved ash *Fraxinus angustifolia*, Royal Botanic Gardens, Kew. (a) Early signs of decline can be seen in the crown. (b) A closer view of the basal sprouts.

tree is still a seedling. These buds grow outwards to maintain their position just underneath the bark. Indeed, if the bark is removed, these dormant buds can be seen on the surface of the xylem as small conical protrusions. Morphologically, these buds are similar to epicormic buds found higher up the stem but are functionally distinct as they produce secondary trunks rather than branches (del Tredici 2001).

As with all epicormic buds, the longevity of basal buds varies tremendously: some species maintain their ability to sprout for decades, whereas other species use them to ensure persistence of saplings in challenging environments but have a diminishing ability to sprout as they mature.

The ultimate fate of released buds depends to a great extent on the growth environment and the point of origin. Sprouts developed on the trunk will be dependent on the original root system. They are also susceptible to pathogens crossing over from the trunk (see Chapter 9). If the sprout develops below the soil, adventitious roots may establish on the lower portion of the new stem and offer independence from the original tree.

While many gymnosperms are capable of basal sprouting at the sapling stage, most lose this ability as they mature. Notable exceptions include species of *Sequoia, Cunninghamia, Taxus* and *Torreya*, and fire adapted pines such as Canary Island pine *Pinus canariensis*.

Branch and Trunk Sprouts

Unless the species is known to produce abundant epicormic sprouts in the absence of stress (e.g. common lime *Tilia x europaea*), the development of epicormic sprouts (*syn.* suckers) on the trunk or branches often signals some form of stress is present. These features may rise vertically from lateral branches within the crown (Figure 2.15a), or may appear towards the end of a branch that has been broken in a storm event (Figure 2.15b) or incorrectly pruned. In any case, further investigation into the cause of epicormic sprouting is warranted as it may represent the onset of tree decline or recent injury.

(a)

(b)

Figure 2.15 (a) Branch epicormic sprouts growing within the crown on common lime *Tilia × europea*. (b) Branch epicormic sprouts growing in response to storm damage on sycamore *Acer pseudoplatanus*.

Partially uprooted trees with horizontal trunks, as well as lateral branches, often develop rows of vertical sprouts, derived from epicormic structures. These reiterative shoots exploit the newly opened canopy gap whilst using the remaining root system. This scenario is particularly frequent along riverbanks where erosion has undermined the root system, leaving the tree vulnerable to wind-throw. Trees in the genera *Alnus, Salix* and *Populus* are often seen with this type of growth but it also occurs in a wide range of species, including conifers such as European larch *Larix decidua* and Douglas fir *Pseudotsuga menziesii*.

In a similar but distinct way, when a living trunk or branch comes into contact with soil, *layered sprouts* may also arise. This can occur in open grown trees where lower side branches touch the ground as a consequence of their own weight; where repeated snow loading may cause lower branches to permanently touch the ground; or where a particular growth form frequently brings stems in contact with soil (Figure 2.16). In such cases, at the point of soil contact adventitious roots develop prior to adventitious shoots arising from the same region of stem. In some cases, the original branch may develop upwards into a new trunk. Subsequent growth may either augment an existing crown or, over time, develop into a fully functional and independent tree, forming a replacement to an ageing or damaged parent stem. This type of tree development is often referred to as 'layering'. While it has the potential to expand the crown area in open-grown situations, it is largely a mechanism to survive suppression and disturbance rather than colonise new territory or increase population size.

Since most trees are capable of layering, plant propagators use this to vegetatively produce new material from established plants by *air layering*. Essentially, young active stem portions are partially buried in soil or an alternative growing substrate held around the aerial stem in a small container. When sufficient adventitious roots have developed

(a)   (b)

Figure 2.16 Layering occurs when a branch comes into contact with the soil. This produces adventitious roots and allows epicormic sprouts to develop from the point of soil contact.
(a) A weeping beech *Fagus sylvatica* 'pendula' has layered and subsequently produced a number of stems around the periphery of the original crown that have the effect of substantially increasing the radial dimensions of this tree. (b) Layered branches on a horse chestnut *Aesculus hippocastanum*. Both examples are from the Royal Botanic Gardens, Kew.

to sustain the distal stem portion, it is removed to produce a new plant capable of autonomous growth. This approach is particularly useful for species that are not easily grown from seed or propagated using other techniques.

Sprouts Originating Underground

Some trees also have specialised underground stems, lignotubers and rhizomes, which are capable of producing sprouts that are spatially separated from the original stem. Since their point of origin is underground, they have the potential to become autonomous from their parent plant through the development of an adventitious root system.

The lignotuber first develops from buds in the axils of the seedling cotyledons. These proliferate to form a large organ (Figure 2.17) holding a bud bank and a carbohydrate store that will help sprouts grow if the main trunk is damaged. Many trees have been found to develop lignotubers, particularly those found widely in fire-prone ecosystems (Paula *et al.* 2016). They include species from the families: *Asteraceae, Bruniaceae, Caprifoliaceae Cupressaceae, Elaeocarpaceae, Ericaceae, Fagaceae, Myrtaceae, Oleaceae, Proteaceae, Rhamnaceae, Rosaceae* and *Tiliacea*.

Specialised underground stems known as rhizomes can be distinguished from roots by their distinct nodes and internodes. These are capable of producing sprouts some distance away from the original primary stem while staying connected to the root system of the parent tree. They may remain connected as part of a clonal colony or produce their own adventitious roots and become fully autonomous of the parent plant. Drought adapted trees, such as Gambel oak *Quercus gambelii* and chokecherry *Prunus virginiana*, can produce rhizomes.

Many angiosperm trees, such as species of *Alnus, Hippophaë, Populus, Pterocarya, Robinia, Ulmus* and *Zelkova* are also capable of producing sprouts from their roots.

Figure 2.17 A large lignotuber formed at the base of a eucalyptus tree, growing to the west of Sydney, Australia.

(a) (b)

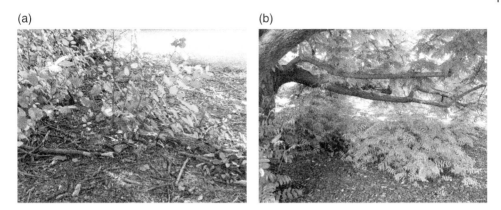

Figure 2.18 Root suckers on: (a) Japanese zelkova *Zelkova serrata*; and (b) Caucasian wingnut *Pterocarya fraxinifolia*, Royal Botanic Gardens, Kew.

Root suckers are not normally considered epicormic structures as they are derived from roots rather than stems, but in many ways their development is analogous to sprouts that form on above-ground parts of the tree. They develop either from *additional buds* that arise from deep within the root tissue (the pericycle, cambium, phellogen or phelloderm) of young uninjured roots or *reparative buds* that form in the superficial tissues (root callus) of injured roots. *Root suckers* are typically produced in response to traumatic injury or disturbance but they can also be part of normal plant development as in species of *Rhus, Banksia* and some *Prunus* (e.g. blackthorn *Prunus spinosa*). Excavation of the upper root system often reveals substantial thickening of the root system away from the parent tree, whereas the root section connecting it with the parent tree shows no further development and eventually dies, leaving an independent secondary stem.

Suckering can have important consequences for managing trees because open ground and high light conditions can also stimulate root suckering in some species (Figure 2.18). This can cause problems in amenity areas where the suckers conflict with other plantings, lawns or other infrastructure. There is no easy remedy for this situation as cutting off the sprouts may only serve to stimulate further sprouting, at least in the short term. Therefore, a species' potential to root sucker can be an important characteristic to consider when selecting trees for amenity situations.

Perhaps the most impressive root suckering is in the quaking aspen *Populus tremuloides* which colonises vast areas (the largest is up to 43 hectares), with a single individual clone comprised of thousands of individual trunks. Aspen *Populus tremula* and white poplar *P. alba* are also very successful at colonising new ground using root suckers.

Practical Considerations of Sprouting

For centuries, the ability of trees to sprout has been utilised to produce stems for charcoal making, fencing materials and fodder. The practices of both pollarding and coppicing rely on the tree's ability to sprout after injury. These sprouts are then allowed to develop for 3–7 years before being harvested. In the case of coppicing, the cuts are made around 15–25 cm above the ground to encourage a large number of sprouts from the remaining trunk. Pollarding is similar but the stems are cut around 2 m above ground to keep new shoots away from browsing animals. When pollarding or coppicing young trees (e.g. 10–15 years), the cyclical pruning helps to form a mechanically strong 'knuckle' that improves the stem attachments.

Sprouting in Response to Mechanical Injury

In managed landscapes, poor pruning technique can stimulate sprouting. For example, a stub cut (caused by internodal cutting) that leaves a short branch projecting beyond the branch collar, stimulates a similar response to a naturally caused branch failure due to wind. In many species, multiple sprouts will grow out from epicormic buds close to the site of injury or adventitious buds formed in callus (Figure 2.15b). These then compete with each other to occupy the space left by the removed branches in order to replace lost leaf area. If left unchecked, this will result in the development of large epicormic stems (>5 cm diameter at their base) that have an increasing risk of failure as the stem grows. Primarily this is a result of drag forces being focused on the point of attachment (see Chapter 3) but numerous epicormic shoots arising from the same area of stem also causes crowding at the point of attachment which reduces its quality. In some cases the attachment may be further compromised by decay that colonised the original wound. Damage to the parent stem during pruning (a flush cut) may also result in the development of epicormic shoots; these too will be vulnerable to failure at the point of attachment if they are allowed to develop into substantial stems. Consequently, unless deliberately coppicing or pollarding, epicormic growth in managed landscapes should be minimised by following best practice pruning guides (e.g. BS3998; International Society of Arboriculture Best Management Practices; see also Tree Pruning in Chapter 3). However, if the management of epicormic growth is necessary to restore a previously damaged crown, careful selection of a single epicormic shoot over the course of several pruning cycles can lead to a more resilient attachment capable of maintaining crown integrity over the longer term.

Sprouting in Response to Disturbance

Epicormic sprouts naturally appear as part of a tree's strategy for persisting through periods of disturbance: their appearance can be an indicator of environmental disturbance, injury or some form of internal physiological problem. Consequently, visual inspection of trees should include assessment of epicormic structures and other forms of sprouting, because their presence can help diagnose stresses in a tree. When doing this, it is important to remember that sprouting response is by no means ubiquitous or uniform; there is wide variation between species in their ability to sprout. For some species (e.g. common lime *Tilia × europaea*), prolific basal and epicormic sprouting is part of their normal development. Other species may not produce epicormic sprouts at all. Therefore, the absence of sprouts does not indicate the absence of site disturbance or tree decline.

Secondary Growth

It is difficult to overstate the importance of secondary growth. In many ways, the history of mankind has been dependent on the evolution of woody plants. Their ability to take energy from sunlight, add a few other compounds, and turn them into wood and bark has provided the material for shelter, fuel, paper, fabrics, medicine and numerous other commodities. However, secondary growth did not evolve to meet the needs of humans, but rather for the plant to grow tall in a quest for light. By expanding the crown upwards, plants with woody stems competed for light more effectively. This evolutionary impetus has been the driving force for the dominance of trees in many of our terrestrial biomes and has led to a number of species reaching over 70m in height. A further ecological consequence of tough secondary growth is that plants have a greater resilience to environmental stress and biological degradation: they can get old.

To allow the plant body to grow, compete for and capture light, trees must be capable of reinforcing and protecting shoots generated by primary growth. In the vast majority of trees this involves the woody skeleton, above and below ground, becoming wider through the action of two lateral meristems known as the *vascular cambium* and the *cork cambium* (*syn. phellogen*). The evolution of these cambia provided the key for the formation of true wood and is one of the fundamental differences between herbaceous and woody plants. Secondary xylem (wood) and secondary phloem (inner bark) originates from the vascular cambium, and bark originates from the cork cambium. In most trees the inner and outer bark are progressively lost as they build up, so the main outcome of secondary growth is a series of xylem layers, added during periodic growth, which are encapsulated underneath the bark.

Wood provides several key functions for the tree: it provides biomechanical support for the crown; conducts materials between the roots and the shoots; and acts as a store for carbohydrates, nutrients, water and defensive compounds (Chave *et al.* 2009). Phloem has a primary role in the transport of sugars and other compounds from their source in the leaves or storage sites to various *sinks* that use carbohydrates for growth, reproduction or defence. The outer bark physically protects the stem from extremes of temperature, desiccation and biological threats, while facilitating the exchange of gases needed for cellular respiration. In young stems, a further role of bark is to increase the mechanical resistance of the stem to bending by wind, snow and/or ice (Niklas 1999).

The Vascular Cambium

The vascular cambium is a permanent secondary meristem that envelops the entire tree with the exception of the very apex of the stems and roots. In most trees, it first develops in the vascular bundles of young stems (Figure 2.4) between the xylem and phloem tissue (called the fascicular cambium). Some of the normal plant cells (ground tissue or parenchyma cells) between these vascular bundles then differentiate into cambial cells (producing the interfascicular cambium) before the two merge to form a continuous cylinder of meristematic cells capable of producing secondary xylem internally (centripetally) and secondary phloem externally (centrifugally) (Figure 2.19). In new twigs, this vascular cambium joins up to the cambium in older growth, thus forming a continuous sheath down the stem and into the roots.

The vascular cambium is composed of meristematic cells known as *initials* that, during the growth period, continuously regenerate to produce a mother cell and another initial. Typically, xylem and phloem cells then divide again before differentiating into their mature tissue types. This thin sheath (a few millimetres at its most active) of immature and meristematic cells is collectively known as the *cambial zone* (Figure 2.20). In the vast majority of species, the xylem mother cells divide more frequently than the phloem mother cells, giving rise to more xylem than phloem in any growth period. Most species have xylem to phloem ratios of between 4 and 10 : 1. However, a 1 : 1 ratio

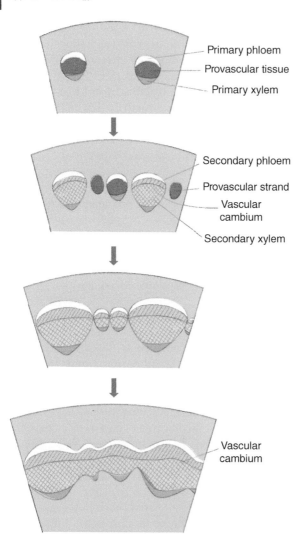

Primary phloem
Provascular tissue
Primary xylem

Secondary phloem
Provascular strand
Vascular cambium
Secondary xylem

Vascular cambium

Figure 2.19 Vascular cambium development in typical trees. The cambium initially forms between the primary xylem and phloem of the vascular bundle, before expanding around the circumference between the vascular bundles to form a continuous cylinder of meristematic tissues. *Source:* Adapted from Beck (2010). Reproduced with permission of Cambridge University Press.

has been observed in the tropical angiosperm *Mimusops elengi* and a 15 : 1 ratio has been observed in the temperate gymnosperm American arborvitae *Thuja occidentalis*, showing that at least a few species operate outside of this range (Fromm 2013).

Even with the naked eye, in many tree species, if you look down upon the cross (transverse) section of a woody stem it is possible to make out two distinct sets of cells within the wood. These are the axial (parallel to the axis of the stem) and radial (perpendicular to the axis of the stem) systems, and each originates in particular portions of the vascular cambium. *Fusiform initials* generate axially orientated cells in the stem (tracheids, vessel elements, fibres, axial parenchyma) and the *ray initials* create those that are radially orientated (ray parenchyma, ray tracheids). Most divisions of the cambium are *periclinal* (radial) to produce new xylem and phloem but, as more xylem is deposited, the cambium must expand circumferentially to maintain continuity by adding cells via *anticlinal* division (Figure 2.20). For example, a young stem of eastern white pine

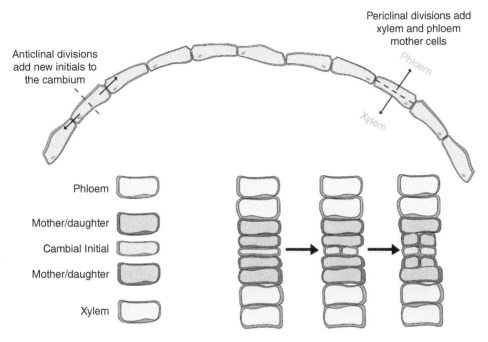

Figure 2.20 The cambial zone of a tree. The cambial initials (the original meristem cells) divide in two distinct ways. Anticlinal divisions (perpendicular to the axis of the tree) produce new cambium cells increasing the circumference of the cylinder, and periclinal divisions (parallel to the axis) produce new xylem and phloem. *Source:* Adapted from Spicer and Groover (2010). Reproduced with permission of John Wiley and Sons.

Pinus strobus was found to have around 800 initial cells in its cambium; after 60 years that number had multiplied 40 times to ~32 000 cells forming the cambium around its circumference (Bailey 1923).

The rate and nature of cell division is under genetic control but it may be modified by environmental cues. The creation of vessels and sieve tube elements (the main transport cells of the phloem) is triggered by auxin, (specifically, indole-3-acetic acid; IAA). However, a host of additional plant growth regulators including cytokinins, abscisic acid, brassinosteroids and ethylene have been implicated in xylem development and growth and are likely to interact in highly complex ways (Aloni 2010). Immediately after the initiation of the xylem mother cell, cell expansion is driven by the uptake of water – thought to be under the control of membrane proteins called aquaporins – which helps regulate cell turgor (the pressure of water inside the cell). A series of 'switches' appearing to be under the control of auxin then determine if the xylem mother cell differentiates into parenchyma, fibres or sap conducting cells.[1] Therefore, secondary xylem consists of several cell types, each performing specific functions that are closely allied to their mature form. Growth regulators also modulate the shape of these cells and their response to ageing and stress (Schweingruber *et al.* 2013).

---

1 Sap-conducting cells such as vessels in angiosperms and tracheids in gymnosperms are collectively referred to as *tracheary elements*.

In seasonal climates, cambial activity is strongly influenced by the environmental conditions. Unfavourable conditions, such as low temperatures or low water availability, lead to the cambium becoming dormant. When the cambium is dormant, it exists in two states or phases: *rest* and *quiescence*. During rest, no cell division is possible even in the presence of favourable environmental conditions; but during quiescence, cell division can resume, providing key environmental conditions are met. The switch from rest to quiescence is triggered when a chilling requirement is met (see Leaf Phenology in Chapter 3). As the days shorten in autumn, the cambium becomes insensitive to auxin and cell division ceases, coinciding with the completion of bud set and the start of leaf senescence in deciduous species. Once the specific chilling requirements are met over winter (again, see Chapter 3), the quiescent phase is reached and the cambium once again becomes sensitive to auxin and warming temperatures. As auxin is produced in developing buds and transported from the top to the bottom of a plant (basipetally), cambial reactivation tends to migrate downwards from the top of the stems to the base of the trunk before moving into the roots. Additional plant growth regulators such as gibberellin, cytokinin and ethylene, as well as nutrients such as calcium, have also been linked to the complex process of cambial reactivation (Fromm 2013). While photoperiod and temperature are the most important environmental signals regulating the growth–dormancy cycle, social status may also have a role in groups of trees. For example, *Abies alba* occupying a dominant part of the canopy has been found to start cambial activity earlier and complete it later than intermediate or suppressed trees within a forest plantation in France, suggesting that an interaction between tree size and environmental factors controls the timing of dormancy (Rathgeber *et al.* 2011).

In tropical regions where there is little annual variation in day-length or temperature, the cambium can be active throughout the year. In seasonal tropics, periods of drought can temporarily halt cambial activity. Equally, long periods of flooding can cause cambial dormancy because of its profound impact on root activity. As a result, the width of tree rings in tropical trees is usually influenced by variable environmental conditions or even some internal regulation. Therefore, growth rings of tropical trees may not be made annually and cannot be used to age the tree.

Thickening of Woody Cell Walls

Both xylem and phloem go through a period of adding new cells before turning these into particular cell types (differentiation). For cells that are required to support the tree physically or resist large negative pressures during water conductance, an important process in their development is cell wall thickening.

Mature xylem cells, as well as other cells such as phloem fibres, have walls made up of multiple layers, each different in its construction. The outermost layer is the middle lamella and is essentially the matrix (up to 5 μm thick) on which the cells are built, made up of pectin and lignin (Figure 2.18). Pectin sticks the neighbouring cells together, while strength is provided by high concentrations of lignin (>80% in some species). A series of walls are then built up inside each other (Figure 2.21). The primary cell wall is made up of very fine strands of cellulose known as cellulose microfibrils (20–30%), scattered throughout a matrix of pectin (30–50%), hemicelluloses (20–25%), glycoproteins (e.g. expansin) and water (5–10%). Together, the middle lamella and the primary cell wall are usually referred to as the compound middle lamella (CML). Inside the CML, three secondary cell walls (S_1, S_2, S_3) are built in sequence while the cell is expanding

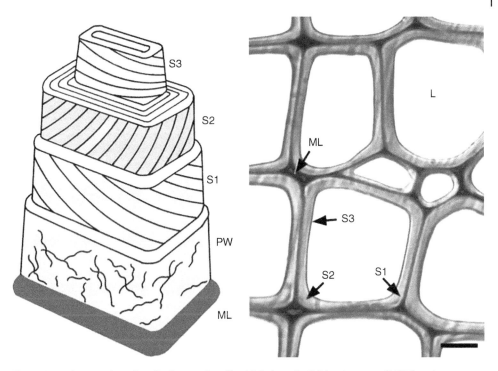

Figure 2.21 Layers of a cell wall of a woody cell: middle lamella (ML), primary wall (PW) and secondary walls (S_1, S_2, S_3) – see text for a description of these. *Source:* Schwarze (2007). Reproduced with permission of Elsevier.

(Figure 2.21). These secondary walls provide the vast majority of the cell's structural strength and differ markedly from the primary cell wall in that they are comprised of numerous helically arranged cellulose microfibrils between 3 and 10 nm thick[2]. In turn, these are sheathed and tethered together by hemicellulose and embedded in a lignin matrix. The lignin gives compressive and bending strength (stiffness) that is integral for support and resistance to cell collapse. However, the water-repelling (hydrophobic) properties of lignin also resist water absorption into the cell wall, helping to keep water flowing up the mature xylem vessels by reducing friction and water loss. Antimicrobial properties of the lignin also resist decay from all but the most specialised microorganisms (see Pathogenic Fungi in Chapter 9). The cellulose component provides the tensile strength; the amount of cellulose (typically 40–50% of the dry weight of the cell wall) and the angle of the microfibrils have a substantial effect on the mechanical properties of wood, particularly the stiffness. Hemicellulose has an important role in integrating and binding the cellulose microfibrils into the lignin matrix.

The precise composition of the secondary cell walls varies among species and its formation can be altered by factors such as the mechanical pressure experienced by the cells. In general terms, the S_1 layer is 0.1–0.4 μm thick[3] and has a fairly flat microfibril angle that is wound helically at 60–80° relative to the cell axis. More significant in

---

2 A nanometre (nm) is a billionth of a metre or a millionth of a millimetre, so 10 nm = 0.00001 mm.
3 A micrometre (μm) is a thousandth of a millimetre, so 1 μm = 0.001 mm.

determining the biomechanical properties of wood is the S_2 layer. This varies from 1 to 10 μm thick but is always proportionally thinner inside vessel elements than in fibres. Microfibril angle varies from 4° to 30°, with the angle becoming less steep as tension is increased. For example, microfibril angle in a poplar *Populus* sp. fibre S_2 layer is typically 4° but under tension was shown to be 13° (Lautner *et al.* 2012). The innermost S_3 layer (i.e. immediately adjacent to the cell lumen) has a similar microfibril angle to the S_1 layer (60–90°) and is marginally thicker at 0.5–1.0 μm. In combination, these secondary cell walls are vital to the structural integrity of the wood. However, the secondary cell wall does not develop in regions of the cell that join neighbouring cells together (perforation plates or pits). This gives an efficient compromise, allowing the cells to conduct water axially and between cells while maximising structural strength.

Programmed Cell Death

Tracheary elements (i.e. vessel elements or tracheids), by the very nature of their main role in water transport, are required to be hollow before they can become functional. To facilitate this, they must die. Further, once cell wall thickening has occurred there is no advantage in keeping the fibres alive – they can function in their structural role just as adequately when they are dead and do not use up valuable carbon for cellular respiration. Even parenchyma cells, which have a range of functions, eventually become moribund and are allowed to die. This *programmed cell death* is genetically controlled (that is, it is 'transcriptionally regulated') and influenced by various internal signals such as from the plant hormone ethylene. In vessel elements, differentiation and death can take place within a few days. Cell wall thickening and lignification are prioritised in vessel elements, because they are essential in maintaining full hydraulic function. It is also important to note that when the tracheary elements die, they are full of water – an important factor for subsequent water transport. Fibres may take about a month to mature, lignify and die, whilst parenchyma cells can be kept alive for decades before they die during their transition from sapwood to some form of heartwood. In this regard, an important concept in the function of trees is that cell death is a key component of successful living.

Bark and Secondary Phloem

Outer bark is analogous to the tree's skin because it provides a protective layer between the internal body and the outside world. If this boundary is compromised, stems can rapidly become dysfunctional. Entry of air into the xylem will disrupt water conduction; microorganisms and/or harmful insects may also rapidly invade exposed xylem surfaces. Therefore, bark integrity is absolutely critical for normal stem function and tree resistance to pathogens.

Although bark works in a similar way in different species, its appearance varies hugely among species (see Vaucher 2003 and Pollet 2010). While this is an example of biological variation, it does help us in species identification, in interpreting abnormalities and can often help us understand the ecological heritage of the tree. For example, the thick bark of the giant sequoia *Sequoiadendron giganteum* and cork oak *Quercus suber* (Figure 2.22) has evolved to resist periodic forest fires and is vital for survival in their native environments.

Figure 2.22 Cross-section of a cork oak *Quercus suber* stem showing the thick, corky bark that offers protection from forest fires.

How Bark Grows

Secondary phloem (*syn.* inner bark or bast), produced on the outside of the vascular cambium, provides a series of open tubes known as *sieve cells* in gymnosperms and *sieve tube elements* in angiosperms. During cell development they maintain a living plasma membrane inside the cell wall, but lose several organelles, including the nucleus, in order to give the hollow structure required to conduct sugars and other compounds around the tree. As a result, sieve elements are unable to control their own genetic and metabolic activities so they form associations with adjacent parenchyma cells, known as *companion cells* in angiosperms and *Strasburger cells* in gymnosperms. These cell complexes actively help to control the transport of materials in and out of phloem sieve elements. Phloem also contains axial and radial parenchyma cells that assist in such things as moving sugars for storage in the rays; and phloem fibres that give rigidity to the tissue. In some trees (e.g. *Calocedrus decurrens, Robinia pseudoacacia, Ulmus* spp. and *Tilia* spp., especially *T. cordata*), the phloem is particularly rich in long, durable, phloem fibres that make it suitable for making ropes and coarse fabrics (Vaucher 2003). Bark cloth is also made in Asia from *Broussonetia papyrifera* and, in Uganda, from *Ficus natalensis*. However, these long phloem fibres make these species vulnerable to bark tearing during pruning operations or natural stem failure. To combat this, undercuts must be made before removing branches during pruning operations (see Chapter 3). Undercutting branches during pruning operations is also essential to prevent the wood from tearing.

While the inner bark is involved in transporting sugars and other compounds around the tree, the outer bark acts as a tough skin for the tree. In gymnosperm and dicotyledonous trees, the outer bark covers all but the very youngest tips of the stems and the fine roots. It is derived from the cork cambium, which forms at approximately the same time as the vascular cambium. This cork cambium can form anywhere within the

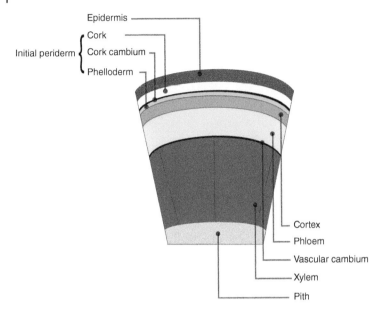

Figure 2.23 Simplified diagram of initial bark or periderm formation in the cortex. Together the cork (phellem), cork cambium (phellogen) and phelloderm make up the periderm.

outer cortical parenchyma (most typical) to deeper within the cortex or the outer secondary phloem (Figure 2.23). Modified parenchyma cells create a cylindrical cork cambium that divides to generate phellem (cork) cells on the outside and phelloderm cells on the inside, collectively known as the *periderm*. Phellem is usually produced in much larger quantities than phelloderm, and its secondary cell walls are impregnated with a hydrophobic and antimicrobial chemical called suberin. This chemical modification of the bark cells helps to prevent water loss from the xylem (sapwood), as well as colonisation by all but the most specialised microorganisms. Phellem is dead at maturity as it does not need to remain alive to fulfil its protective role. As the phelloderm ages, some component cells may develop secondary cell walls (i.e. become sclerids, *pl.* sclerenchyma) that provide additional mechanical support for the stem, and resist compressive forces.

In many species, the phelloderm is also useful as it contains chlorophyll (forming chlorenchyma) and therefore has the ability to utilise light penetrating the bark for photosynthesis. In *Fraxinus excelsior*, for example, 50% of daylight reaches the phloem and 0.4% reaches the pith in 1-year-old stems, 6 mm in diameter (Thomas 2016). This stem chlorenchyma has been found to absorb CO_2 before bud break (Langenfeld-Heyser *et al.* 1996) and so must be photosynthesising. Chlorenchymal photosynthesis may not help growth that much, but it can substantially offset (60–90%) the respiratory losses of adjacent stem cells and help to provide oxygen inside the stem (Pfanz *et al.* 2002). In winter, stem photosynthesis can also help reduce the depletion of stored carbohydrates used to maintain respiration of living tissues. Chlorenchyma can easily be found by gently removing the outermost bark to reveal the green layer. In dormant trees, identifying this green layer beneath the surface of the bark can be used to verify if the stem is still living.

Variation in Bark

In some species, for example within the genera *Acacia, Acer, Citrus, Ilex* and *Eucalyptus*, the cortical tissue and epidermis continues to develop throughout the life of the tree and a periderm is never formed. In such cases the 'bark' is very thin and can be susceptible to damage. However, in most trees the initial periderm replaces the epidermis as the external protective tissue. While this can take over 50 years in some trees (e.g. some species of *Abies, Picea, Betula* and *Prunus* – and explains why many old cherries lose their characteristic bark), in most temperate trees it is typically formed in less than 10 years. In species that form a succession of continuous cylindrical internal periderms within the cortex (Figure 2.24a), the outside of the bark is smooth – as in beech *Fagus sylvatica* and hornbeam *Carpinus betulus*. In these species, the cork cambium undergoes anticlinal (sideways) division to keep pace with the expansion of the stem. In contrast, other species form a series of overlapping crescent-shaped periderms (lenticular periderms) between the initial periderm and the secondary phloem. It is the arrangement of these successive periderm layers and the ongoing radial expansion of the stem that creates the external pattern of bark that results in more textured bark surfaces (Figure 2.24b). Often, the later formed periderms produce more phellem (cork) than the initial one, enhancing its protective quality. This is exemplified in cork oak *Quercus suber*, where the initial periderm is thin and of little value but the successive periderms form much higher quantities of cork, protecting the stem from fire and making it a commercially valuable product.

As the formation of a periderm effectively isolates the external tissues from nutrition from the phloem, almost all cells external to the periderm die. However, prior to death at

(a)

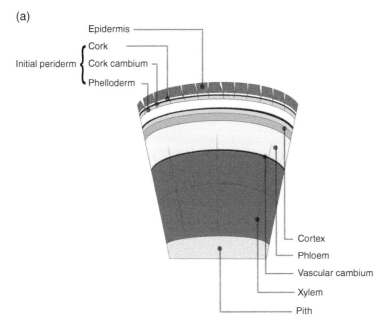

Figure 2.24 (a) Continuous internal periderms formed immediately underneath the initial periderm gives the smooth bark of trees such as beech *Fagus sylvatica* and hornbeam *Carpinus betulus*.

(b)

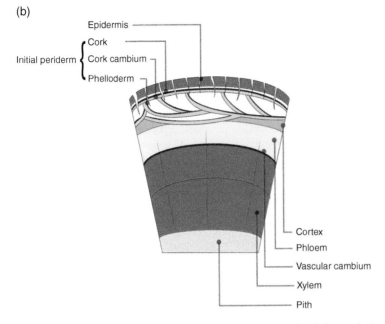

Epidermis

Initial periderm { Cork
Cork cambium
Phelloderm

Cortex
Phloem
Vascular cambium
Xylem
Pith

Figure 2.24 (Cont'd) (b) Lenticular periderms formed immediately beneath the initial periderm produces a more textured bark, typical of many temperate trees.

least some of these cells may be impregnated with substances such as resin (e.g. *Pinus*), mucilage (e.g. *Ulmus*) or tannins (e.g. *Alnus, Betula, Picea, Pinus, Quercus, Salix* and *Tamarix*), and/or deposited with waste materials (e.g. calcium oxalate or silicate crystals in *Tilia* and *Fagus*). The result is a highly complex region of periderm and non-living tissue that makes up the *rhytidome*, more generally referred to as outer bark.

Bark is shed (exfoliated) when unsuberised phellem cells (phelloid cells) function as excision layers. In the case of *Betula* spp. these phelloid cells also contain betulin, a white substance that gives birch bark its characteristic colour. In other species (e.g. *Platanus* spp., *Lagerstroemia* spp. and *Parrotia persica*), excision layers are more developed and bark is excised in larger plates.

While many of the cells within woody stems are dead when functioning, the remaining living tissues still need oxygen. Lenticels are specialised regions of loosely arranged cells derived from the phellogen, which allow gas exchange with the atmosphere. On young and thin-barked species, lenticels can be seen across the surface of the stem; thicker bark often obscures them but they can also become a characteristic feature (Figure 2.25).

Secondary Xylem – Wood

Even to the untrained eye, the sight of woody plant stems under magnification reveals an array of cell types and wide variation in their arrangement across species. Knowledge of the component cell types, their configuration and key features can be used to identify wood, understand the ecological heritage of a plant, explore past climate and disturbance

(a) (b)

Figure 2.25 (a) Lenticels can often be seen in the form of light flecks on the younger (green) portions of shoots, as in field maple *Acer campestre*. (b) However, lenticels can also be seen in many older stems, as seen in white poplar *Populus alba*. In more deeply fissured bark, the lenticels cannot be easily seen but they will be present.

events, as well as help predict how species' distribution may be modified by a changing climate. Therefore, features within wood can provide information that is relevant for a wide range of applications within wood processing, trade, conservation, ecology, forensic science, archaeology and paleobotany. Before considering the impact of alternative wood structure on tree performance, it is first necessary to identify the component parts of wood. Box 2.2 will be useful to help clarify some important principles used when looking at wood.

Box 2.2 Looking at Wood

Three views are helpful in visualising the three-dimensional structure of wood: transverse or cross-section; tangential section; and radial section (Figure B2.2). In the transverse section, for example a cross-section of a tree trunk, the axial elements are revealed end-on and the radial cells are seen in longitudinal cross-section. Tangential sections show the axial elements in longitudinal cross-section and the radial elements end-on. If you imagine peeling away the bark to reveal the wood, this will be the tangential view. If you look at the side face of a perfectly split log, this is the radial section. Radial sections from the middle of the tree to the outside expose both the axial and radial systems in longitudinal cross-section. At first, it can be difficult to visualise these contrasting views but given enough time spent pondering a block of wood, or indeed splitting firewood, the different views become clear.

In some species, the structure of the xylem varies substantially through the year. Even with little or no magnification, this creates distinctive rings (layers) that in temperate or seasonal climates are likely to reflect one year's growth. However, in tropical climates

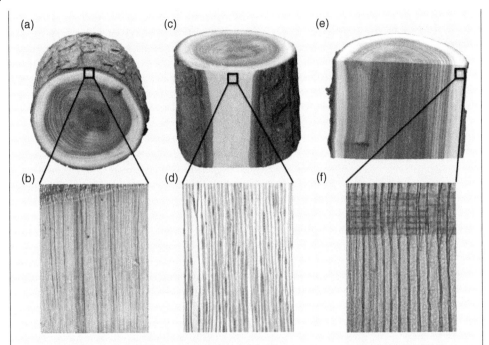

Figure B2.2 Three-dimensional structure of wood: (a,b) transverse or cross-section (×100); (c,d) tangential section (×200); and (e,f) radial sections (×400). *Source:* Wood micrographs © Fritz Schweingruber, used with permission.

where favourable conditions for growth exist throughout the year, boundaries between annual growth layers are often not prominent or, if they are, do not correspond to calendar years. Further, in some climates, multiple growth rings may be observed in any single year.

Where conspicuous rings are evident, each growth ring can often be divided into early-wood (*syn.* spring-wood) and late-wood (*syn.* summer-wood). In gymnosperms, wider cells with thinner cell walls predominate in the early-wood; narrower cells with thicker cell walls form the late-wood. Boundaries between each ring are apparent because of the contrast between the late-wood and the following growth period's early-wood (Figure 2.31). Similarly, in angiosperms, annual rings are quite easy to spot where big differences exist between the late-wood of one growth period and the early-wood of the next, as in ring porous species (see text for details); but they can be harder to distinguish in diffuse porous species (again, see text) that have a more uniform wood structure.

Different Cell Types Found in Wood

Wood must provide a secure conduit for water and mineral nutrients, as well as biomechanical support for the plant body. Various cells have evolved with different degrees of specialisation to meet these fundamental requirements (Table 2.1).

Important differences occur between the structure of dicotyledonous wood (angiosperm) and gymnosperm (conifer) wood. In common parlance, the terms

Table 2.1 Xylem cell types and principal functions.

Cell type	Principal function(s)
Axial system	
Vessel members (mainly in angiosperms)	Conduction of sap
Tracheids (mainly in gymnosperms)	Conduction of sap and support
Fibres	Support; sometimes storage
Parenchyma cells	Storage, support, transport and defence
Ray system	
Parenchyma cells	Storage, support, transport and defence
(Tracheids in some conifers)	Support and transport

'hardwood' and 'softwood' are frequently used to separate these two types of wood respectively; however, this is not very helpful as 'hardwoods' may be physically soft and 'softwoods' may be physically hard, depending on the particular species. For example, poplars *Populus* and willows *Salix* are classed as hardwoods but have relatively soft wood, whereas yew *Taxus* has strong wood but is classed as a softwood. It is therefore preferable to use the botanically accurate dicotyledonous or gymnosperm wood.

In gymnosperms, biomechanical support for the stem is provided by *tracheids*, particularly those of the late-wood that have thicker cell walls. In the more evolutionarily advanced wood of dicotyledonous angiosperms, a division of labour has occurred and *fibres* have specialised to provide support for the stem, whilst *vessel elements* specialise in the conduction of sap. Fibres are relatively long (typically 0.8–2.3 mm) cells with thick lignified cell walls and a very narrow cell lumen. While the secondary cell wall provides most of the fibres' structural strength, some fibres develop thin transverse walls known as septa across their cell lumen. These structures within *septate fibres* are analogous to cavity wall ties used in buildings. In addition to their structural role, some fibres (as in genera such as *Acer* and *Robinia*) also maintain their protoplasts (cell contents) in mature sapwood and may therefore be involved with storage of reserve materials in a similar way to the xylem parenchyma cells.

Living Cells in the Wood – Parenchyma

With the possible exception of a few living fibres, the radial and axial parenchyma make up the living component of xylem. A combination of radial and axial parenchyma forms a highly interconnected, three-dimensional lattice of living cells throughout the xylem and phloem, known as the symplast. Extensive symplastic connections between the parenchyma cells, in the form of simple pits and plasmodesmata, give trees the ability to store and redistribute materials throughout these living tissues (Spicer 2014). This extensive symplast is crucial for the movement of carbohydrates, nitrogenous compounds and water. Xylem parenchyma is also vital for the storage of carbohydrates (mainly in the form of starch) and its subsequent reallocation to fuel growth or other active responses within the tree. These include the defence of stems (Morris *et al.* 2016a) and the repair of transport networks following embolism (Brodersen and

(a) (b)

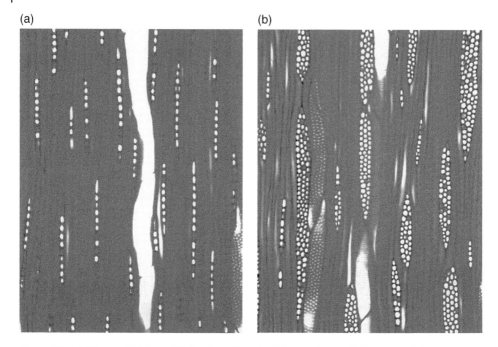

Figure 2.26 (a) Tangential view of *Salix alba* with uniseriate rays (one cell thick) containing heterocellular cells (upright cells on upper and lower margins); and (b) *Acer pseudoplatanus* with multiseriate rays (multiple cells thick) with homocellular cells (all procumbent cells). Further details on the definitions of these terms can be found in the text. *Source:* Schoch *et al.* (2004). Reproduced with permission of Wood anatomy of central European Species (www.woodanatomy.ch).

McElrone 2013). Additionally, xylem parenchyma is involved with the synthesis and storage of a range of compounds such as oils, phenols, tannins and terpenes.

Axial parenchyma is made up of axially orientated living cells that are highly diverse in their arrangement. They form connections with the radial parenchyma as well as with vessels and/or tracheids, which they may surround in partial or complete sheaths (paratracheal). Axial parenchyma may also be found embedded amongst the fibres with no contact with conducting elements, or, in species where it is abundant, in tangential bands running through the xylem (apotracheal).

Rays, predominantly made up of radial parenchyma, vary a great deal among different tree species, especially in their relative size, as they may be uniseriate (one cell thick), biseriate (two cells thick) or multiseriate (multiple cells thick, can be up to 30 cells wide) (Figure 2.26). Inevitably, it is much easier to see the large multiseriate rays of many *Quercus* spp. than it is to pick out the thread-like uniseriate rays of *Salix* spp. Rays also vary in height from one cell to hundreds of cells high.

Cells within the rays occur in different shapes and arrangements, which can be helpful in identifying wood: *procumbent* cells are arranged radially with their longest axis perpendicular to the stem axis; *upright* cells are arranged with their longest axis parallel to the stem axis; and *isodiametric* (square) cells have no definite orientation. As procumbent cells are elongated from the centre of the stem to the outside of the tree, they are designed to conduct materials radially within the stem. As demand for carbohydrates increases with stem age, procumbent cells become more profuse. By contrast, upright cells within the rays are usually associated with axial parenchyma and so have a vital role in integrating the two parenchyma systems into the elaborate three-dimensional network of living tissue. Rays made up exclusively of procumbent cells are called *homocellular* whilst rays comprised of different cell types are termed *heterocellular*. The lignified secondary cell walls also have a significant mechanical role within the stem, particularly in resisting shear failure and preventing delamination between tree rings.

As parenchyma possess an ability to re-differentiate at maturity, both axial and radial parenchyma cells provide a vital service of regenerating the cambium after damage.

Non-Living Cells in the Wood – Vessels, Tracheids and Fibres

Vessels and tracheids are the most highly specialised cells of the xylem, responsible for the conduction of water and other dissolved substances, such as nutrients and hormones. At functional maturity they are dead and have lignified cell walls with a variety of connections to other conducting cells. In tracheids, *pits* through the secondary cell walls of two adjacent tracheids allow water to move between them. In vessel elements, *perforation plates* (openings) at each end of the cell, and occasionally in the side-wall, connect adjacent cells; these may either be simple, with only one hole or perforation, or compound (multiperforate), if more than one perforation exists. Compound perforation plates can be described as scalariform, if the perforations are arranged parallel to each other; reticulate, if the perforations form a net-like pattern; or foraminate, if observed as a group of circular holes (Figure 2.27). Pits may also be found on the radial cell walls of vessel elements. A longitudinal series of vessel elements form *vessels* that may be up to 0.5 m or longer in some species.

In most angiosperm trees, fibres make up 30–80% of the xylem by volume (Spicer 2016). Fibres are specialised for mechanical support as they have well-developed secondary cell walls, a very narrow cell lumen and are relatively long, averaging between 1 and 2 mm. Two types of fibre are important in non-reaction wood and are distinguished by differences in their cell wall pitting: *fibre-tracheids* have bordered pits, whereas, *libriform-fibres* have simple pits (without borders). In the reaction wood of angiosperms (i.e. tension wood; see Chapter 3), a third type of fibre is found in some species. *Gelatinous fibres* develop a modified gelatinous cellulose-rich cell wall layer (G-layer). These specialised fibres are thought to be responsible for generating a tensile force that pulls the stem upward (Groover 2016).

The nature and composition of pits varies widely among species. Characteristically, a minute cavity (up to ~200 nm) is partially enveloped by the secondary cell wall border which has a narrow aperture leading to a wider pit chamber inside (Figure 2.28). These *bordered pits* maximise the structural support offered by the secondary cell wall, and also the surface area available to transfer water between tracheary elements (Carlquist 2001). However, the extent of the border can vary and lead to half-bordered and even simple pits (i.e. without the overarching portion of a secondary cell wall). At the base of the cavity, the two adjacent primary cell walls and the intervening middle lamella make up the pit membrane, typically composed of a closely woven mesh of

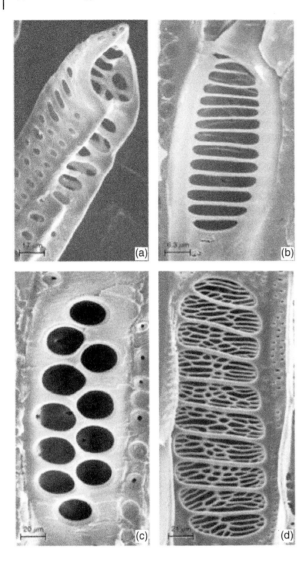

Figure 2.27 Scanning electron micrographs of perforation plates between vessel elements. (a) Simple perforation plates with a single large opening from *Pelargonium*; (b) the parallel bars of a scalariform perforation plate from *Rhododendron*; (c) foraminate perforation plate with circular perforations from *Ephedra*; (d) contiguous scalariform and reticulate perforation plates from *Knema furfuracea*. From Ohthani *et al.* (1992). *Source:* Evert (2006). Reproduced with permission of John Wiley and Sons.

cellulose and hemicellulose microfibrils held in a matrix of pectin. Most gymnosperms (and a very few angiosperms; see Jansen *et al.* 2007) have two distinct regions within the pit membrane: the *torus* (*pl.* tori) and the *margo*. The outer margo region is porous to allow the movement of water through the pit, while the central torus is thickened and lignified to create an impermeable plate that can be moved sideways to block the aperture and prevent the movement of gas and pathogens between xylem cells (closed pits are termed *aspirated*). In contrast to the torus-margo pits, most angiosperms have a uniform membrane that facilitates the low-resistance transfer of water between vessels (Choat *et al.* 2008). Usually, numerous pits occur between adjoining tracheary elements; few or no pits occur between tracheary elements and fibres; and half-bordered or simple pit-pairs occur between tracheary elements and parenchyma cells.

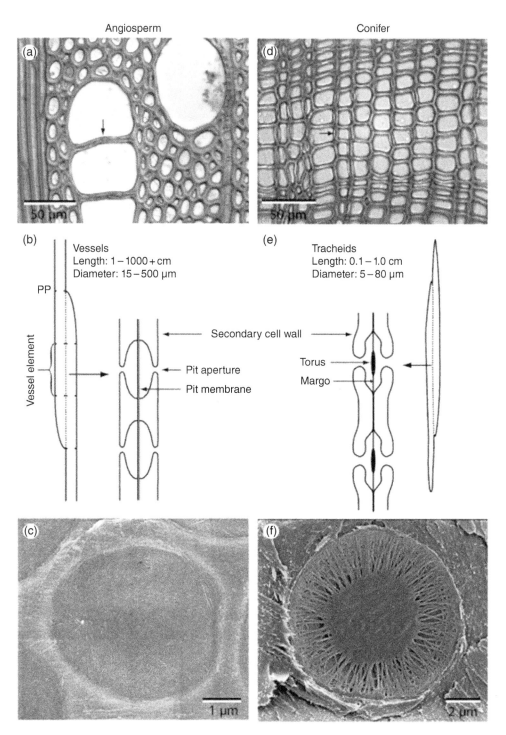

Figure 2.28 Structure of pits in dicotyledonous and gymnosperm wood. (a) Transverse section of dicotyledonous xylem tissue showing vessels connected through pitted walls. (b) Each vessel is made up of multiple vessel elements joined end-on-end through a perforation plate, but vessels are connected sideways through bordered pit-pairs with a pit membrane consisting of two primary cell walls and a middle lamella. (c) Electron microscope scan showing 'homogeneous' pit membrane of angiosperms, with a uniform deposition of microfibrils across the surface of the membrane. (d) Transverse section of typical gymnosperm xylem tissue made up of tracheids with bordered pits located in radial walls. (e) The architecture of bordered pits is similar to that of vessels although the pit membrane structure is different, (f) because it has a central thickening (torus) and very porous outer region (margo). *Source:* Choat (2008). Reproduced with permission of John Wiley and Sons.

Variation in Wood Structure

Gymnosperm Wood

From an evolutionary perspective, gymnosperm wood is less advanced than dicotyle-donous wood and has a rather homogeneous wood structure (homoxylous). It lacks vessels, and is dominated by tracheids that act both as conduits and to physically support the stem (Figure 2.29). Tracheids in most species are less than 5 mm in length but several members of the Araucariaceae have shown tracheid lengths of around 10 mm.

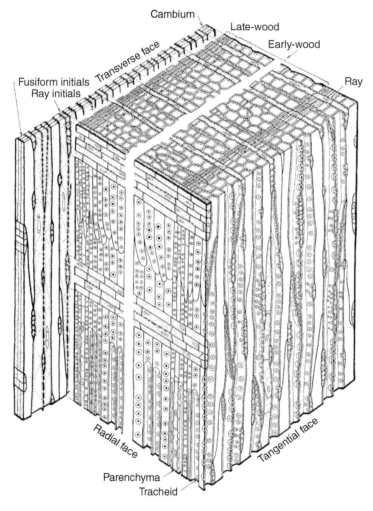

Figure 2.29 A block of gymnosperm wood, American arbor-vitae *Thuja occidentalis*. The centre of the tree is to the bottom right. The block shows one complete growth ring of early-wood and late-wood, and the cambium just lifted off the surface of the xylem. The axial system is composed of tracheids and some parenchyma cells. The rays only contain parenchyma cells. *Source:* Esau (1960). Reproduced with permission of John Wiley and Sons.

Tracheid diameters range from 10 to 65 µm; their tapered ends overlap and share numerous bordered pit-pairs of the torus-margo type, which create a low resistance to water movement.

In the early-wood (xylem produced early in the growth season), tracheids have a relatively large lumen and thinner cells walls to improve the water conduction of the cells. In the late-wood (xylem produced later in the growth season), the lumen is diminished in favour of much thicker cell walls that confer greater mechanical support. These late-wood tracheids provide most of the mechanical strength in gymnosperm wood.

While the majority of axial elements in gymnosperms are tracheids, axial parenchyma may be present. Where this is the case, axial parenchyma will make up less than 10% of the xylem (Plavcová and Jansen 2015). Most often, the axial parenchyma is restricted to the cells associated with resin ducts that can run along and across the xylem, forming a three-dimensional network capable of responding to pathogens or injury. Resin ducts are a constant feature of some gymnosperm trees (e.g. *Pinus, Pseudotsuga, Larix* and *Picea*) but in others (e.g. *Sequoia, Cedrus* and *Abies*) they are only produced in response to a traumatic event such as mechanical wounding or insect damage; then, sometimes only in the bark (e.g. *Abies* and *Cupressus*). Resin ducts are unknown in *Cupressus* xylem.

Generally, rays are made up of parenchyma cells; however, they may contain ray tracheids that closely resemble ray parenchyma, but have no living cell contents at maturity and have secondary walls with bordered pits. Some species contain both axial and radial resin ducts to enable the transport of resin throughout the sapwood.

Dicotyledonous Wood

Wood from the evolutionarily more advanced dicotyledonous plants displays much more variation in structure than gymnosperm wood (for further details see Schweingruber *et al.* 2011, 2013). It may contain vessels and/or, more rarely, tracheids; one or more fibre types, as well as axial and radial parenchyma of more than one kind (Figure 2.30). Further, cellular enlargement, particularly from vessel elements, displaces neighbouring cells causing quite a variable appearance across species.

One of the most striking sources of variation in dicotyledonous wood is the number and distribution of the vessels within the wood. On a cross-section of a tree, these vessels look like holes and so are referred to as *pores* by wood anatomists. The distribution of these vessels, which make up about 25% of the xylem, fall into three broad patterns. In *diffuse porous* wood, the vessels are distributed fairly uniformly across each growth ring (Figure 2.31b). *Semi-ring porous* wood has vessels in the early-wood that are three to five times larger in diameter than those in the late-wood. *Ring porous* wood (Figure 2.31a) has early-wood vessels that are six to more than ten times larger than those in the late-wood (Schweingruber *et al.* 2011). Globally, diffuse porous trees make up over 90% of all woody species but in northern hemisphere temperate regions this drops to just over 50%, with ring and semi-ring porous species making up the remainder in approximately equal proportion (Wheeler *et al.* 2007). For simplicity, as semi-ring and diffuse porous woods are variable even within individuals, it is normally adequate to classify trees as either

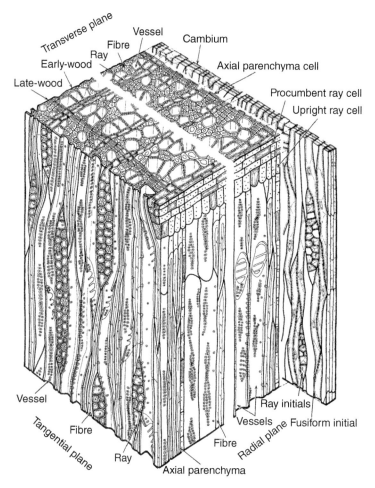

Figure 2.30 A block of wood from a dicotyledon, tulip tree *Liriodendron tulipifera*. The axial system consists of vessel members joined by scalariform perforation plates, fibre-tracheids and parenchyma strands. The centre of the tree is to the bottom left and, as in Figure 2.26, the cambium is shown lifted off the xylem. *Source:* Esau (1960). Reproduced with permission of John Wiley and Sons.

diffuse or ring porous (sometimes called microporous and cycloporous, respectively). Because most dicotyledonous trees are diffuse porous, it is likely that ring porous structure is a more recent innovation in the evolution of xylem.[4]

Mean vessel length varies from approximately 0.01 to 0.45 m. Maximum vessel length (possessed by relatively few vessels within the tree) is about five times the mean length, suggesting that individual vessels in some trees species can exceed 2 m (Jacobsen *et al.* 2012). In trees (but not in shrubs and vines), vessels tend to be longer in

---

4 Ring porous wood has developed across a wide range of plant families and in multiple species of the following genera: *Acacia, Ailanthus, Carya, Castanea, Castanopsis, Catalpa, Celtis, Cotinus, Dalbergia, Elaeagnus, Fraxinus, Gleditsia, Gymnocladus, Koelreuteria, Laburnum, Lagerstroemia, Malclura, Morus, Paulownia, Phellodendron, Pistacia, Platycarya, Prunus, Quercus, Robinia, Rhus, Sassafras, Senegalia, Ulmus, Vachellia,* and *Zelkova.* Based on information from the InsideWood database (2004–onwards, Wheeler 2011).

(a)   (b)

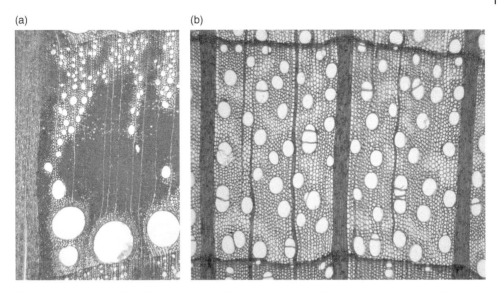

Figure 2.31 Cross-section of wood showing: (a) ring porous vessel arrangement in pedunculate oak *Quercus robur*; and (b) diffuse porous vessel wood in sugar maple *Acer saccharum*. In each case, the centre of the tree is below the diagram, the vertical dark lines are rays, and the horizontal lines are the breaks in growth caused by winter.

ring porous species (0.19 m) than with diffuse porous wood (0.05 m). Further, where large differences occur between the early-wood and late-wood, shorter vessels are found in the late-wood. There is also evidence that vessel length increases with stem diameter, with the longest vessels found in more mature stems, and shorter vessels found in younger (smaller diameter) stems.

Although we often think of trees as having either tracheids (gymnosperms) or vessels (dicotyledonous angiosperms), a number of trees have both. This combination offers better security to conduction because tracheids are much less vulnerable to embolism (the blocking of tracheary elements by gas bubbles). In good times, the vessels will be responsible for the vast majority of hydraulic conduction, but the tracheids provide a certain degree of resilience to total hydraulic failure by offering a subsidiary conducting system if times get tough. For this reason, having some tracheids to support vessels seems quite widespread in temperate and Mediterranean trees with examples of this type of xylem found in multiple species in a wide range of genera.[5]

The southern hemisphere family, Winteraceae, are botanical misfits in the sense that, whilst they are dicotyledonous, they are vesselless and only contain tracheids. Five genera (*Drimys, Pseudowintera, Takhtajania, Tasmannia* and *Zygogynum*) and approximately 65 species of trees and shrubs seem to have been exposed to selective pressure against vessels during their evolutionary past. The most likely impetus for this is the greatly improved hydraulic conductance of tracheid-based xylem after freeze–thaw

---

5 Vessels and tracheids can be found in a number of species in the following genera: *Alnus, Arbutus, Betula, Castanea, Celtis, Cercis, Corylus, Cotinus, Fagus, Ligustrum, Lithocarpus, Olea, Osmanthus, Pistacia, Prunus, Quercus, Rhamnus, Rhus, Sambucus, Ulmus* and *Zelkova*. Based on information from the InsideWood database (2004–onwards, Wheeler 2011).

events (see Trade-offs in Wood Design later in this chapter). This enables them cope with the cold nights in the wet temperate and tropical alpine habitats in which they occur (Feild *et al*. 2002). The small family of Trochodendraceae native to Southeast Asia are another rare angiosperm group with vesselless wood (Carlquist 2012).

Parenchyma fractions in dicotyledonous wood can vary a great deal across species. In temperate angiosperm trees, on average, parenchyma makes up 26% of the xylem, but a wide range of 7–62% has been observed. Tropical angiosperm trees have, on average, 38% parenchyma in their xylem with a range of 6–92%. Although exceptions do occur, in both groups, radial parenchyma tends to make up the greater portion of xylem parenchyma (Morris *et al*. 2016b). Analysis of xylem from different climates shows that mean annual temperature drives a change in parenchyma fractions, with an increase in axial parenchyma being particularly apparent as the temperature increases (Morris *et al*. 2016b). The reasons for this positive shift in parenchyma fractions with temperature warrant further investigation.

Maintaining a high proportion of living cells within the xylem is energetically expensive so there must be a substantive advantage to the tree where this occurs. Given the variety of functions associated with xylem parenchyma, the specific advantages of developing high parenchyma fractions are likely to vary somewhat by climate and habitat. In temperate environments, long-lived trees, such as oaks *Quercus* spp., have relatively high proportions of parenchyma: this aids tree longevity by enhancing xylem defensive capabilities and increasing the ability to store carbohydrates within the xylem. These characteristics will, no doubt, be advantageous in tropical climates as well.

In other cases, the increase in xylem parenchyma appears to be much more closely associated with survival during periods of water shortage. For example, baobabs *Adansonia* spp. (see Figure 10.12), have about three-quarters of their xylem made up of parenchyma (mostly axial parenchyma) (Chapotin *et al*. 2006). This enables these specialist trees to store large volumes of water in their stems. During periods of water shortage, a process known as *hydraulic capacitance* releases the stored water into the transpiration stream to help maintain photosynthesis. The disadvantage of this survival strategy is that the high parenchyma fraction reduces the proportion of fibres in the xylem; in turn, this reduces the stiffness of the stem. To compensate, the girth of baobabs has had to increase to ensure their stability: some specimens have even been known to exceed 10 m in diameter.

Although it is not widely considered by those managing landscape trees, knowledge of xylem anatomy can greatly aid our understanding of how trees respond to their environment and tree management practices. It can be used to help characterise the way a tree species responds to stress, gives insight into the strength of the wood and helps determine the tree's response to wounding, pathogens and pests. Therefore, understanding the functional significance of variation in wood structure is hugely valuable for those managing trees.

Sapwood and Heartwood

Dissection of a mature tree stem often reveals a contrast between the younger *sapwood* in the outer portion of the wood and the older, age-altered *heartwood* in the centre of the stem (Figure 2.32). Whilst all the component cells of the xylem are present in both,

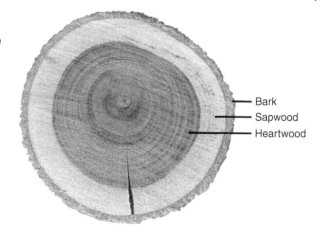

Figure 2.32 Dark heartwood surrounded by lighter sapwood in common laburnum *Laburnum anagyroides*.

Bark
Sapwood
Heartwood

it is the presence of living parenchyma cells that defines sapwood. These living cells typically make up somewhere between 5% and 40% of the sapwood by volume, and provide the capacity to store and synthesise materials needed by the tree.

Perhaps unsurprisingly, there is a huge variation in the width of sapwood between species and in the nature of its coordinated transition to heartwood. For example, just one to two growth rings make up the sapwood in northern catalpa *Catalpa speciosa*, while black tupelo *Nyssa sylvatica* has been described with 100 growth rings of sapwood. A number of conifers have been found to contain well in excess of that (Hillis 1987). A complicating factor is that sapwood and heartwood are not always different in colour, and so can be difficult to distinguish. In these cases, the presence or absence of live parenchyma cells gives useful guidance. As a general rule, gymnosperms show a gradual loss of living parenchyma, and so a gradual transition from sapwood to heartwood. This is also true of angiosperms that produce an irregular form of heartwood known as *ripewood*. However, many angiosperms produce *regular heartwood* that shows quite an abrupt transition from sapwood to heartwood. By assessing the proportion of living parenchyma cells at different depths within the xylem, it is possible to determine how sapwood varies within and between species. Figure 2.33 shows that variation within a number of gymnosperm and angiosperm species can be high. This is likely to be because of a range of genetic and environmental differences but the mechanisms are not yet fully understood.

Sapwood and Water Movement

In theory, all of the sapwood should be capable of conducting water up the tree but, in reality, conduction is unlikely to occur across the full width of sapwood because inner, and thus older, parts of the sapwood are likely to progressively accumulate embolisms (gas blockages). For this reason, it is worth making the distinction between *sapwood* that contains living parenchyma and *conducting sapwood* that can transport water. Many species rely on relatively few rings for their entire hydraulic needs. This is particularly apparent in ring porous species that often rely on only one ring of xylem to transport sap to the entire crown despite having sapwood of 40 to 50 rings (for example, see *Fraxinus* in Figure 2.33). Diffuse porous trees tend to have a greater proportion of sapwood involved in conduction,

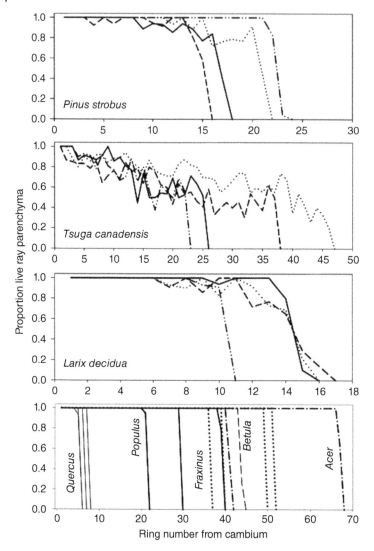

Figure 2.33 The width of sapwood in different species of gymnosperms and angiosperms determined by the proportion of rays that contain living cells across a radial transect through the sapwood. There can be much variation among individuals in the same species. Each line represents an individual (i.e. four individuals are shown for each gymnosperm). Gymnosperms shown are eastern white pine *Pinus strobus*, eastern hemlock *Tsuga canadensis* and European larch *Larix decidua*. Angiosperms shown are red oak *Quercus rubra*, bigtooth aspen *Populus grandidentata*, white ash *Fraxinus americana*, paper birch *Betula papyrifera* and red maple *Acer rubrum* (each with a designated line pattern, e.g. dotted for *Fraxinus* and dash-dotted for *Acer*). *Source:* Spicer (2005). Reproduced with permission of Elsevier.

as shown by the contrast between white ash *Fraxinus americana* and red maple *Acer rubrum* in Figure 2.34, where the ash only uses the outer 1 cm of sapwood, unlike the maple which is still conducting some water across 10 cm of sapwood. Gymnosperms, by contrast, can keep their very safe tracheids functional for more than 40 years. However,

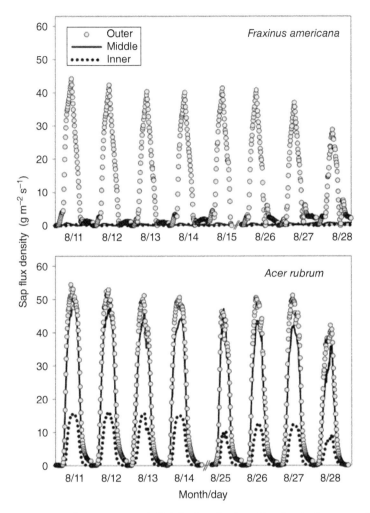

Figure 2.34 Sap movement (officially described as a flux density in grams of water conducted per square metre of cross-section, per second) measured at different radial depths in two angiosperms with wide sapwood: red maple *Acer rubrum* (11 cm sapwood width, 35 years old) and white ash *Fraxinus americana* (9 cm sapwood width, 30 years). The *inner* depth in each species represents sapwood within 1 cm of the heartwood–sapwood boundary; *outer* depth represents the outermost 1 cm of sapwood and *middle* is equidistant between the two. *Source:* Spicer (2005). Reproduced with permission of Elsevier.

comparison of only a few species probably obscures the variation between species and between individuals, because of the range of environmental conditions (e.g. drought and freezing temperatures) that can influence water transport.

Heartwood

Xylem that does not conduct water and contains few or no living cells can be called heartwood. Although this largely dead, central core of age-altered wood may have no role in the daily working of the tree, it is nonetheless a vital component in the development and maturation of the tree. It is also commercially important in many species because its aesthetic properties and increased durability means that it makes up the majority of wood sold commercially.

A number of different heartwood types have been recognised. In *regular heartwood* species, heartwood development keeps pace with that of sapwood, resulting in a uniform band of sapwood around a heartwood core that shows a consistent colour, often darker than the sapwood (Figure 2.32). It is generally quite durable to decay as it is chemically modified as it forms. In *irregular heartwood* species (*syn.* facultative heartwoods), as in some *Acer, Betula, Carpinus, Fagus, Fraxinus* and *Olea* species, the heartwood can be eccentrically placed, often has a very irregular shape (e.g. cloud-like or stellate) and frequently contains bands of varying colour intensity. A number of sub-categories of irregular heartwood have also been described: for example, *ripewood* is lighter in colour and has a lower moisture content than the sapwood, as frequently seen in European beech *Fagus sylvatica*. This is not chemically modified to any great extent so is much less durable than regular heartwood and tends to decay more rapidly, particularly if it becomes well aerated (Lonsdale 2013a). In some species, the modifications to the heartwood are so minor that it is very difficult to distinguish from sapwood. These can be termed *delayed heartwood* species, as many only seem capable of producing heartwood in response to wounding or in response to other kinds of damage, such as that caused by extremes of temperature. Where this type of response is seen, it should be termed *false heartwood* as it is frequently discontinuous and does not necessarily follow growth rings.

Heartwood, in all its forms, is a fundamental feature of mature trees as maintaining a three-dimensional living network of parenchyma over decades (centuries even) is energetically expensive. Good economics suggests that once sufficient new xylem is produced to maintain water conduction and to store carbohydrates, then the maintenance of older cells only serves to divert potential resources away from current growth, reproductive and defensive demands. Hence, we see the senescence of inner (older) sapwood to heartwood: a 'shedding' of internal wood that is no longer required for daily life, analogous to the loss of branches that no longer positively contribute to carbon gain. This process can be fairly abrupt in regular heartwood species but can be quite gradual in species that develop an irregular form of heartwood. Further, the transformation that occurs in heartwood formation implies an intentional transition of programmed cell death; the old cells do not just wither and die: they are intentionally killed. Whilst heartwood no longer functions to conduct water, store or synthesise materials, it does still have a structural role – particularly in those species with narrow sapwood – as the greatest physical loads are borne by wood in the outer part of the stem.

The formation of heartwood occurs in a transition zone between the sapwood and heartwood. In regular heartwood species, secondary metabolites (chemical compounds distinct from the products of primary metabolism, occasionally referred to as *heartwood extractives*) such as phenolic compounds, are synthesised in the parenchyma, often filling the cell lumen in addition to being deposited in the walls of adjacent cells. It is the addition of these compounds that gives some species' heartwood its distinctive colour and smell, making it much more resilient to decay and desirable for wood-workers. In gymnosperms, heartwood also has a lower moisture content than

(a) (b)

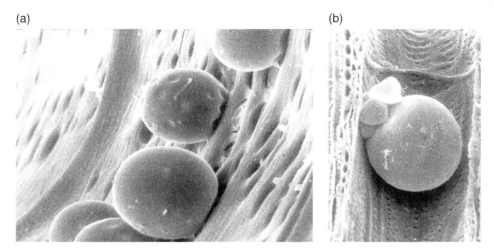

Figure 2.35 Scanning electron micrographs of *Quercus* xylem. (a) Tyloses are visable as balloon-like outgrowths from parenchyma cells that extend into the vessels via pits. (b) After tylosis formation is complete, the cell lumen is fully blocked. *Source:* Schwarze *et al.* (2000). Reproduced with permission of Springer.

sapwood but in angiosperms, the moisture content is highly variable and species dependent. Typically, heartwood begins formation in summer; reaches a maximum in early autumn (fall); and ceases during winter (Kampe and Magel 2013).

As vessel elements and tracheids cease to carry water, they become blocked through a number of different mechanisms. Pits become aspirated and/or have secondary metabolites deposited within their membranes, effectively sealing them. In some species, *gels* (often referred to as gums) formed in the axial or radial parenchyma are secreted into the lumen of vessels via pits. In others, *tyloses* (*sing.* tylosis) are formed primarily in radial parenchyma by an outgrowth of the parenchyma cell that protrudes into the vessel lumen (Figure 2.35). For most vessels, multiple tyloses will need to combine to form an effective blockage. Whether a species uses tylosis or gels seems to be governed by the size of the vessel and its pits. Species with wider vessels and larger diameter pits are more likely to have tyloses than species with narrower vessels and small diameter pits (Bonsen and Kučera 1990; De Micco *et al.* 2016). However, such is the variation in xylem that many exceptions occur to this trend. In all, tyloses are common in about 17% of the world's woods (Wheeler *et al.* 2007) and are known to form in a number of the species in the temperate world.[6] Gum and resin canals found in some species may similarly be blocked; in this case by tylosoids that form from the epithelial cells lining the canal, rather than external parenchyma as in tyloses.

---

6 Tyloses can be found in a number of species from the genera: *Carya, Castanea, Castanopsis, Catalpa, Celtis, Cercidiphyllum, Corymbia, Eucalyptus, Fagus, Ficus, Fraxinus, Juglans, Lithocarpus, Magnolia, Morus, Nothofagus, Paulownia, Pistacia, Platanus, Quercus, Rhus, Robinia, Sambucus, Sassafras* and *Ulmus*. Based on information from the InsideWood database (2004–onwards, Wheeler 2011).

Mechanical injury, as a result of extreme weather or pruning, can be serious for the tree as it usually results in the loss of leaf area and some level of dysfunction in the stem. Small pruning wounds on branches that have not yet developed any heartwood are always preferable to larger pruning wounds. Although even small wounds can cause disruption to water transport (hydraulic dysfunction) as a result of embolism, the relatively high proportion of living cells associated with the wound makes some form of active defence possible across the cut surface. It also reduces the wound area that needs to be occluded before it is sealed by new growth. In contrast, pruning wounds that expose heartwood (as well as sapwood) expose a larger fraction of the xylem to hydraulic dysfunction and limit the wound response because few, if any, living cells occur in the heartwood. Thus, the tree is much more reliant on its inherent resilience to decay (see Chapter 9), rather than active responses. Consequently, species with ripewood (e.g. hornbeam *Carpinus betulus* and beech *Fagus sylvatica*) are particularly vulnerable to large wounds as their heartwood is not durable (Lonsdale 2013a). Larger pruning wounds also require much longer to fully occlude so remain potential sites for decay fungi colonisation for extended periods. These factors should be taken into account when deciding on pruning points in the crown, particularly on mature and old trees. It is always preferable (from a tree health perspective) to have a series of smaller branches removed rather than a single larger branch (see also Tree Pruning in Chapter 3).

Trade-offs in Wood Design

The tree habit has evolved in a vast number of different species to allow greater capture of light. Woody stems provided the structural support necessary to arrange leaves above neighbouring plants and ultimately make the tree life-form more competitive. Consequently, trees have become the dominant form of vegetation in most terrestrial biomes where environmental conditions allow.

The woody stem has a number of functions: it supports the leaves so they receive light; it conducts water up to these leaves; and it stores a variety of compounds such as carbohydrates, defensive compounds and water. How the woody skeleton of the tree develops is driven by three main factors: competitive ability, resistance to stress (such as drought) and resistance to disturbance caused by wind, snow and ice. Each woody stem represents a pragmatic compromise between these major driving forces that can be visualised by the 'wood economics spectrum' trade-off triangle (Figure 2.36).

Competitiveness is at least partly governed by growth rate, so a highly efficient system of water transport (conductive efficiency in Figure 2.36) is a prerequisite to ensure sap is rapidly delivered where and when it is needed. Maximising the lumen area of tracheary elements and minimising resistance to sap flow can achieve this. However, there is little value in having a very efficient hydraulic system if a minor disturbance (e.g. light wind) causes stem failure. Equally, a tree that sacrifices all hydraulic efficiency to maintain structural integrity is likely to become over-topped and lose the battle to capture light. A balance must be struck between the ability to conduct sap and wood strength, as shown on the right-hand side of Figure 2.36.

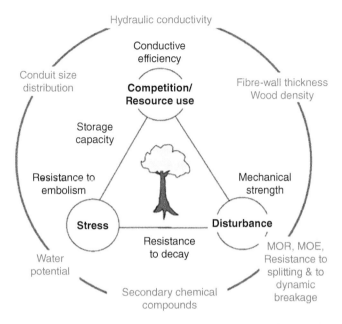

Figure 2.36 The trade-offs between the major ecological driving forces (competitive ability, resistance to stress and disturbance). Important wood properties to cope with these are reported around the triangle and the outer circle identifies key wood traits associated with these wood properties. MOR (modulus of rupture) and MOE (modulus of elasticity) are widely used measures of wood mechanical strength. *Source:* Chave *et al.* (2009). Reproduced with permission of John Wiley and Sons.

Trees in more stressful environments, such as those growing in dry or saline areas, where water is in short supply, will have a greater emphasis on maintaining their hydraulic conductivity and the ability to resist xylem dysfunction from gas blockage (embolism). Trees in very windy areas or on unstable slopes would in turn put more emphasis on structural strength. Increased resistance to stress implies a trade-off, leading to slower growth. Species have resolved these tensions over millennia, through adaptations that have led to their survival in a way that maximises their fitness for the environmental conditions faced throughout their history. By understanding the nature of these trade-offs, it is possible to see some of the key reasons underlying a species' ecological performance (i.e. how effective it is at competing within a particular environment). Further, these observations help anticipate the performance of a species to future climate scenarios and/or challenges presented by managed landscapes.

Trade-offs and the Movement of Water

Vessel structure and configuration exert a fundamental influence on the hydraulic performance of xylem, and, by implication, the entire tree.

Hydraulic conductivity is greatly influenced by perforation plates and pits. Although the holes through perforation plates can appear quite large (look back at Figure 2.27), the circular or bar-shaped holes are bridged by cellulose strands, leaving holes <0.02 μm wide. These are present to help hold back gas bubbles from jumping between vessel elements (and will do so down to −15 MPa), but will also slow water flow. In gymnosperms, where tracheids tend to be 1.5–5 mm long and are joined by pits, water will have to travel through >20 000 pits in a 100-m tall Douglas fir *Pseudotsuga menziesii*. It is perhaps not surprising that these pits account for >50% of total xylem hydraulic resistance (Choat *et al.* 2008).

While it is clear that compound perforation plates do increase resistance to flow (think about how a mesh filter slows the drainage of a sink), they also have other qualities. Bubbles are formed in winter when gas is released from solution by freezing (see Freezing-Induced Cavitation). These bubbles become trapped by the minute bars of the perforation plates, preventing them from rising up the vessels to form one large bubble that would be difficult to re-dissolve in the spring. They are also likely to add a certain degree of rigidity to the end wall of the vessel element, helping to prevent collapse under negative tension. Clearly, the resistance offered by a compound plate is reduced as the number of bars is reduced towards a simple perforation plate. For this reason, some species (e.g. of the genus *Styrax*) have simple plates in the early-wood to help water movement, but develop scalariform plates in the late-wood to strengthen the wood and improve the security of the transport system.

Vessel length has an important influence on a number of aspects of xylem performance. As perforation plates represent resistance in the hydraulic pathway, species with long vessels have lower resistance to sap-flow than species with shorter vessels (all other factors being equal). However, in shorter vessel elements, any hydraulic dysfunction caused by breaking the water column (cavitation) is more localised and more readily reversed (Holbrook and Zweiniecki 1999).

Even what would appear to be extremely minor modifications in xylem anatomy have been shown to influence performance. For example, in some species, tiny helical ridges or grooves (helical sculpturing) appear on the lumen surface of the innermost of the secondary cell walls (S_3) to increase its surface area. By increasing surface wettability, this helps to both prevent and remove gas bubbles caused by cavitation which in turn helps to explain why helical sculpturing on the S_3 layer is more common in colder and drier habitats (Carlquist 2012).

Above all else, speed of water movement is governed by the size of lumen in the xylem cells. In the same way that wider pipes move water around more quickly than narrow pipes, on a microscopic scale the diameter of the vessel or tracheid lumen makes a huge difference to the conductive properties of xylem. The Hagen–Poiseuille equation states that the hydraulic conductivity of cylindrical tubes increases with the radius to the fourth power (Reiner 1960), so doubling the width of a tube allows it to carry 16 times the amount of water. This is primarily a result of reducing friction between the water and the tube wall – a wider tube touches less of the water so there is less friction, and faster flow. As a result, any slight increase in lumen diameter through evolution has had a profound impact on the number of tracheary elements and amount of wood needed to conduct water at a given rate. It also affects speed: in the narrow tracheids in European larch *Larix decidua* sap moves at 10 cm per hour while in the wide vessels in the ring porous *Fraxinus americana* it moves at 125 cm per hour.

Following this relationship, one vessel 40 μm wide is as conductive as 16 vessels at 20 μm or 256 vessels at 10 μm wide (Figure 2.37). Further, if all three of these vessels were found in the same area of wood, the smallest one (10 μm) would carry 0.4% of the water, the middle one (20 μm) 5.9% and the largest one (40 μm) 93.8% of the water

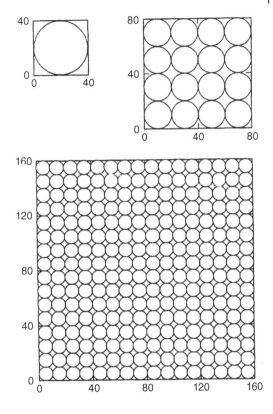

Figure 2.37 The effect of the Hagen–Poiseuille equation on the hydraulic conductivity of cylindrical channels. Water flow is much faster through a wider tube so each of these three groups of tubes would conduct the same amount of water per unit of time. *Source:* Tyree and Zimmermann (2002). Reproduced with permission of Springer.

(Tyree and Zimmermann 2002). In real life, the internal structure of the cell walls with their numerous pits, combined with the resistance offered by perforation plates or pits, means that this relationship is not strictly true inside a tree; but the advantage of increasing the lumen diameter is still substantial. To illustrate this further, consider two 'model trees' that are required to shift the same volume of water through their stems but have widely contrasting vessel size. The tree with relatively narrow vessels (50 μm diameter) would require 625 times as many vessels as a tree with large vessels (250 μm). When this is coupled with the potential difference in vessel length, the difference in the number of vessels required to shift the same volume of water (sap) can vary by several orders of magnitude. However, in the event that the conduit is damaged or becomes dysfunctional, the significance of the lost vessel to the hydraulic conductivity of the stem may be orders of magnitude greater for trees with large vessels than for those with narrow vessels. Furthermore, wider tubes are also more at risk of being blocked (embolised). A trade-off between hydraulic efficiency and safety must be made.

To understand the nature of this efficiency–safety trade-off further, it is important to consider the key mechanisms of a process known as cavitation, where conduits become gas-filled (embolised). Cavitation may occur as a result of three circumstances: freeze–thaw cycles, water deficits (drought-induced) or physical injury to the xylem. The embolised xylem is still important in structurally supporting the tree, but every embolised tube reduces the crucial hydraulic conductivity between the roots and the crown of the tree.

Freezing-Induced Cavitation

When sap in the xylem freezes, any gases that are ordinarily held in solution come out and form bubbles. As the sap thaws and is put under negative pressure when the tree begins pulling water up the xylem, these bubbles can create an embolism. Small bubbles can re-dissolve before the negative pressure gets too large but large bubbles can lead to permanent embolism. The likelihood of this occurring is thus closely related to the conduit diameter (Davis *et al.* 1999; Pitterman and Sperry 2003): as the conduit diameter increases, so too does the potential size of the bubbles that can form within it and therefore the likelihood of embolism upon thawing.

This relationship between conduit size and susceptibility to freeze–thaw induced embolism was elegantly demonstrated by Davis *et al.* (1999), who froze stem segments from 12 species with a wide range of conduit diameters. Negative pressure (−0.5 MPa[7]) was produced by centrifugal force in newly thawed stems. Those species with a mean conduit diameter <30 μm showed almost no loss of conductivity, while those species with mean conduit diameters >40 μm lost most of their conductivity as a result of the freezing event (Figure 2.38).

As narrow conduits allow species to avoid any significant loss of conductivity as a result of freezing-induced embolism, they occur more often in those species naturally

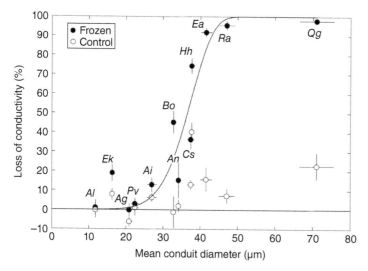

Figure 2.38 Percentage loss of conductivity for frozen and thawed (•) and control (o) stems plotted against mean conduit diameter. Stems were spun in a centrifuge to simulate a moderately negative pressure of −0.5 MPa (Ag, *Acer grandidentatum*; Ai, *Alnus incana*; Al, *Abies lasiocarpa*; An, *Acer negundo*; Bo, *Betula occidentalis*; Cs, *Cornus sericea*; Ea, *Elaeagnus angustifolia*; Ek, *Euonymus kiautschovicus*; Hh, *Hedera helix*; Pv, *Prunus virginiana*; Qg, *Quercus gambelii*; Ra, *Rhus aromatica*). *Source:* Davis *et al.* (1999). Reproduced with permission of the Botanical Society of America.

---

7 Pressure is universally measured in pascals (Pa) and in living things, millions of pascals (MPa) is most common. 0.1 MPa is equivalent to 1 bar or 1 atmospheric pressure. Pressure normally has a plus sign, so if we are dealing with suction (which goes below atmospheric pressure) then it is given a negative sign. The more negative the number, the more suction is being applied, so −1 MPa is a greater suction than −0.5 MPa.

found in cold regions and at high elevation (Sperry *et al.* 1994). Indeed, conifers, with their narrow tracheids, have remained highly competitive in these challenging environments despite the dominance of angiosperms in many of the world's more favourable environments. Species with larger vessels, such as those found in the early-wood of ring porous species (see *Elaeagnus angustifolia*, *Quercus gambelii* and *Rhus aromatica*, in Figure 2.38), lose almost all of their conductivity as these vessels become gas-filled during the first freeze–thaw cycle: only the narrow late-wood vessels (and/or tracheids depending on the species) retain their conductive ability. As a consequence, ring porous trees must grow new early-wood vessels before leaves appear the following spring, and most of the water used by the crown is conducted up just this one ring of new wood. New xylem must be produced from carbohydrates stored from the previous growth season. This causes a pronounced decline in stored carbohydrates in stems of ring porous species in early spring (Barbaroux *et al.* 2003; Hoch *et al.* 2003), and tends to reduce the duration of the growing season as they expand their leaves later in the year than their diffuse porous neighbours. A combined effect of this fundamental trade-off between hydraulic efficiency and the preservation of hydraulic function through freezing events is that vessel diameters tend to be greater in temperate regions that do not experience very low temperatures (Baas *et al.* 2004).

Species with narrower vessels carry fewer embolised vessels into spring and, by using positive stem (and/or root) pressure, can refill embolised vessels. The advantages of refilling conduits are that hydraulic conduction can resume prior to the development of new xylem, growth can occur over a longer period and stored carbohydrates do not need to be used to develop new functional xylem. Trees that have the ability to produce positive pressure in the xylem (e.g. many species in the genera *Acer*, *Alnus*, *Betula* and *Juglans*) tend to 'bleed' if they are wounded (pruned) during early spring. While, in most cases, this is not a real problem for the tree, it can cause rather unsightly sap-runs down the stem(s), especially when these are colonised by various microorganisms that turn the nutritious sap black.

Drought-Induced Cavitation

Where the demand for water by the crown exceeds the water supply from the roots, tension within the water column becomes more and more negative (a greater suction). When this tension passes a particular threshold, a gas bubble from outside the water-filled conduit is drawn in and *gas-seeding*[8] occurs. This causes the water column to break, creating a gas-filled embolism. As the major entry route for gases through the conduit wall is via the pits, it is the construction of the pit membranes, the shape of the pit complex and the pits' distribution across the conduit cell wall that most strongly determines a species' vulnerability to cavitation via this mechanism (Hacke *et al.* 2006; Sperry *et al.* 2006; Choat *et al.* 2008). For example, in angiosperms, the thickness of the pit membrane varies enormously (70–1900 nm), as does the size of the pores within the membrane (10–225 nm); both these affect the species' vulnerability to gas-seeding (Lens *et al.* 2013) as thinner membranes and bigger holes are less resistant at keeping gas out.

---

8 This is sometimes referred to as air-seeding but, because it is often made up mainly of carbon dioxide rather than air which is comprised of a number of different gases, it is more accurate to refer to the process as gas-seeding.

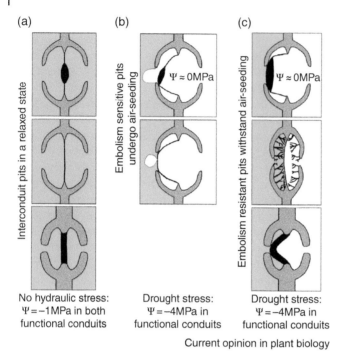

(a)   (b)   (c)

Interconduit pits in a relaxed state

Embolism sensitive pits undergo air-seeding

Embolism resistant pits withstand air-seeding

$\Psi \approx 0MPa$

$\Psi \approx 0MPa$

No hydraulic stress: $\Psi = -1MPa$ in both functional conduits

Drought stress: $\Psi = -4MPa$ in functional conduits

Drought stress: $\Psi = -4MPa$ in functional conduits

Current opinion in plant biology

Figure 2.39 Adaptations within the pits between xylem cells that help regulate gas-seeding and embolism. (a) The top row shows the bordered pits of gymnosperms with their torus-margo pit membranes. The middle and bottom rows show the more homogeneous pit membranes in angiosperms. The pit membranes are in a relaxed state facing no hydraulic stress as the suction (shown as −1 MPa) is equal on both sides. (b) A prolonged period of drought increases the pressure difference between the water-filled conduit on the left and the air-filled, embolised conduit on the right, causing the pit membranes to deflect. At a critical pressure difference, gas-seeding occurs with air being pulled through into the water-filled conduit. (c) However, various adaptations in pit design can prevent gas-seeding at the same pressure difference as in (b): (top) an increase in the size of the central torus in comparison to the pit aperture diameters in gymnosperms, creating a better seal; (middle) the growth of *vestures* inside the pit which limit how far the membrane can be stretched; and (bottom) thicker pit membranes with reduced pores in angiosperms. *Source:* Lens *et al.* (2013). Reproduced with permission of Elsevier.

Some species, such as the Japanese pagoda tree *Styphnolobium japonicum* (formerly *Sophora japonica*), possess outgrowths from the secondary cell wall to the pit membrane. These 'vestured pits' (Figure 2.39c) have been shown to reduce gas-seeding by limiting the defection of the pit membrane under tension (Choat *et al.* 2004). Other adaptations acting to reduce gas-seeding are also shown in Figure 2.39. These may act to impede water flow, but variations in pit structure represent important trade-offs between the efficiency of sap conduction and the safety of the hydraulic pathway.

Generally, evolutionary change over time has resulted in the cavitation-resistance of a species being well adjusted to the typical levels of water deficit that trees experience within their natural habitat. A standard measure for quantifying the resistance of a species to air-seeding is to calculate the water potential at which 50% loss of conductivity occurs (Ψ_{50}[9]). Huge variation in Ψ_{50} has been found in trees from very low resistance (−0.07 MPa in Euphrates poplar *Populus euphratica*; Hukin *et al.* 2005) to very high resistance (−18.8 MPa in mallee pine *Callitris tuberculata* from western Australia; Larter *et al.* 2015). The fact that *C. tuberculata* can operate at such low water potentials is remarkable because it is likely that some tracheids are functioning at the absolute physical limit for water transport during drought (Larter *et al.* 2015). Given the range of habitats that woody plants grow in it is perhaps not surprising that such large differences in species-response to declining water availability have been found. However, a global analysis (Choat *et al.* 2012) revealed that 70% of woody species (trees, shrubs and vines) operate within a rather narrow safety margin (<1 MPa between the minimum water potential observed in the field (Ψ_{min}) and Ψ_{50}). This suggests that trees in more moist regions are just as much at risk from a drying climate as trees that currently grow in drier areas.

Vulnerability curves describe the relationship between the drying of stems (technically the declining stem water potential), usually a result of declining amounts of soil water, and the loss in conductivity within the stem resulting from embolism (Figure 2.40). These curves are very useful in showing which species rapidly lose conductivity over a narrow range of water potential (low safety factor) and those species that lose conductivity over a much wider range of soil moisture (high safety factor). Clearly, those with a high safety factor are less at risk from cavitation during periods of water shortage. Even in closely related species, the vulnerability curve can differ widely across a range of habitats, from mesic (moist) to xeric (dry). For example, in European oaks, pedunculate oak *Quercus robur* and the Pyrenean oak *Quercus pyrenaica* show very steep vulnerability curves with a rapid loss of conductivity over a narrow range of water potentials; whereas, the kermes oak *Quercus coccifera* and holm oak *Quercus ilex* show a much more gradual loss of conductivity. By comparison, downy oak *Quercus pubescens*, Portuguese oak *Quercus faginea* and cork oak *Q. suber* were intermediate in their response (Figure 2.40a). Perhaps not surprisingly, the safety of their hydraulic system was closely related to the duration of the dry period (Figure 2.40b).

The construction of xylem can therefore reveal a great deal about the strategies used to cope with a particular environmental stress. Ring porous species are specialists in seasonal climates: they have highly efficient early-wood vessels which makes them very competitive during late spring and summer, but the loss of conductivity should cavitation occur can be substantial. This is so likely during winter months with multiple freeze–thaw events that the entire hydraulic infrastructure requires a re-build (a new ring of early-wood) before leaves can emerge in spring. Diffuse porous species that generally display narrower, more evenly distributed vessels operate with greater resilience to freezing events, giving them an advantage earlier in the year as they have a functioning hydraulic system immediately after winter that can facilitate leaf expansion earlier in spring. This extends the growth season and reduces the significance of

---

9 The Greek letter psi (Ψ) is used to represent *water potential*. Strictly speaking, this is a measure of the potential energy of water but it is easier to think of it as a measure of suction. It is also used to measure the degree of water stress a plant is experiencing.

(a)

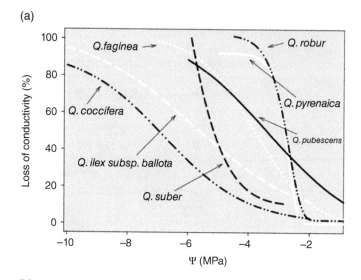

(b)

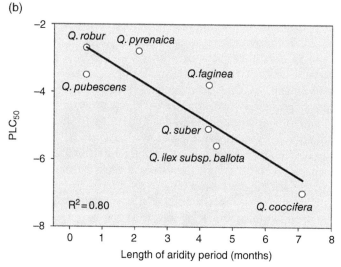

Figure 2.40 Vulnerability to drought-induced cavitation in several European oak (*Quercus*) species from different habitats. (a) Loss of conductivity is plotted against water potential (Ψ); and (b) the relationship between the length of the dry (aridity) period and the water potential that gives a 50% loss of hydraulic conductivity (PLC$_{50}$). (a) Redrawn from Tognetti *et al.* (1998) for Q. *pubescens*; Tyree and Cochard (1996) for Q. *robur* and Q. *suber*; Vilagrosa *et al.* (2003) for Q. *coccifera*, Corcuera *et al.* (2005) for Q. *ilex* subsp. *ballota*; Esteso-Martínez *et al.* (2006) for Q. *faginea*; and Corcuera *et al.* (2006) for Q. *pyrenaica*. *Source:* Vilagrosa *et al.* (2012). Reproduced with permission of Springer.

embolism in any one vessel as it is responsible for a lesser proportion of the overall stem hydraulic conductance. Gymnosperms have xylem that is very resilient to freezing events because of their narrow tracheids. This has given them a distinct advantage in cold regions, and is one of the reasons that conifers are still so dominant in the boreal and alpine environments.

The hydraulic network of vessels and/or tracheids may be either: well integrated, so that water may move around the whole tree; or sectored, where sapflow from any one portion of the root system is restricted to discrete 'sectors' of the xylem. It is important to note that *vascular sectorality* is really a continuous spectrum from highly integrated to highly sectored, and the relative position of each species is determined by a suite of traits such as vessel grouping (tight groups tend to keep water within themselves), vessel dimensions, inter-vessel pitting (the more pitting the more water will move between tubes) and the tortuosity of the axial pathway (the more twisted the tubes, the greater the number of other tubes they are likely to be linked to by pits). This hydraulic network is ecologically significant as it strongly influences the sap's pathway up the stem, and ultimately determines whether a discrete portion of the root system supplies a specific part of the crown or a wider proportion of it. Sectored species, including oaks (*Quercus* spp.) and ashes (*Fraxinus* spp.), may do best in temporally variable environments, whereas integrated species may be more competitive in spatially variable but temporally predictable environments (Ellmore *et al.* 2006). For example, highly sectored species (e.g. many ring porous) often dominate xeric (dry) sites that are likely to experience very variable water availability during the growth season. Their large early-wood vessels exploit periods of high water availability in spring, while limited or no integration between early-wood and late-wood vessels limits the potential spread of embolism, should this occur as a result of environmental conditions or pathogens. More integrated species (e.g. many diffuse porous) are likely to do better in highly disturbed, wet or understorey conditions (spatially patchy resource environments). Flooded soils, for example, restrict root access to oxygen and so some roots will be better able to take up water than others; by sharing water around the tree, the whole structure is more likely to survive. Of course, the potential limitation of this strategy is that xylem dysfunction is not always easily contained. This sectorality also works for other processes within the tree. For example, light conditions in the understorey are governed by gaps in the forest canopy and periodic sunflecks that may only reach a small portion of an individual crown at any one time. More integrated species cope well with this because they can share sugars produced by the patchy light resource, ensuring impoverished sectors of the tree are able to draw upon resources from enriched sectors (Zanne *et al.* 2006). As trees become old, sectorality within the xylem can become particularly pronounced and can even begin to divide the stem into semi-autonomous functional units (Lonsdale 2013b; see also Expert Box on Ancient and Veteran Trees by David Lonsdale in Chapter 9).

There are a number of practical implications of vascular sectorality. In highly sectored species, damage or depletion of resources in one portion of the root system is likely to have an impact upon a particular portion of the crown. The exact nature of the impact may vary from reduced growth and premature senescence of leaves, through to branch mortality. Moreover, plant health care products, such as fertilisers and pesticides, that rely on the transpiration stream to distribute the product to the crown can also be affected by vascular sectorality. Application of root drenches, soil injections or stem (trunk) injections that do not take account of xylem structure risk partial treatment of the crown; increased coverage of the drench (rooting) area or injection sites usually mitigates this risk.

References

Aloni, R. (2010) The induction of vascular tissue by auxin. In: Davies, P.J. (ed.) *Plant Hormones: Biosynthesis, Signal Transduction, Action!* 3rd edition. Springer, Berlin, Germany, pp. 485–518.

Baas, P., Ewers, F.W., Davis, S.D. and Wheeler, E.A. (2004) Evolution of xylem physiology. In: Hemsley, A.R. and Poole, I. (eds) *The Evolution of Plant Physiology: From Whole Plants to Ecosystems*. Elsevier, Academic Press, Amsterdam, The Netherlands, pp. 273–295.

Bailey, I.W. (1923) The cambium and its derivative tissues. *American Journal of Botany*, 10: 499–509.

Barbaroux, C., Bréda, N. and Dufrêne, E. (2003) Distribution of above-ground and below-ground carbohydrate reserves in adult trees of two contrasting broad-leaved species (*Quercus petraea* and *Fagus sylvatica*). *New Phytologist*, 157: 605–615.

Beck, C.B. (2010) *An Introduction to Plant Structure and Development: Plant Anatomy for the Twenty-first Century*, 2nd edition. Cambridge University Press, Cambridge, UK.

Bonsen, K.J.M. and Kučera, L.J. (1990) Vessel occlusions in plants: Morphological, functional and evolutionary aspects. *IAWA Bulletin*, 11: 393–399.

Brodersen, C. and McElrone, A. (2013) Maintenance of xylem network transport capacity: A review of embolism repair in vascular plants. *Frontiers in Plant Science*, 4: 1–11.

Brown, C.L., McAlpine, R.G. and Kormanik, P.P. (1967) Apical dominance and form in woody plants: A reappraisal. *American Journal of Botany*, 54: 153–162.

Carlquist, S. (2001) *Comparative Wood Anatomy*, 2nd edition. Springer, Heidelberg, Germany.

Carlquist, S. (2012) How wood evolves: A new synthesis. *Botany*, 90: 901–940.

Chapotin, S.M., Razanomeharizaka, J.H. and Holbrook, N.M. (2006) A biomechanical perspective on the role of large stem volume and high water content in baobab trees (*Adansonia* spp.; Bombacaceae). *American Journal of Botany*, 93: 1251–1264.

Chave, J., Coomes, D., Jansen, S., Lewis, S.L., Swenson, N.G. and Zanne, A.E. (2009) Towards a worldwide wood economics spectrum. *Ecology Letters*, 12: 351–366.

Choat, B., Cobb, A. and Jansen, S. (2008) Structure and function of bordered pits: New discoveries and impacts on whole plant hydraulic function. *New Phytologist*, 177: 608–626.

Choat, B., Jansen, S., Brodribb, T.J., Cochard, H., Delzon, S., Bhaskar, R., *et al.* (2012) Global convergence in the vulnerability of forests to drought. *Nature*, 491: 752–755.

Choat, B., Jansen, S., Zwieniecki, M.A., Smets, E. and Holbrook, N.M. (2004) Changes in pit membrane porosity due to deflection and stretching: the role of vestured pits. *Journal of Experimental Botany*, 55: 1569–1575.

Cline, M.G. (1991) Apical dominance. *The Botanical Review*, 57: 318–358.

Cline, M.G. (2000) Execution of the auxin replacement apical dominance experiment in temperate woody species. *American Journal of Botany*, 87: 182–190.

Corcuera, L., Camarero, J.J., Sisó, S. and Gil-Pelegrin, E. (2006) Radial-growth and wood-anatomical changes in overaged *Quercus pyrenaica* coppice stands: Functional responses in a new Mediterranean landscape. *Trees*, 20: 91–98.

Corcuera, L., Morales, F., Abadia, A. and Gil-Pelegrin, E. (2005) Seasonal changes in photosynthesis and photoprotection in a *Quercus ilex* subsp *ballota* woodland located in its upper altitudinal extreme in the Iberian Peninsula. *Tree Physiology*, 25: 599–608.

Crawley, M.J. (1994) Life history and environment. In: Crawley, M.J. (ed.) *Plant Ecology*, 2nd edition. Blackwell Science, Oxford, UK, pp. 73–131.

Critchfield, W.B. (1971) Shoot growth and heterophylly in *Acer. Journal of the Arnold Arboretum*, 52: 248–266.

Davis, S.D., Sperry, J.S. and Hacke, U.G. (1999) The relationship between xylem conduit diameter and cavitation caused by freezing. *American Journal of Botany*, 86: 1367–1372.

del Tredici, P. (2001) Sprouting in temperate trees: a morphological and ecological review. *Botanical Review*, 67: 121–140.

Di Micco, V., Balzano, A., Wheeler, E.A. and Baas, P. (2016) Tyloses and gums: A review of structure, function, and occurrence of vessel occulsions. *IAWA Journal*, 37: 186–205.

Ellmore, G.S., Zanne, A.E. and Orians, C.M. (2006) Comparative sectorality in temperate hardwoods: hydraulics and xylem anatomy. *Botanical Journal of the Linnean Society*, 150: 61–71.

Esau, K. (1960) *Anatomy of Seed Plants*. John Wiley, New York, USA.

Esteso-Martínez. J., Camarero, J.J. and Gil-Pelegrin, E. (2006) Competitive effects of herbs on *Quercus faginea* seedlings inferred from vulnerability curves and spatial-pattern analyses in a Mediterranean stand (Iberian System, northeast Spain). *Ecoscience*, 13: 378–387.

Evert, R.F. (2006) *Esau's Plant Anatomy: Meristems, Cells, and Tissues of the Plant Body: Their Structure, Function, and Development*, 3rd edition. John Wiley, New York, USA.

Feild, T.S., Brodribb, T. and Holbrook, N.M. (2002) Hardly a relict: Freezing and the evolution of vesselless wood in Winteraceae. *Evolution*, 56: 464–478.

Fink, V.S. 1980. Anatomische Untersuchungen über das Vorkommen von Sproß- und Wurzelanlagen im Stammbereich von Laub- und Nadelbäumen. I. Proventive Anlagen. *Allgemeine Forst und Jagdzeitung*, 151: 160–180.

Fromm, J. (2013) Xylem development in trees: From cambial divisions to mature wood cells. In: Fromm, J. (ed.) *Cellular Aspects of Wood Formation*. Plant Cell Monographs 20, Springer, Berlin, Germany, pp. 3–39.

Groover, A. (2016) Gravitropisms and reactions woods of forest trees: Evolution, functions and mechanisms. *New Phytologist*, 211: 790–802.

Hacke, U.G., Sperry, J.S., Wheeler, J.K. and Castro, L. (2006) Scaling of angiosperm xylem structure with safety and efficiency. *Tree Physiology*, 26: 689–701.

Hillis, W.E. (1987) *Heartwood and Tree Exudates*. Springer, Berlin, Germany.

Hoch, G., Richter, A. and Körner, C. (2003) Non-structural carbon compounds in temperate forest trees. *Plant Cell and Environment*, 26: 1067–1081.

Holbrook, N.M. and Zwieniecki, M.A. (1999) Xylem refilling under tension. Do we need a miracle? *Plant Physiology*, 120: 7–10.

Hukin, D., Cochard, H., Dreyer, E., Le Thiec, D. and Bogeat-Triboulot, M.B. (2005) Cavitation vulnerability in roots and shoots: does *Populus euphratica* Oliv., a poplar from arid areas of central Asia, differ from other poplar species? *Journal of Experimental Botany*, 56: 2003–2010.

Jacobsen, A.L., Pratt, R.B., Tobin, M.F., Hacke, U.G. and Ewers, F.W. (2012) A global analysis of xylem vessel length in woody plants. *American Journal of Botany*, 99: 1583–1591.

Jansen, S., Sano, Y., Choat, B., Rabaey, D., Lens, F. and Dute, R.R. (2007) Pit membranes in tracheary elements of Rosaceae and related families: new records of tori and pseudotori. *American Journal of Botany*, 94: 503–514.

Kampe, A. and Magel, E. (2013) New insights into heartwood and heartwood formation. In: Fromm, J. (ed.). *Cellular Aspects of Wood Formation*. Plant Cell Monographs 20, Springer, Berlin, Germany, pp. 3–39.

Kikuzawa, K. and Lechowicz, M.J. (2011) *Ecology of Leaf Longevity*. Springer, Berlin, Germany.

Langenfeld-Heyser, R., Schella, B., Buschmann, K. and Speck, F. (1996) Microautoradiographic detection of CO_2 fixation in lenticel chlorenchyma of young *Fraxinus excelsior* L. stems in early spring. *Trees*, 10: 255–260.

Larter, M., Brodribb, T.J., Pfautsch, S., Burlett, R., Cochard, H. and Delzon, S. (2015) Extreme aridity pushes trees to their physical limits. *Plant Physiology*, 168: 804–807.

Lautner, S., Zollfrank, C. and Fromm, J. (2012) Microfibril angle distribution of poplar tension wood. *IAWA Journal*, 33: 431–439.

Lens, F., Tixier, A., Cochard, H., Sperry, J.S., Jansen, S. and Herbette, S. (2013) Embolism resistance as a key mechanism to understand adaptive plant strategies. *Current Opinion in Plant Biology*, 16: 287–292.

Lonsdale, D. (2013a) *Ancient and Other Veteran Trees: Further Guidance on Management*. The Tree Council, London, UK.

Lonsdale, D. (2013b) The recognition of functional units as an aid to tree management, with particular reference to veteran trees. *Arboricultural Journal*, 35: 188–201.

Meier, A.R., Saunders, M.R. and Michler, C.H. (2012) Epicormic buds in trees: A review of bud establishment, development and dormancy release. *Tree Physiology*, 32: 565–584.

Morris, H., Brodersen, C., Schwarze, F.W. and Jansen, S. (2016a) The parenchyma of secondary xylem and its critical role in tree defense against fungal decay in relation to the CODIT model. *Frontiers in Plant Science*, 7: 1–18.

Morris, H., Plavcová, L., Cvecko, P., Fichtler, E., Gillingham, M.A., Martínez-Cabrera, H.I., *et al.* (2016b) A global analysis of parenchyma tissue fractions in secondary xylem of seed plants. *New Phytologist*, 209: 1553–1565.

Niklas, K.J. (1999) A mechanical perspective on foliage leaf form and function. *New Phytologist*, 143: 19–31.

Ohtani, J., Saitoh, Y., Wu, J., Fukazawa, K. and qun Xiao, S. (1992) Perforation plates in *Knema furfuracea* (Myristicaceae). *IAWA Journal*, 13: 301–306.

Paula, S., Naulin, P.I., Arce, C., Galaz, C. and Pausas, J.G. (2016) Lignotubers in Mediterranean basin plants. *Plant Ecology*, 217: 661–676.

Pfanz, H., Aschan, G., Langenfeld-Heyser, R., Wittmann, C. and Loose, M. (2002) Ecology and ecophysiology of tree stems: Corticular and wood photosynthesis. *Naturwissenschaften*, 89: 147–162.

Pitterman, J. and Sperry, J. (2003) Tracheid diameter is the key trait determining the extent of freezing-induced embolism in conifers. *Tree Physiology*, 23: 907–914.

Plavcová, L. and Jansen, S. (2015) The role of xylem parenchyma in the storage and utilization of non-structural carbohydrates. In: Hacke, U. (ed.) *Functional and Ecological Xylem Anatomy*. Springer, Berlin, Germany, pp. 209–234.

Pollet, C. (2010) *Bark: An Intimate Look at the World's Trees*. Frances Lincoln, London, UK.

Rathgeber, C.B.K., Rossi, S. and Bontemps, J.D. (2011) Cambial activity related to tree size in a mature silver-fir plantation. *Annals of Botany*, 108: 429–438.

Reiner, M. (1960) *Deformation, Strain and Flow: An Elementary Introduction to Theology*. Lewis, London, UK.

Schoch, W., Heller, I., Schweingruber, F.H. and Kienast, F. (2004) *Wood Anatomy of Central European Species*. Available from www.woodanatomy.ch (accessed 22 June 2017).

Schwarze, F.W.M.R. (2007) Wood decay under the microscope. *Fungal Biology Reviews*, 2: 133–170.

Schwarze, F.W.M.R., Engels, J. and Mattheck, C. (2000) *Fungal Strategies of Decay in Trees*. Springer, Berlin, Germany.

Schweingruber, F.H., Börner, A. and Schulze, E.D. (2011) *Atlas of Stem Anatomy in Herbs, Shrubs and Trees; Volume 1*. Springer, Berlin, Germany.

Schweingruber, F.H., Börner, A. and Schulze, E.D. (2013) *Atlas of Stem Anatomy in Herbs, Shrubs and Trees; Volume 2*. Springer, Berlin, Germany.

Sillet, S.C. and Van Pelt, R. (2007) Trunk reiteration promotes epiphytes and water storage in an old-growth redwood forest canopy. *Ecological Monographs*, 77: 335–359.

Sperry, J.S., Hacke, U.G. and Pittermann, J. (2006) Size and function in conifer tracheids and angiosperm vessels. *American Journal of Botany*, 93: 1490–1500.

Sperry, J.S., Nichols, K.L., Sullivan, J.E.M. and Eastlack, S.E. (1994) Xylem embolism in ring-porous, diffuse-porous and coniferous trees of northern Utah and interior Alaska. *Ecology*, 75: 1736–1752.

Spicer, R. (2005) Senescence in secondary xylem: Heartwood formation and an active developmental program. In: Holbrook, N.M. and Zwieniecki, M.A. (eds) *Vascular Transport in Plants*. Elsevier, Amsterdam, The Netherlands, pp. 457–475.

Spicer, R. (2014) Symplasmic networks in secondary vascular tissues: Parenchyma distribution and activity supporting long distance transport. *Journal of Experimental Botany*, 65: 1829–1848.

Spicer, R. (2016) Variation in angiosperm wood structure and its physiological and evolutionary significance. In: Groover, A.T. and Cronk, Q.C.B. (eds) *Comparative and Evolutionary Genomics of Angiosperm Trees, Plant Genetics and Genomics: Crops and Models*, Springer, Berlin, Germany, pp. 1–42.

Spicer, R. and Groover, A. (2010) Evolution of development of vascular cambia and secondary growth. *New Phytologist*, 186: 577–592.

Spicer, R. and Holbrook, N.M. (2005) Within-stem oxygen concentration and sapflow in four temperate tree species: Does long-lived xylem parenchyma experience hypoxia? *Plant Cell Environment*, 28: 192–201.

Sterck, F. (2005) Woody tree architecture. *Annual Plant Reviews, Plant Architecture and its Manipulation*, 17: 209–237.

Thomas, P.A. (2014) *Trees: Their Natural History*, 2nd edition. Cambridge University Press, Cambridge, UK.

Thomas, P.A. (2016) Biological Flora of the British Isles: *Fraxinus excelsior*. *Journal of Ecology*, 104: 1158–1209.

Tng, D.Y.P., Williamson, G.J., Jordan, G.J. and Bowman, D.M.J.S. (2012) Giant eucalypts: Globally unique fire-adapted rain-forest trees? *New Phytologist*, 196: 1001–1014.

Tognetti, R., Longobucco, A. and Raschi, A. (1998) Vulnerability of xylem to embolism in relation to plant hydraulic resistance in *Quercus pubescens* and *Quercus ilex* co-occurring in a Mediterranean coppice stand in central Italy. *New Phytologist*, 139: 437–447.

Tyree, M.T. and Cochard, H. (1996) Summer and winter embolism in oaks: Impact on water relations. *Annals of Forest Science*, 53: 173–180.

Tyree, M.T. and Zimmermann, M.H. (2002) *Xylem Structure and the Ascent of Sap*, 2nd edition. Springer, Berlin, Germany.

Vaucher, H. (2003) *Tree Bark: A Color Guide*. Timber Press, Portland, USA.

Vilagrosa, A., Bellot, J., Vallejo, V.R. and Gil-Pelegrín, E. (2003) Cavitation, stomatal conductance, and leaf dieback in seedlings of two co-occurring Mediterranean shrubs during an intense drought. *Journal of Experimental Botany*, 54: 2015–2024.

Vilagrosa, A., Chirino, E., Peguero-Pina, J.J., Barigah, T.S., Cochard, H. and Gil-Pelegrín, E. (2012) Xylem cavitation and embolism in plants living in water-limited environments. In: Aroca, R. (ed.). *Plant Responses to Drought Stress*. Springer, Berlin, Germany, pp. 63–109.

Ward, H.M. (1904) *Trees, Volume 1: Buds and Twigs*. Cambridge Biological Series, Cambridge University Press, Cambridge, UK.

Ward, H.M. (1909) *Trees, Volume 5: Form and Habit*. Cambridge Biological Series, Cambridge University Press, Cambridge, UK.

Wheeler, E.A. (2011) InsideWood: A web resource for hardwood anatomy. *IAWA Journal*, 32: 199–211.

Wheeler, E.A., Baas, P. and Rodgers, S. (2007) Variations in dicot wood anatomy: A global analysis based on the Insidewood database. *IAWA Journal*, 28: 229–258.

Wilson, B.F. (2000) Apical control of branch growth and angle in woody plants. *American Journal of Botany*, 87: 601–607.

Zanne, A.E., Sweeney, K., Sharma, M. and Orians, C.M. (2006) Patterns and consequences of differential vascular sectoriality in 18 temperate tree and shrub species. *Functional Ecology*, 20: 200–206.

Zimmerman, M.H. and Brown, C.L. (1971) *Trees: Structure and Function*. Springer, New York, USA.

3

Leaves and Crowns

Leaves, and the conglomeration of leaves that make up the tree crown, have the fundamental task of intercepting light. However, leaves must also be supplied with water and have access to carbon dioxide (CO_2) in the atmosphere so that photosynthesis can occur. The driving force for water uptake is evaporation of water from the leaf (transpiration; see Chapter 6), so water is unavoidably lost from pores in the leaf (known as stoma, *pl.* stomata) in order to supply water for photosynthesis. This system works very effectively when water is plentiful, as open stomata permit CO_2 diffusion into the leaf from the atmosphere. However, as water deficits within the tree begin to develop, a fundamental challenge is exposed. Supply of carbon for photosynthesis necessarily involves substantial water loss: plants must 'lose water to fix carbon'. When water becomes less available, the leaf must resolve (with the help of various hydraulic and chemical signals) whether to risk dehydration in order to maintain photosynthesis, or to close the stomata to save water and thereby limit the supply of CO_2. This most ancient of problems can never be solved in the complete sense, but compromises have been reached that enable trees to perform in a wide range of contrasting environments.

In addition to the 'water-loss, carbon-gain' compromise, leaves have further challenges to overcome. Seasonal climates produce marked changes in temperature and sunlight throughout the year that are highly relevant for leaves. For most trees, photosynthesis works best between 20 and 35 °C but the sunlight that is needed to drive photosynthesis can easily push leaf temperature 10–20 °C above the ambient air temperature, an asset during cooler periods but a liability during warmer periods. Either through their physical design or physiological processes, leaves must be able to keep themselves cool or at least prevent overheating. At the other extreme, special measures need to be employed if leaves are to survive freezing temperatures (see Chapter 10). In fact, many trees avoid the problem of coping with temperature extremes by being deciduous: they simply lose their leaves during the period of greatest environmental limitation. In higher latitudes, this relates to cold winters (winter-deciduous) but in lower latitudes, seasonal climates relate to hot and dry periods (summer-deciduous or drought-deciduous).

Exposing leaves for maximum light interception and growing tall to maintain a competitive advantage for light creates a set of mechanical problems for the tree to 'solve'. The sideways force (referred to as drag) caused by the wind poses much more of a problem than the downward force caused by gravity, despite the massive size of some trees.

Applied Tree Biology, First Edition. Andrew D. Hirons and Peter A. Thomas.
© 2018 John Wiley & Sons Ltd. Published 2018 by John Wiley & Sons Ltd.

The force of the wind on the leaves within the crown acts to bend the whole tree. If the drag generated by wind loading overcomes the strength of a component part of the tree's aerial structure, whether this is the leaf petiole, branch, stem or root, then the tree, or part of it, will fail. Leaves and branches must be able to reduce drag to the extent that biomechanical failure only occurs in exceptional circumstances.

The life of a leaf therefore represents a series of compromises. They must successfully capture light, move water and nutrients to the sites of photosynthesis whilst preventing dehydration; maintain a supply of sugars for the tree; regulate temperature; and cope with drag caused by wind. If these constraints are not problematic enough, many organisms view the leaf as a food source. Like so many complex problems, solutions are available but most come at a cost. Positive selection pressures over time have refined the design of leaves to meet the most important demands of its given environment, the outcomes of which are seen in the extraordinary diversity of leaves.

Angiosperm Leaves

Angiosperm trees are often referred to as *broadleaved* trees because they normally produce broad leaf blades (*lamina*) for the interception of light. Leaves are typically attached to a stem via a *petiole* (leaf stalk), either singly (alternate), in pairs (opposite) or in groups of three or more (whorled). However, some leaves lack petioles and are termed *sessile*. At the base of the petiole there may also be outgrowths (superficially similar to tiny leaves) known as *stipules*. These often appear not to have any major purpose, but in some species seem to help protect the expanding axillary bud as in the tulip tree *Liriodendron tulipifera*. In other species (e.g. false acacia *Robinia pseudoacacia*), the protective role of the stipules goes further because they develop into woody thorns.

Although leaves all have the same basic purpose, they come in a vast array of sizes and forms. They may either be *simple* (Figure 3.1), if they are undivided, or *compound*, if a number of smaller leaflets make up a single leaf (Figure 3.2). Compound leaves are further divided into *palmate* (or *digitate*), where all leaflets originate from the same point (e.g. *Aesculus*), or pinnate where many leaflets are attached to a common rachis (e.g. *Ailanthus*; Figure 3.2). Occasionally, pinnate leaves subdivide again to form bipinnate leaves (e.g. *Aralia* and *Gymnocladus*). Amongst angiosperm trees, leaves tend to vary from around 10 mm in length, in species such as Mountain beech *Fuscospora cliffortioides* (Figure 3.3) to comparatively massive fronds of several metres found in some palm trees (Figure 3.4). Indeed, some palms such as *Raphia farinifera*, native to tropical regions in Africa, can have leaves up to 20 m long. The issue of drag is so substantial that larger leaves tend to be divided so they reduce resistance to the wind. In most cases, this means they are compound leaves, but some species, such as the Madagascan traveller's palm *Ravenala madagascariensis*, grow an entire leaf that tears easily in the wind, creating a 'compound' leaf that reduces drag quite effectively without any compromise in leaf area (Figure 3.5) – a process known as *tattering*.

In very hot, dry environments, leaves may be reduced to small, thick leaves (microphylls) in order to cope with the high temperatures, intensity of light and limitation in water supply (as seen in acacias). Some trees have reduced their leaves to an even

(a)

(b)

(c)

(d)

Figure 3.1 Simple leaves of (a) gutta-percha *Eucommia ulmoides;* (b) mountain camellia *Stewartia ovata;* (c) moose-bark maple *Acer pensylvanicum;* and (d) tulip tree *Liriodendron tulipifera.*

(a)

(b)

Figure 3.2 Compound leaves. (a) Palmate leaves of Ohio buckeye *Aesculus glabra* and (b) pinnate leaves of tree of heaven *Ailanthus altissima.* The pinnate leaf has a petiole (leaf stalk) that continues up between the leaflets where it is referred to as the rachis.

(a) (b)

Figure 3.3 (a) Mountain beech *Fuscospora cliffortioides*, a native of New Zealand, seen here at the timberline. (b) A close-up of a lower branch showing the small leaves of about 7–10 mm in length.

(a) (b)

Figure 3.4 (a) *Beccariophoenix madagascariensis*, a rare forest palm found in forest around Sainte Luce in southeast Madagascar, has massive leaves. (b) A single leaf about 6–7 m long; note the gecko on the rachis towards the bottom of the picture. Other palms can have even bigger leaves.

greater extent. The she-oaks *Casuarina* spp. of Australia and Asia have minute scale-leaves that surround the modified green stems that make up the crown of these dryland specialists (Figure 3.6).

Variation in leaf shape has led botanists to come up with a vast array of terms to describe every minute detail of the leaf: from the overall shape; the way it is arranged on

Figure 3.5 The traveller's palm *Ravenala madagascariensis* in southeast Madagascar. Younger leaves in the centre of the crown (or on younger plants in the foreground) are less torn, whereas older leaves are progressively torn by the action of the wind. This process of tattering provides an effective way of reducing drag without any substantive reduction in leaf area.

(a)   (b)

Figure 3.6 (a) She-oak *Casuarina equisetifolia*, Noosa Heads National Park, Queensland, Australia, showing photosynthetic stems. (b) The diminutive leaves of this species can be seen as yellow bands that are formed at the nodes of these modified stems.

the shoot; the form of the tip and base; the nature of the margin; the network of veins; and even the microstructure of leaf hairs. All these features help to identify individual species. However, some species remain fiendishly difficult to distinguish from each other by the leaf alone, and other features such as the flowers, fruit, bark and crown form will need to be consulted before positively identifying a tree. Specialist texts are thus often invaluable as identification aids, particularly if they are well illustrated.

Angiosperm Leaf Anatomy

Layers of epidermal cells cover the outer surfaces of the leaf and provide a waterproof (but still just about gas permeable) layer that protects the inner leaf tissue (mesophyll) from dehydration. A waxy layer made up largely of cutin is secreted from the upper epidermal cells to form a *cuticle* that confers further resistance to leaf dehydration (Figure 3.7). The thickness of this cuticle can vary quite considerably depending on the leaf microclimate in which the leaf has developed. Shade-grown leaves tend to have thinner cuticles, approximately 1 µm thick, as water loss is much lower in shaded environments. Leaves exposed to direct sunlight tend to have much thicker cuticles, up to around 15 µm, to help them cope with the high-light environment.

Both the upper and lower epidermis may have stomata. These are surrounded by two guard cells that control the aperture of the pore opening, regulating gas exchange and water loss from the leaf. Most of the carbon dioxide used in photosynthesis diffuses into the leaf through the stomata, and water lost by transpiration diffuses out. In temperate angiosperm trees, stomata are between 17 and 50 µm in length and have a density of between 100 and 600 mm^{-2} (Pallardy 2008). As a general rule, those species with the lowest density of stomata also have the largest stomata, and vice versa. Size and frequency can vary greatly, even between species of the same genus. Most angiosperm trees have relatively few or no stomata on their upper surfaces to reduce water loss in direct sunlight. However, a number of willow *Salix* and poplar *Populus* species have stomata in both the upper and lower surfaces. Where this does occur, most of the stomata are still found the lower epidermis.

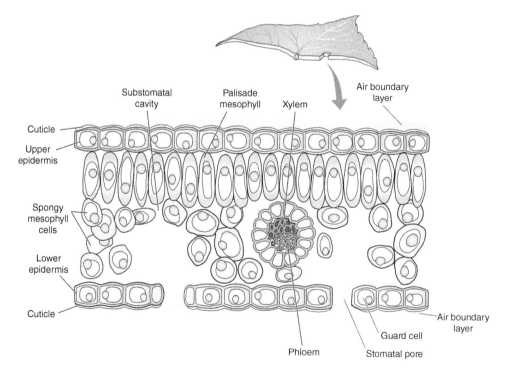

Figure 3.7 Dicotyledonous leaf seen in cross-section. *Source:* Adapted from Taiz and Zeiger (2010). Reproduced with permission of Oxford University Press.

The inside of the leaf, the mesophyll, is predominately made up of parenchyma cells that form two distinctive tissue regions (Figure 3.7). *Palisade mesophyll* consists of regimented columnar cells with high numbers of chloroplasts (the cellular organelles containing chlorophyll that are fundamental for photosynthesis). These tightly packed cells are found in the upper portion of the leaf and are arranged perpendicular to the upper epidermis. In high light environments, palisade mesophyll may be two or three layers thick to maximise light interception. Some trees, including many *Eucalyptus* spp., have palisade mesophyll in the upper and lower portions of the leaf. This allows them to hold their leaves vertically to reduce the heat load and water loss, whilst still allowing photosynthesis in both sides of the leaves. *Spongy mesophyll* is located immediately below the palisade mesophyll and is much more irregular in shape. This creates numerous spaces between the cells to allow effective gas exchange.

Embedded within the mesophyll is a network of veins designed to transport water and minerals into the leaf, and carbohydrates out. The upper portion of the vein is made up of xylem that delivers water, minerals and other chemical signals to the leaf cells. The lower portion of the vein consists of the phloem and is predominately involved with the export of carbohydrates from the leaf to the rest of the plant, although it is also involved with the redistribution of carbohydrates within the leaf.

Gymnosperm Leaves

Under favourable tropical and sub-tropical conditions, gymnosperms, such as *Podocarpus, Agathis, Ginkgo* and *Araucaria*, have quite broad leaves; however, needle-like leaves are typical in the family Pinaceae and scale-like leaves are typical of Cupressaceae (Figure 3.8). Compared with angiosperm leaves, those of gymnosperms tend to be much smaller in size and most feel more robust than angiosperm leaves. This structural toughness is, at least in part, a consequence of protective tissues that make these leaves capable of enduring up to several years on the tree, often whilst being exposed to some of the most challenging environments on the planet. Thus, most keep their leaves all year (i.e. are evergreen), which also makes them well adapted to the short growing seasons typical of hot arid regions and very cold environments where conifers are most abundant. Deciduous or semi-deciduous conifers (*Glyptostrobus, Larix, Metasequoia, Pseudolarix, Taxodium*) have a different strategy in that they save energy by not developing protective tissues, and so tend to have softer-textured leaves. Reasons for this change in habit are discussed in Evergreen and Deciduous Leaves.

Needle-like leaves are common in gymnosperms (Figure 3.8c,d). True fir *Abies* and spruce *Picea* needles are arranged systematically along stems in a similar way to most angiosperm trees, although the precise phyllotaxy may be quite different between species. In *Picea*, the very base of the needles have evolved swollen bases (*pulvini*) which contain cells with large vacuoles that help regulate needle gas exchange and water supply; they also increase the flexibility of the leaf when frozen, so that it does not break so easily under heavy snow loads (Debreczy and Rácz 2011). Close observation of *Pinus* needles reveals that they are actually borne in terminal clusters on extremely dwarfed shoots, encapsulated at the base by a needle sheath – this arrangement is known as *fascicled*. The entire unit tends to have two, three or five needles within the cluster although, as its name suggests, *Pinus monophylla* bucks the trend by only having

(a)

(b)

(c)

(d)

(e)

(f)

Figure 3.8 Gymnosperm leaves come in various forms. (a) Bunya pine *Araucaria bidwillii* and (b) *Podocarpus macrophyllus* have relatively broad, flat leaves. Note the new younger leaves of the *P. macrophyllus* are a much lighter green than the older leaves. (c) Scots pine *Pinus sylvestris* and (d) eastern hemlock *Tsuga canadensis* have needle-leaves. (e) Giant sequoia *Sequoiadendron giganteum* and (f) hiba *Thujopsis dolabrata* have scale-leaves.

one needle per cluster (a few other exceptions also exist). On the twigs of genera such as *Cedrus, Larix* and *Pseudolarix*, after 1 year's growth with a simple alternate needle arrangement, needle clusters form on older twigs with very tight spirals (whorls), giving the needle arrangement a rosette appearance.

Nearly half of the conifer genera and about one-fifth of conifer species have *scale-like leaves* (Figure 3.8e,f). While scale-leaves are varied, telling them apart is often very challenging. They vary from around 1 to 10 mm in length and each scale is typically longer than it is wide. Size, shape and features such as resin blisters, colouring and the smell of freshly crushed sprigs can help resolve the differences between species.

In a few gymnosperms (*Phyllocladus* and *Sciadopitys*), true leaves are much diminished and cladodes, flattened stems that take on a leaf-like form, carry out photosynthesis. Other anomalies include *Glyptostrobus*, *Metasequoia* and *Taxodium* which have small simple leaves borne on short-lived shoots. These are often shed in response to cold or water stress, and so, superficially, give the appearance of a pinnately-compound leaf.

Gymnosperm Leaf Anatomy

The internal anatomy of gymnosperm leaves is different from that of angiosperms, although they do have features in common. In many gymnosperms, needle rigidity is increased by a layer of lignin-rich *hypodermal sclerenchyma cells* below the epidermis (Figure 3.9). These effectively increase the depth of the epidermis and result in sunken stomata (particularly in *Pinus*) that are somewhat protected from the outside environment. Such structures can help increase the resistance to water loss and help protect the delicate stoma from damage. Inside the needles, the xylem and phloem are surrounded by *transfusion tissue* made up from tracheids and parenchyma that help control the movement of substances into, and from, the vascular bundle. Many species will also have specialised resin ducts (or canals) that help transport resin to sites of

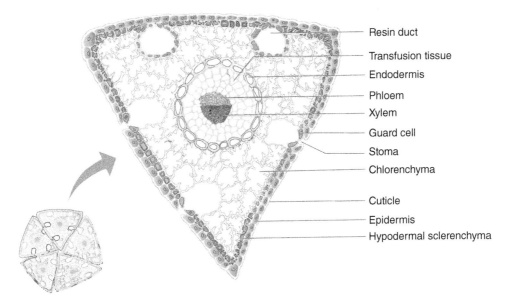

Figure 3.9 Cross-section of a typical pine needle from eastern white pine *Pinus strobus*. In this five-needled pine, the triangular cross-section represents a segment of the original needle cylinder to emerge from the bud, as indicated in the lower left of the figure. *Source:* Adapted from Kramer and Kozlowski (1979). Reproduced with permission of Elsevier.

injury. In *Pinus*, the mesophyll is not as organised as in angiosperms; it is made up of chlorophyll-containing parenchyma cells (*chlorenchyma*) that have a cloud-like appearance as a consequence of their deeply infolded cell walls (Figure 3.9). This is likely related to the fact that pine needles absorb light from all leaf surfaces rather than being designed to intercept light through their upper surface, as in most broadleaves. However, in the flatter leaves of genera such as *Abies, Araucaria, Dacrydium, Ginkgo, Podocarpus, Pseudotsuga, Sequoia, Taxus, Torreya* and *Tsuga*, the mesophyll is differentiated into palisade and spongy parenchyma as it is in angiosperms.

Juvenile Leaves

A number of tree species show *leaf dimorphism*: a distinct difference between juvenile and adult leaves. Leaves produced on young shoots (not necessarily young plants, because new epicormic shoots that form in response to injury can have juvenile leaves) have a totally different form from their adult counterparts. Occasionally, this change is quite gradual and includes intermediate leaf forms (*homoblastic*), whilst in other species it is abrupt with no apparent transitional forms (*heteroblastic*) (Pallardy 2008). Most scale-leaved gymnosperms have needle-like juvenile leaves that give them a greater photosynthetic area to help them compete in the more shaded environments of their early life. In fact, juvenile leaves may also be present in adult scale-leaved trees that find themselves in shaded environments. However, the expanded leaf area increases the tree's vulnerability to water deficits, so it may be a disadvantage if environmental conditions suddenly change (Debreczy and Rácz 2011); it certainly will be when the tree is bigger and in full sunlight. Many angiosperm trees also have starkly contrasting juvenile and adult foliage. For example, a number of *Populus* spp. (*P. euphratica, P. ilicifolia, P. mexicana* and *P. monticola*) show heteroblastic leaf development, where foliage on shoots less than about 10 years old is distinctly different in venation and shape compared to foliage on older stems (Eckenwalder 1980). *Eucalyptus* species also generally form very distinctive juvenile foliage (a favourite of florists; Figure 3.10) that is normally sessile and much rounder than the adult foliage (Wrigley and Fagg 2010). As if these

(a)   (b)

Figure 3.10 (a) Juvenile and (b) adult leaves of *Eucalyptus* species.

differences were not pronounced enough, some species such as *Raukaua edgerleyi* and *Weinmannia racemosa*, both native to New Zealand, have compound juvenile leaves but simple adult leaves (Dawson and Lucas 2011).

Sun and Shade Leaves

In large tree crowns and in woodlands, it is inevitable that some leaves grow in a shaded environment. There are many adaptations to enable trees to survive in high and low light levels (see Chapter 7); perhaps the most important are *sun* and *shade leaves*. Shade leaves tend to be thinner and have a larger surface area than their sun-leaf counterparts (Figure 3.11). For example, in European beech *Fagus sylvatica*, shade leaves are about 50% thinner and 60% larger in area than sun leaves (Larcher 2003). Most of the difference in thickness results from the fact that sun leaves have two or three layers of palisade mesophyll (Figure 3.12; see also Figure 3.7) to help them make the most of higher light levels; the epidermis and cuticle also tend to be thicker. Additional palisade mesophyll results in more chloroplasts per unit of leaf area and higher rates of photosynthesis in sun leaves. The corollary of this is that respiration is higher and therefore the light compensation point is also higher (this is the light level at which photosynthetic fixation of CO_2 equals respiratory production of CO_2): a disadvantage if light levels drop. However, as sun leaves are adapted to high-light environments, the light saturation point (the level above which light is no longer limiting to photosynthesis) is also higher (Lambers *et al.* 2008).

Where the leaf is lobed, sun leaves also tend to be more deeply lobed than shade leaves. This increases the light penetration through the crown so that the outer leaves do not excessively shade more internal leaves, but is also related to the need to keep the sun leaves from overheating: deeper lobes have been shown to have lower average and peak temperatures by dissipating heat more effectively, because most evaporation of water is from the edge of the leaf and lobes increase the length of that edge (Vogel 2012).

Figure 3.11 Sun and shade leaves can markedly differ in size. Here, the relative size of sun (left) and shade (right) leaves are shown for holm oak *Quercus ilex*.

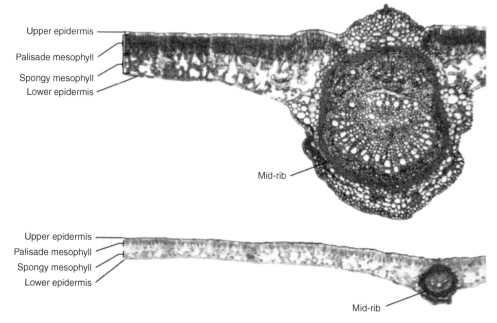

Upper epidermis

Palisade mesophyll

Spongy mesophyll

Lower epidermis

Mid-rib

Upper epidermis
Palisade mesophyll
Spongy mesophyll
Lower epidermis

Mid-rib

Figure 3.12 Partial cross-section of a sun (upper) and shade (lower) leaf from European beech *Fagus sylvatica*. The sun leaf has thicker epidermal layers, two layers of densely packed palisade mesophyll cells, a deeper spongy mesophyll layer and a larger mid-rib (leaf vein) hosting the xylem and the phloem. The shade leaf has thinner epidermal layers, only a single layer of palisade mesophyll, a thinner layer of spongy mesophyll and a much smaller mid-rib. Shown at relative size.

Consequently, even relatively simple modifications in leaf shape and size can improve the overall photosynthetic efficiency of the whole crown by reducing self-shading and leaf overheating.

Most trees have both sun and shade leaves that make the most of the contrasting light environment across the crown; unsurprisingly, more shaded trees tend to have more shade leaves. Taken to its extreme, it is possible that a tree will have only shade leaves in very shaded conditions, such as those found beneath a forest canopy. Conversely, in high-light environments, some species only produce sun leaves. This is seen in trees with very open canopies where lots of light reaches even the deepest leaves within the tree's crown (as in European ash *Fraxinus excelsior*).

At the whole tree level, trees can be divided into *multilayer* and *monolayer* species (Horn 1971). Multilayer trees tend to be composed of sun and shade leaves and are at their best in high-light environments. As light intensity becomes higher, the top leaves may soon reach their light saturation point and will not be able to increase their photosynthetic rate; however, the extra light penetrating into the crown may still lead to an overall increase in photosynthesis. At the other extreme, in low-light environments (typically less than 25% of full sunlight), trees can operate with a monolayer of leaves, this reduces self-shading so each leaf is exposed to as much light as possible. Monolayer trees are typically: pioneers in very shady gaps in rainforests (Figure 3.13); and tropical forest canopy trees growing in very cloudy environments or heavily shaded positions. Some temperate trees, such as sycamore *Acer pseudoplatanus*, also have a monolayer crown.

Figure 3.13 *Cecropia peltata*, a monolayered tree growing in a gap in tropical cloud forest in Honduras. Growing in the shade, all the leaves are aligned in a single layer to reduce self-shading and maximise the amount of light they receive.

The fact that leaves are effectively designed for the light environment in which they grow means that sudden changes to the light environment can be problematic. Whilst it is rare that sun leaves are suddenly plunged into low-light conditions, it is possible when trees are transplanted from the nursery to a very shaded site. More common is the removal of an overstorey, neighbouring trees or the peripheral crown on a single tree. This is a particular problem where the crown of a tree has been drastically reduced, and in over-clipped hedges. These sudden increases in light levels can easily lead to leaf scorch and physiological damage in hitherto shaded leaves. Removal of internal shade leaves appears to be less of a problem for the tree because the remaining shade leaves are still shaded by the sun leaves. In fact, the tree will not usually miss them much. In a study using 6–7 m tall wild cherry *Prunus avium*, removing the bottom 22% of the tree crown by pruning all but the top five whorls of branches did not affect height growth, and only reduced diameter growth by 4% in the subsequent year. Removing 50% of the crown by pruning all but the top three whorls of branches had no impact on total crown height and only reduced diameter growth by 5% in the first year and 9% in the following year (Springmann *et al.* 2011). Therefore, removal of lower branches predominantly carrying shade leaves would seem to have only a very modest impact on growth, providing the upper branches with their sun leaves remain intact and with plenty of light. However, whilst shade leaves may not be particularly productive, they are likely to help trees in woodland situations by reducing the amount of light that is available to the seedlings of potential competitors growing below (Margalef 1997). They may also help the tree's stability. Movements of branches within the crown help dissipate wind energy through a process known as (mass) damping. By reducing the wind energy transferred to the tree, internal branches can have an important role in improving tree stability (James *et al.* 2006; see Biomechanical Design of Tree Crowns). Thus, it is important to consider the crown as a whole, rather than just from the perspective of photosynthesis, as relatively unproductive leaves and branches may have important roles in the reducing available light resources to competitor plants or by increasing the biomechanical integrity of the tree.

Leaf Arrangement

Even with shade leaves, self-shading is a challenge in the low light levels within a larger crown or under a forest canopy. This is why trees in the understorey often have leaves on a branch displayed in a flat, non-overlapping layer (a *planar array*) that minimises self-shading and maximises light interception (Figure 3.14). Many maples *Acer* and beech *Fagus* species show this particularly well.

The spiral arrangement of leaves (*phyllotaxy*) along a shoot prevents successive leaves emerging directly above each other. If you look down the length of a shoot, the different orientation of each leaf is usually very clear (Figure 3.15). Adjustments of the distance along the stem between successive leaves (internode length), as well as using long and short shoots (in some species), can also help reduce self-shading. Petiole length can also be adjusted to prevent leaves overlapping and, in some species (e.g. aspens *Populus tremula* and *P. tremuloides*), flattened petioles encourage leaf fluttering to allow flecks of unfiltered light (*sunflecks*) to reach their internal leaves. The combined effect of these developmental characteristics results in highly organised and efficient interception of light.

Conversely, even using efficient sun leaves, the high light intensity at the top of the crown can be too much for the leaves and, ironically, lead to reduced photosynthesis

Figure 3.14 Flat, non-overlapping layers of leaves on a branch (planar arrays) can often be observed in trees growing in shaded environments. This sugar maple *Acer saccharum* in the understorey demonstrates how precise arrangement of leaves can minimise self-shading and maximise light interception.

Figure 3.15 Leaves are arranged in spiral patterns around shoots to help reduce self-shading: (a) white oak *Quercus alba*; (b) sweetgum *Liquidambar styraciflua*; (c) old man banksia *Banksia serrata*; and (d) monkey puzzle *Araucaria araucana*.

(*photoinhibition*), reducing the efficiency of photosynthesis at the leaf and crown scale. Under such circumstances, spilling some light by adjusting the leaf angle or even encouraging self-shading of leaves can actually increase photosynthetic output of the whole crown (Cescatti and Niinemets 2004). In fact, some eucalypt forests are known as 'shadeless forests' because their leaves hang vertically to reduce the radiation load (Figure 3.16), which also has the advantage of keeping the leaves cooler.

Compound Leaves

Trees with compound leaves (Figure 3.2) are successful in a range of different environments, particularly in arid regions and seasonal rainforests, but they are also common in some habitats of temperate regions. A single compound leaf and a series of individual simple leaves on a twig would appear to be very similar; so what underlies the apparent success of compound leaves in some habitats? The answer seems to relate to the fact that compound leaves consist of leaflets arranged on a deciduous twig called the *rachis*.

Figure 3.16 Some eucalypt forests are sometimes known as 'shadeless forests' because they hang their leaves vertically to reduce the radiation load. This means that the environment underneath these trees can seem quite light, as shown here in a karri *Eucalyptus diversicolor* forest of Leeuwin-Naturaliste National Park, Western Australia.

These disposable branches can make a real difference to plant performance in a number of contrasting environments. In regions prone to seasonal drought, anything that helps conserve water will be important. Once the leaves are shed, the thinnest branches, with their high surface to volume ratios, can still transpire significant amounts of water. In compound leaved trees, this vulnerability to further water loss is mitigated when the rachis is also shed. Although this is only one strategy to cope with water deficits, small margins can make big differences in such challenging environments. At least partly for this reason, trees with compound leaves, such as baobab *Adansonia digitata*, *Vachellia* spp. (previously members of *Acacia*) and *Erythrina* spp., are often found in African savannahs. North American trees found in similar drought-prone areas include desert ironwood *Olneya tesota*, and species of *Parkinsonia* and *Prosopis*.

In cooler, moister climates, compound leaves give a different advantage. The rachis is cheaper to grow than an equivalent sized branch because the rachis is held up primarily by turgor (water) pressure and only a small amount of fibrous tissue: they increase the leaf area with much less investment in woody material. As with young green stems in general, the green rachis will also recoup at least some of its production cost through photosynthesis. An excellent example of this use can be seen in the eastern deciduous forest of North America. Devil's walking stick *Aralia spinosa* has large bipinnate leaves and by using compound leaves avoids the need to extensively branch. In comparison, a co-occurring tree, flowering dogwood *Cornus florida*, has simple leaves and needs to invest 7–15 times more in wood than *A. spinosa* to support the same leaf area (White 1984). Although not all comparisons are likely to be as marked, compound leaves help conserve energy that can then be used for rapid vertical growth (Givnish 1978). High disturbance habitats, such as floodplain forests, where rapid opportunistic

Figure 3.17 Staghorn sumac *Rhus typhina* colonising a gap on the edge of woodland in Ithaca, New York. This is a good example of a compound-leaved tree being suited to invading gaps by investing less in a compound leaf than in a branch with simple leaves.

growth is an advantage, frequently have a number of compound-leaved trees such as ashes *Fraxinus* spp. and Manitoba maple *Acer negundo*. Gap colonisers and early-successional species such as tree of heaven *Alianthus altissima*, Japanese angelica tree *Aralia elata*, Kentucky coffee-tree *Gymnocladus dioica*, false acacia *Robinia pseudoacacia*, some *Fraxinus* spp. and many species of *Sorbus* and *Rhus* (Figure 3.17) are good examples of this strategy.

Whilst compound-leaved trees is a strategy that has emerged again and again in habitats where a fast growth gives an advantage, it is important to note that rapid growth is certainly not the preserve of trees with compound leaves.

Evergreen and Deciduous Leaves

Leaves are precious to trees. Their construction, emergence and longevity must be controlled so that their inevitable senescence does not occur before they are able to repay the cost of their construction, maintenance and defence (Chabot and Hicks 1982). Achieving this is no easy feat as it may involve avoiding freezing temperatures, water deficits, heavy rains that can leach-out minerals, herbivores, competition from neighbouring trees, or any combination of these. If leaves do not avoid such hazards, they must be built to endure them. For this reason, the type of leaf a tree produces is closely associated with the environment in which it grows.

No single leaf and crown design can be effective in all environments, so there are a wide range of *foliar strategies*. Perhaps the most conspicuous difference in foliar habit relates to the duration of leaves on a tree. *Evergreen* trees retain at least some leaves

Figure 3.18 Holm oak *Quercus ilex*, an evergreen broadleaved species from the warm-temperate regions of Europe and Central Asia. This species is now widely planted across Europe.

throughout the year and include most conifers (notable exceptions to this are *Glyptostrobus, Larix, Metasequoia, Pseudolarix* and *Taxodium* species), a wide variety of warm-temperate and Mediterranean climate species (such as holm oak *Quercus ilex;* Figure 3.18), and most tropical trees. *Deciduous* trees lack leaves for some part of the year and include most trees from seasonal climates. Despite these general characterisations, in reality there are a whole range of different foliar habits on the evergreen–deciduous spectrum that have developed in response to particular environmental constraints, as shown in Table 3.1.

The definition of 'evergreen' and 'deciduous' is based on looking at the entire crown rather than individual branches (with the exception of heteroptosis species; see Table 3.1). How long individual leaves live for (*leaf longevity*) and how often they are replaced (*leaf turnover*) are important in these definitions. While some leaves on winter-green deciduous species only last for a few months, at the other extreme, conifers such as monkey puzzle *Araucaria araucana* and bristlecone pine *Pinus longaeva* have leaves that can last 20–30 years. As well as variation across species, leaf longevity may vary within the same tree, with the more active leaves at the upper margins of the crown being replaced more frequently than those in more shaded positions.

Value of Evergreen and Deciduous Leaves

Deciduous leaves have evolved in a vast range of species so they must have their advantages. In general, higher leaf nitrogen and thinner, wider leaves facilitate higher levels of photosynthesis per gram of leaf. However, this advantage may be eroded in cold, dry or more nutrient-deprived conditions that require tougher leaves to persist throughout the year. Leaf loss during an unfavourable time of year gets round this problem by eliminating the running cost of the leaf, reducing water loss (transpiration) and lessening the requirement for active roots during this period. Even though the uptake of water to the roots does not require energy, any reduction in root activity helps to conserve energy.

Evergreen leaves can, potentially at least, photosynthesise over a longer period each year than deciduous leaves. In cold-temperate, boreal and taiga ecosystems the growing season is very short, but evergreens can begin photosynthesis as soon as conditions allow without wasting valuable time having to grow new leaves. This is also important

Table 3.1 Variations in the evergreen and deciduous habits of trees.

Foliar habit	Definition
Subdivisions of the evergreen habit	
Leaf exchanger	Leaves that are exchanged within the year. Leaf longevity is shorter than a year but there are always viable leaves in the crown
Semi-evergreen	Immediately after new leaf emergence, old leaves fall. Leaf longevity is essentially a year and a leafless period is neither very apparent nor very short
Brevideciduous	Some leaves are shed during part of the year, but never more than 50% of the crown so the tree appears evergreen
Semi-deciduous	More than 50% of the leaves are lost at some time during the year but the crown is never completely bare
Heteroptosis	Some branches of a tree become completely leafless during unfavourable periods but others retain leaves throughout the year
Subdivisions of the deciduous habit	
Summer-green	Leaves are shed in autumn (fall) and the crown is completely bare during winter. This is typical of the deciduous habit in temperate regions
Winter-green	Leaf emergence occurs at the end of summer, leaves are retained through winter and shed at the onset of summer. Tree crowns are bare through summer. This is typical in Mediterranean climates
Drought-deciduous	Leaves are shed through the dry season. This habit is common in tropical forests with a pronounced seasonal dry period
Spring-ephemeral	Leaves emerge in early spring and are shed by summer. This habit is usually found in herbaceous plants but it has been recorded in a shrub, painted buckeye *Aesculus sylvatica*, native to southeast North America

Source: Kikuzawa and Lechowicz (2011). Reproduced with permission of Springer.

in understorey trees of temperate areas, such as holly *Ilex aquifolium*, which, being brevideciduous, can start photosynthesis very early in spring before the deciduous trees above have grown leaves and the canopy closes. The same also happens in the autumn (fall) when evergreen species can keep photosynthesising long after deciduous trees have shed their leaves. Without these productive 'shoulder periods' in spring and autumn, they are less likely able to compete and survive within that habitat. In temperate regions with milder winters, photosynthesis is also possible during warm spells during the winter. Even on what appear to be relatively cold days, sunlight can warm broad leaves (needle and scale-leaves to a lesser extend) to around 10–20 °C above the ambient air temperature and so provide decent conditions for photosynthesis (Vogel 2009) providing the roots can supply water.

In general, evergreen leaves survive for several years before they are shed so they have a lower leaf turnover rate and longer leaf longevity. This means repayment of the capital 'carbon cost' of leaf construction (technically the grams of carbon per gram of leaf, per year) can be spread over several years, rather than just one (Givnish 2002). A leaf that survives for 4 years only needs to fix about one-quarter of the carbon construction cost each year before a net 'carbon profit' for that leaf becomes possible. Therefore, they can afford to pay back the initial 'carbon cost' of the leaf over a longer period of time than deciduous leaves that must repay the construction costs and make a carbon profit in one growing season. This means that evergreens have what is referred to as a lower 'amortised (carbon) cost' of leaf production, compared to deciduous leaves.

In moist tropical environments, trees tend to be evergreen as warm temperatures, high rainfall and little variation in day length (light) favours photosynthesis all year round. In these ideal growing conditions, no great advantage is found by synchronous whole-crown leaf loss. Evergreens also tend to grow in areas where soil nutrients are in short supply (such as the northern conifer forests) or are strongly competed for (as in tropical forests) and so are difficult to acquire. Fortunately, evergreens also have a lower amortised cost of replacing nutrients because the nutrient investment in new leaves can also be recouped over several years. Further, evergreens tend to have lower nutrient concentrations per gram of leaf (Reich *et al.* 1997), which further reduces the nutrient investment in each leaf. Nutrient resorption in trees appears to be about 50%, so only about half of the nitrogen and phosphorus (measured as milligrams of nutrient per gram of dry leaf tissue) are taken back into the parent plant prior to leaf abscission (Aerts 1996; for more details see Nutrient Cycling in Chapter 8). However, this figure does vary between species and different sites. For example, aspen *Populus tremula* has been shown to draw back 60% of nitrogen and phosphorus during leaf senescence (Keskitalo *et al.* 2005). Any nutrients not saved in this way must be replaced when the new leaves are produced. In impoverished soils, this can require additional costs in root production and even more nutrient acquisition in order to manufacture new leaves. For these reasons, evergreens have a competitive advantage on nutrient-poor sites. Indeed, this advantage is increased, because short-lived deciduous leaves require higher levels of nitrogen and phosphorus to support a higher photosynthetic capacity that is, in turn, required to make the leaf 'profitable' in one growth season. Consequently, species with low leaf nutrient concentrations and high leaf longevity tend to perform well in low-nutrient environments (Craine 2009).

The longer life of evergreen leaves requires them to be more robustly constructed. In the boreal, temperate and Mediterranean areas, they have to survive periodic frost and/or water deficits without suffering lasting damage (Chabot and Hicks 1982). This has resulted in evergreen leaves having a higher leaf mass per unit of leaf area (typically measured as *specific leaf area*, SLA: $m^2 \, kg^{-1}$, or *leaf mass per area*, LMA: $kg \, m^{-2}$). These tougher leaves make them less attractive to herbivores, especially as they also tend to have good chemical defences and, because of relatively low nutrient levels, be poor quality food: crucial when there are few alternative food sources in unfavourable growth periods.

Although the increase in leaf toughness and provision of chemical defences increases the construction costs of each leaf, the actual cost difference between evergreen and deciduous leaves is usually <10%. This is primarily because evergreen leaves are typically small with high costs in making them tough, while deciduous leaves expend almost as much energy building a larger flat leaf blade.

Leaf exchangers, such as the camphor tree *Cinnamomum camphora*, southern magnolia *Magnolia grandiflora*, California live oak *Quercus agrifolia* and cork oak *Q. suber* (Figure 3.19), tend to occupy regions between those environments suited to evergreen and deciduous species. They would appear, superficially at least, not to have any great advantage over those species that keep their leaves for longer than 1 year. However, in impoverished soils, photosynthesis is limited by low nutrients. This diminishes the

Figure 3.19 Cork oak *Quercus suber* growing in Andalucia, Spain, is a leaf exchanger (see Table 3.1) that drops one set of leaves and immediately grows the next set. The commercially valuable corky bark on the main trunk has been removed.

difference between different foliar habits in the amount of sugars produced by photo-synthesis in a growing season. In some regions, between the highly favourable tropical environments and seasonally dry regions, relatively poor soils, fluctuating soil moisture and consistently suitable temperatures favour some form of leaf exchanging or semi-evergreen habit.

Seasonal climates with a period when growth is difficult favour deciduous leaves. In seasonally dry sub-tropical forests, trees tend to be 'rain-green': they flush their leaves during the wet season and lose their leaves as water becomes unavailable throughout the dry season. In seasonally cold environments with relatively high levels of soil fertil-ity, deciduous broadleaved trees are favoured as sufficient nutrients are available to replace the leaves every year. Therefore, highly seasonal environments, whether warm-cold or wet-dry, tend to favour deciduous habits, as leaves can only be productive dur-ing a favourable period. An excellent review of the adaptive significance of different foliar habits is presented by Givnish (2002).

Leaf Phenology

Phenology is the study of the timing of naturally recurring life-history events in relation to climate (Larcher 2003), such as when new leaves emerge in spring or flowers open. Close observation of leaf emergence and other phenological events, such as flowering and leaf senescence, can also be instructive when evaluating any change in climate.

(a)

(b)

Figure 3.20 Early leaf emergence benefits understorey species. (a) Photosynthesis is much easier in spring for understorey species if they expand their leaves before the overstorey. The understorey is predominantly hawthorn *Crateagus monogyna*, in this shelterbelt. (b) By summer the canopy has closed, shading the understorey and reducing the ability to photosynthesise effectively.

Indeed, insights into the relationship between the phenological events and climate have informed much of the traditional wisdom of farmers around the world.

In temperate environments, warmer spring temperatures are the primary trigger for leaf emergence on deciduous trees. Although the precise physiological mechanisms are still quite poorly understood, it is clear that most temperate species are very sensitive to temperature and often its interaction with day length: some, such as European ash *Fraxinus excelsior*, respond almost entirely to temperature; others, such as European beech *Fagus sylvatica*, respond much more to day length. Contrasting responses to temperature and day length account for the observed variation in leaf emergence between years (Lechowicz 1984). It is also a major factor influencing the way different species are responding to climate change. Control over leaf emergence is required so that leaves are produced as early as possible, minimising the likelihood of leaves being damaged by frost. Early emergence is particularly important for understorey species, as they can be highly productive in the higher light conditions prior to canopy closure and rely on these periods to see them through the more shaded days of summer (Figure 3.20). Even saplings of overstorey species, such as sugar maple *Acer saccharum*, usually produce leaves a week or two earlier than the canopy trees above them to capitalise on the increased light availability prior to canopy closure later in spring (Augspurger and Bartlett 2003; Kwit *et al.* 2010).

Two components of temperature influence the timing of leaf emergence in temperate trees: sufficient chilling in winter and warm temperatures in spring. In order to break dormancy, most temperate deciduous trees have a chilling requirement that must be met. The threshold temperature varies between species but temperatures above ~12 °C seem not to contribute to chilling; freezing is not necessary: 6 °C seems to be the optimal chilling temperature with 0–10 °C being the most important range (Polgar and Primack 2011). If most temperate trees are given warm conditions before their chilling requirements are met, they are very reluctant to start spring growth. This makes good sense because it prevents the tree from bursting into spring growth during a warm spell in the middle of winter. After chilling requirements are met, the accumulation of heat

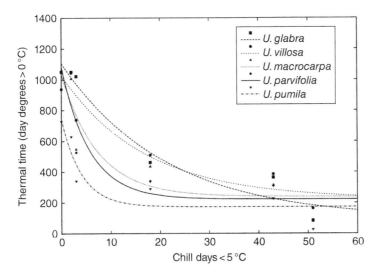

Figure 3.21 The thermal time to budburst (see text for definition) decreases for five elm species (*Ulmus* spp.) as they are exposed to more days with mean temperatures below 5 °C. With increased chilling, buds need less spring heat to trigger their opening, but despite the fact that these elms are related (members of the same genus), the exact relationships between thermal time, chill days and budburst are quite varied. *Source:* Ghelardini *et al.* (2010). Reproduced with permission of Oxford University Press.

energy becomes an important factor in the timing of leaf emergence. This is often referred to as a *forcing requirement* or *thermal time* and quantified in *degree-days*: a cumulative measure of time above a critical temperature threshold representing the minimum threshold temperature for growth. However, there are various approaches to calculating thermal time so care is needed when interpreting different studies (a useful discussion can be found in Jones 2014). The precise forcing requirement is under strong genetic control so is difficult to change and helps explain why some species are slower in getting going in the spring than others.

Crucially, leaves emerge sooner if given a longer chilling period. As an example, Figure 3.21 shows that for elms *Ulmus* spp. the thermal time to budburst is reduced as the number of chill days is increased. In other words, cold winters require less spring warming to trigger leaf emergence than milder winters, but it also shows that trees will eventually start growing even when chilling has been insufficient. This means that if there has been a very mild winter without sufficient chilling (which beech *Fagus sylvatica*, with a high chilling requirement, is currently experiencing in the UK), some species will only start growing when the accumulated spring warmth overcomes the dormancy; this comes at the cost of early spring growth though. In a study of 36 woody species, lack of chilling led to a considerable delay in budburst and substantial changes to the order of those species' budburst (Laube *et al.* 2014). This interplay of winter and spring temperatures can have important implications because milder winters are likely to favour species with lower chilling requirements, providing they do not succumb to spring frosts.

In some species, day length (more accurately referred to as 'photoperiod' because plants are really measuring night-time length) has an important role in regulating leaf emergence. In these species, photoperiod sensitivity is an additional mechanism that prevents them from being duped by abnormal warm spells in winter. Perhaps because temperature cues are less reliable inside forests (i.e. they do not reflect the temperature extremes of adjacent open ground), species that grow up under a canopy of trees seem to be more sensitive to photoperiod than early successional species where it plays only a minor part (Basler and Körner 2012).

The interaction of temperature and photoperiod therefore results in three broad types of temperate trees:

1) those with minor winter chilling requirement and the need for warm spring temperatures (forcing requirement);
2) those with a chilling requirement and a forcing requirement;
3) those with a chilling requirement, a forcing requirement and a photoperiod requirement (Polgar and Primack 2013).

The first category includes species from warmer climates; the second category is most widespread in temperate trees; and the third category includes mostly late successional species that grow under the canopy of other trees.

Autumn senescence of leaves, in contrast, is controlled primarily by day length with just a few species, such as sweetgums *Liquidambar* spp., responding mainly to temperature. The reason for this is that temperate trees have to get ready for winter before they are stopped by freezing conditions, allowing the orderly resorption of nutrients from leaves before they are deliberately shed (for more detail see Chapter 8). Thus, trees need to be warned of approaching winter before it arrives; photoperiod is the most reliable signal. Certainly, the genes that control cold hardiness in eastern cottonwood *Populus deltoides* are switched on by short days before low temperatures are experienced (Park et al. 2008). The truth of this is readily seen in trees growing near streetlights (Figure 3.22), where leaves with artificially longer days (in reality, it is the shorter nights) retain their leaves for a much longer period. However, temperature does have a role because it alters the speed of preparation for winter: in a warmer autumn, it takes longer as trees hedge their bets and keep growing just a little longer to make the most of the growing season. This is particularly true of the younger leaves produced by indeterminate growth (see Chapter 2).

It is worth pointing out here that the shedding or *abscission* of leaves is a deliberate process that costs the tree energy; it is not the leaves just dying and falling off. Looking at a branch that has been snapped off can easily prove this: the leaves wither and die but are remarkably hard to pull off. Once the useful components of the leaf have been resorbed back into the tree, an abscission zone forms at the base of the leaf (Figure 3.23). This consists of layers of corky cells lacking in lignin, similar to those in the bark. Under the control of plant hormones (mostly ethylene and auxin), the cells loosen their contact with each other and eventually the leaf falls by tearing through the corky tissue. The remaining corky cells on the branch then cover the wound (leaf scar) left by the falling leaf, effectively sealing it against water loss and the entry of pathogens.

Knowledge of phenology can be useful when trying to predict what climate change is likely to do to both managed and natural landscapes. Over time, shifts in the timing of leaf emergence may result in the expansion, or contraction, of the natural range of different species. For example, species that require little winter chilling may leaf-out progressively earlier in spring and have a longer growing season; these may grow

Figure 3.22 Most of the leaves of this Norway maple *Acer platanoides* have been shed in response to shortening day length in the autumn, but the leaves immediately around the streetlight on the left are still experiencing long days and are fooled into keeping going for longer.

better and outcompete those species that have more stringent chilling requirements. Where budburst is delayed because of unmet chilling requirements, the growing season is shorter.

Tree Crowns

When we think of trees, we often conjure up images of tall, graceful giants, but in the 'race for light' the competitors set the ultimate distance run: the goal of a successful tree is not to grow tall per se but to grow taller than neighbouring trees (King 1990). Of course, the alternative to getting tall is to find a way to survive in low-light conditions. Nevertheless, the arboreal giants of the world would never have grown so big if they were not trying to keep up with their neighbours. Growing tall and holding the leaves out to catch the maximum amount of sunlight requires a woody skeleton. Forces from gravity and wind act on the woody framework, so a significant amount of biomass is dedicated to ensuring it provides sufficient support. Water and nutrient resources must

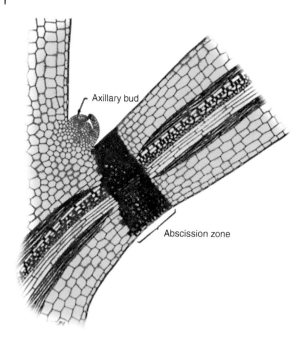

Figure 3.23 An abscission zone forming at the base of a leaf. *Source:* Weier *et al.* (1982). Reproduced with permission of Thomas Rost.

Axillary bud

Abscission zone

be carried up to the leaves through this framework: an extensive vascular system (xylem and phloem) is required. This creates a problem for trees because larger frameworks of stems and branches are increasingly less efficient to construct. Vertical height confers an advantage in capturing light, but it disproportionately extends the cost of structural support and conductive pathways required to hold the leaves and maintain their function. Shorter trees reduce the investment required for support and the length of the conductive pathway, but they become vulnerable to excessive shading by taller neighbours. These contrasting requirements mean that each crown always represents a compromise (Valladares and Niinemets 2007).

This compromise in crown design becomes more complicated when environmental stress is added to the equation. Species adapted to environments with a high degree of mechanical stress tend to lack competitive ability in less stressful environments; species adapted to the less stressful habitats tend to be unsuccessful on mechanically stressful sites (Givnish 1995). This concept is elegantly illustrated by a study on the distribution of woody plants within and around avalanche tracks in the Canadian Rocky Mountains. Johnson (1987) found that short, shrubby stems of dwarf birch *Betula glandulosa* and grey willow *Salix glauca* were able to bend to the ground without snapping, but the taller, thicker stems of Engelmann spruce *Picea engelmannii* and lodgepole pine *Pinus contorta* were less flexible and snapped during avalanches, even though they were stronger. However, without avalanches, *Picea* and *Pinus* were able to overtop *Betula* and *Salix* in about 15–20 years. As a result, *Betula* and *Salix* dominated the high, steep avalanche tracks, whilst *Picea* and *Pinus* dominated the surrounding lower elevation areas with gentler slopes and fewer avalanches. Therefore, the balance between mechanical safety and competitive ability for light can be crucial in determining the success of a species within any given environment.

The environment in which trees grow influences this compromise between gaining access to light and the cost of growing a larger woody skeleton. Some trees do not even bother trying to grow up above their neighbours. In any forest, some tree species grow up to form the canopy while others form distinctive layers (strata) of understorey trees. A key insight into how these trees can gain enough light is given by Terbough (1985) who looked at how light penetrates into a tree canopy. As the sun moves across the sky, gaps in the forest canopy lead to sunlight penetrating to the forest floor over a wide range of angles which, over the course of the day, results in a triangular area of light under the uppermost forest canopy (Figure 3.24). If you imagine the sun sweeping like a searchlight across these 'light triangles' during the day, you will see that taller trees in one of these triangles receive more hours of sunlight than shorter trees. However, where there are multiple gaps in the canopy, the light triangles overlap, and short trees may receive direct light many times a day (corresponding to the number of gaps supplying light to any particular area), even if only for a short time on each occasion. The outcome of these complex gaps in the forest canopy is more spatially uniform light near the

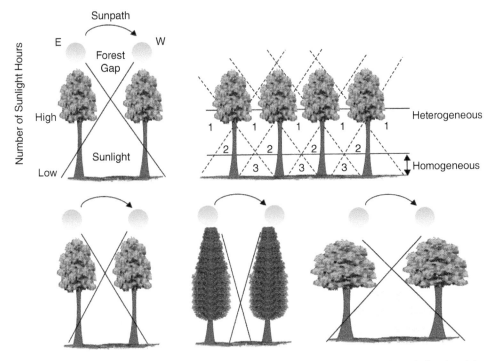

Figure 3.24 A gap in the forest canopy allows direct sunlight to reach the ground in a 'light triangle' (upper left). The number of sunlight hours a tree experiences within a single triangle increases from the ground to the canopy. When the sunlight passing through more than one gap is considered, however, a more complex pattern is found (upper right), with understorey areas affected by one, two, three or more neighbour gaps (indicated by numbers). Where the cones of several gaps intersect, a uniform or homogeneous light field is produced. Both the distance between the trees and the shape of the crown of these trees determine the duration of direct sunlight in the understorey (lower figures). Pyramidal crowns allow little sunlight to reach the understory, whereas the reverse is true for flat and broad crowns. Adapted from Terborgh (1992). *Source:* Valladares (2007). Reproduced with permission of Taylor and Francis Group, LLC, a division of Informa plc.

ground (the homogeneous light field in Figure 3.24) and more variable light just underneath the main canopy (the heterogeneous light field in Figure 3.24). This theoretical analysis predicts that some specialist species grow to the upper limit of the lower homogeneous light field and no higher, because growth above this point may prove too expensive given that at least part of the tree's crown would be in constant shade; it would not therefore receive enough light to pay for its construction and maintenance costs. This helps to explain why some forest species only seem to grow to a certain height, even when provided with unlimited light: their ecological heritage has dictated the upper range of height and adjusted the species' capacity for vertical growth accordingly. These predictions have been found to be true in mature temperate forests of North America (and presumably will be true in other temperate regions) but tropical forests seem to be too complex for such a simple solution to getting enough light (Terbough 1992). Support for this idea also comes from a global analysis of around 1700 woody plants that showed that woody plants tend to be either around 2.5 m or around 25 m in height (Scheffer *et al.* 2014). The shape and depth of the crown can also influence the extent of the light triangle that reaches the ground. Shallower, rounder crowns, typical of lower latitudes, allow more light through gaps than the deeper, conical crowns found in higher latitude boreal environments (Figure 3.24). This may help to explain why the biodiversity of the boreal forest understorey is poor and the tropical understorey is rich (Valladares and Niinemets 2007). Low solar angles at high latitudes are also likely to influence this effect.

For understorey trees growing on slopes, the equation for gaining the maximum amount of light is more complex. Light gradients underneath the canopy run parallel to the slope, so the most effective direction of growth will be at right angles to the ground (Figure 3.25). The natural consequence of this is that understorey trees are required to lean, and this requires further investment in support material to compensate for the leaning trunk: trunks that deviate from the vertical need to be stronger because of the additional loading to the trunk. The resolution tends to be that understorey trees on slopes grow neither vertically nor perpendicular to the slope, but somewhere in between depending on their shade tolerance (Alexander 1997), as shown in Figure 3.25. An alternative option to cope with the horizontally irregular light gradients is with an asymmetrical crown that has more branches on the downhill side (Sumida *et al.* 2002; Matsuzaki *et al.* 2006). Although this solution does not require the whole trunk to change direction, it will still require additional construction costs, including the production of reaction wood, to protect the longer branches from snapping.

Whilst taller trees compete more effectively for light, they also require more light to sustain a greater structural biomass. Growing tall will certainly allow the tree to intercept more light but the gain must be more than the extra costs in reaching the light. Moreover, although getting enough light to grow is an important factor, tree height can also be limited by competition with other vegetation for nutrients and water, and by environmental factors such as high winds and instability of the ground.

Shape of Tree Crowns

Tree crowns are characterised by much more than their height. It is often possible to identify a particular species from its shape and subtleties of form. Having said this, the shape of the crown can be highly variable between species and may even vary within

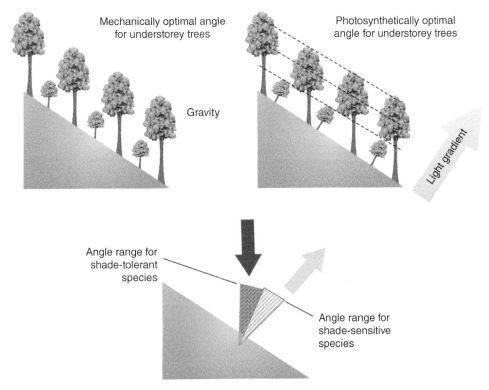

Figure 3.25 Understorey trees that grow on a slope are faced with the choice of growing vertically, which is mechanically optimal (top left), or with their trunks inclined downward according to the light gradient that occurs from the ground to upper canopy, the photosynthetically optimal angle of growth (top right). Depending on their light requirement or shade tolerance, species are expected to exhibit two ranges of trunk angle, as shown in the lower figure. Adapted from: Alexander (1997) and, Ishii and Higashi (1997). *Source:* Valladares (2007). Reproduced with permission of Taylor and Francis Group, LLC, a division of Informa plc.

species across environmental gradients. In forests, the shape of a crown is largely determined by the space that it fills and genetic control of shoot development. Conversely, in open-grown situations, shape is much less encumbered by competing vegetation and its true shape is likely to be seen. Knowledge of crown form is important when specifying trees for particular planting sites, as it can have a big impact on the tree's use (e.g. shade, wind interception or aesthetic impact) within the landscape. Crown form may also have a large impact on the future management of the site: it may be unwise to plant a broad-spreading oak right next to a road.

Variation in the natural shape of tree crowns largely reflects different strategies to 'capture' as much light as possible. The most efficient crown is where all leaves receive enough light to photosynthesise effectively. In complex tree crowns this is virtually impossible to achieve, partly as a consequence of the sheer number of leaves but also because of the changing daily and seasonal patterns of light. In highly diverse tropical forest canopies, a seminal study on tree morphology by Hallé *et al.* (1978) identified about 25 different architectural forms of tree. This suggests that the fundamental

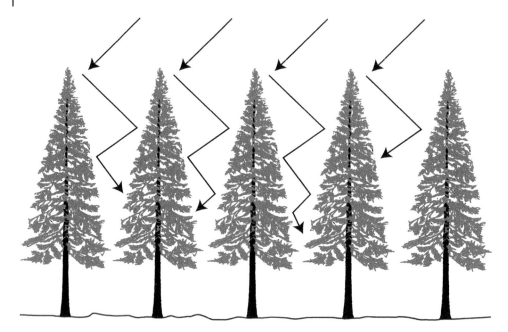

Figure 3.26 The conical shape of boreal conifers assists these trees in intercepting light when the sun is low in the sky, and foliage properties help to scatter light deep within the crowns to maximise light interception. *Source:* Thomas (2014). Reproduced with permission of Cambridge University Press.

trade-off between light capture and crown structure can only be resolved in a relatively few ways. It is therefore possible to make some generalisations on crown form as crown shape can often be linked to the environment in which the trees are found.

Conifers with a pyramidal or conical shape dominate the boreal forest canopy. This shape helps shed snow easily by having relatively short, flexible branches. However, even more importantly, this crown shape facilitates greater light interception when the sun is at a lower angle in the sky, as it is for much of the growing season in the far north. In addition, when these crowns occur in stands, light tends to be scattered downwards within the canopy (Figure 3.26), resulting in a very efficient system for capturing light at high latitudes (Walker and Kenkel 2000). This is the main reason why conifer forests look much darker in aerial pictures than the colour of the foliage would suggest – they reflect very little light. Such efficient absorption of radiant energy has also been shown to raise the inner crown temperatures by 5–10 °C. In cold environments, this is particularly valuable as it can extend the length of the effective growing season (Smith and Carter 1988). Other factors, such as their evergreen leaf habit (with the exception of larches *Larix* spp.) and a greater resilience to freeze–thaw cavitation, also contribute to the dominance of conifers in this region. This conical shape is also found in conifers growing in hot, arid areas further south. In this environment, the shape helps the tree intercept the least light when the sun is overhead; this reduces the heating of foliage and so helps reduce the amount of water the tree needs to transpire in order to keep cool.

At low latitudes, the sun traverses the sky at much higher angles and flatter-topped trees become more efficient for light capture (Kuuluvainen 1992). This may also have other benefits. By creating an aerofoil form, winds tends to pass above and below the

(a)

(b)

Figure 3.27 (a) The aerofoil form of a savannah *Vachellia sp.* (formally members of *Acacia*) in the Maasai Mara national reserve, Kenya. This crown shape helps reduce the impact of drying winds; helps to shade the stem; and keeps valuable foliage out of reach from most browsers. (b) The same tree is also seen at distance showing the scarcity of other trees. Giraffes can obviously still reach the foliage but because they help pollinate the trees by pollen sticking to their faces, some loss of foliage is a low price to pay.

crown rather than through it, so leaves can shelter each other from drying winds and provide shade for their stem. In this case, cooling is mostly by convection upwards of hot air drawing cooler air in to the bottom of the canopy. A number of Mediterranean trees, such as the stone pine *Pinus pinea*, and savannah trees, such as *Vachellia* spp. (Figure 3.27), have this crown shape. In the African savannah there is also an advantage in keeping the leaves above the height of most hungry browsers. Despite this, there are other crown shapes in Mediterranean and savannah environments, so not all species adopt this strategy.

In mid-latitudes dominated by more temperate environments, a more rounded crown is typical. Cloudy skies in these humid environments result in diffuse light from the entire sky, rather than light from a point source as in the clear skies of hotter climates. Great domes of foliage (Figure 3.28) capture light most effectively in the cloudier temperate regions – even if the sun does come out occasionally!

Role of Branches in Tree Crowns

When plants first evolved, a modest increase in height was possible using tissues reinforced with lignin that gave them sufficient advantage in light interception to ensure success. However, this was not adequate for long (relative to geological time), and fossil evidence suggests progymnosperms, such as *Archaeopteris*, from ~370 million years ago were amongst the first 'trees' to produce a crown with perennial branches (Meyer-Berthaud *et al.* 1999). Branching crowns subsequently evolved in many plant groups as a way to intercept more light and compete with neighbouring plants. At least in terms of light capture, branching has been one of the most important innovations in crown design. While it is clear that numerous trees remain successful without complex branching (e.g. palm trees and tree ferns; Figure 3.29), it remains a vital characteristic of more recently evolved trees.

Branching increases the number of growing points (apical meristems) of the shoot system and reduces the impact of damage to any one apical meristem. Injury to the

Figure 3.28 An open-grown chestnut-leaved oak *Quercus castaneifolia* in the Royal Botanic Gardens, Kew. This great dome of foliage captures diffuse light from a cloudy sky very efficiently.

apical meristem in the simple crowns of palms and tree ferns can be catastrophic for the whole plant. Trees with branched crowns will normally recover from injury to one or a few apical meristems as they are less reliant on an individual growing point for success. However, it was probably the photosynthetic advantage gained from more effective light interception that led to the evolutionary development of branches in trees, but, as with so many aspects of plant design, trade-offs must be made. With the advantage of a larger crown comes the disadvantage of additional support requirements.

Biomechanical Design of Tree Crowns

Trees develop wide trunks and thick branches to support their crown, in many cases, tens of metres above the ground. All this woody material can weigh thousands of kilograms, but despite wood being about twice as strong in tension as it is in compression, its ability to resist compression still way exceeds the gravitational load exerted on trees. Assuming that the density of wood is half that of water ($0.5\,\mathrm{g\,cm^{-3}}$ or $500\,\mathrm{kg\,m^{-3}}$),[1] wood has a compressive strength of around 50 megapascals or 50 meganewtons per square metre. This means that trees could reach around 10 km in height before they were at

---

1 Wood actually varies enormously in density from about 0.1 grams per cubic centimeter ($\mathrm{g\,cm^{-3}}$) to about $1.4\,\mathrm{g\,cm^{-3}}$ (i.e. more dense than water, so some woods actually sink).

Figure 3.29 Tree ferns, such as this *Cyathea* sp. in Sherbrooke Forest, Victoria, Australia, have very simple, unbranched crowns that are held on a single stem. Despite this rudimentary form, they remain an important part of the understorey in many forest ecosystems.

risk of crushing themselves by their own weight – even without accounting for taper (Vogel 2012). Trees simply do not operate anywhere near their limit with regards to resisting their own weight under gravitational loading.

A far more significant risk of failure comes from *Euler buckling*. This occurs when loading on the tree causes the trunk (column) to deflect from vertical, and bend beyond some critical limit: failure can then occur without any additional downward force (Niklas and Spatz 2014). It is easy to illustrate this if you hold a piece of dry spaghetti at its base vertically on a table, and apply a downward force on its top end: once the spaghetti bends beyond a certain point, it will break very easily. If the concept of Euler buckling is applied to trees, more realistic maximum heights can be predicted.[2] By making a few assumptions – that the tree is rooted in unyielding ground; the top of the tree is able to freely move; the load forces are concentrated towards the top of the trunk; the cross-sectional shape is cylindrical with no taper; and average values are used for wood density and stiffness – the theoretical height limit for trees would be around 115 m (Vogel 2012). This seems a much more realistic limit. However, factors such as stem taper and loading more evenly along the column would increase this theoretical maximum height of trees. Therefore, for most trees, height varies with trunk diameter

2 For detailed descriptions of the mathematics of Euler buckling, see Vogel (2012) or Niklas and Spatz (2014).

in such a way that there is a considerable margin of safety against buckling in high winds (Niklas and Spatz 2014). For open-grown trees, where competition for light is small and an increase in height would be of no major advantage, trees are often only about 25% of their theoretical buckling height limit (McMahon 1975). In forest environments, where there are large advantages of height in light interception and trees are sheltered from the wind, safety margins tend to be smaller. In the tropical lowland forests of Malaysia, tree heights of 91 different species were assessed and evaluated against their theoretical buckling height. On average, these forest trees reached 65% of their theoretical buckling height although some trees reached 88%, suggesting that in sheltered environments they were operating close to biomechanical limits (King *et al.* 2009).

Gravity is the overwhelmingly dominant force once a tree has failed, but it has little to do with the initiation of toppling in tall trees, as most trees will not grow close to their biomechanical limit in terms of height. A more significant force acting on the tree is provided by wind. When looking at how a tree reacts, it is much more helpful to think of a tree (and its branches) as a cantilever beam(s) (i.e. a beam that is attached at just one end). Tree crowns are loaded by the weight of their branches, leaves and fruits, but *drag*, caused by the wind pulling the crown in a leeward direction, is the most important force at play. The combination of drag (generated by wind moving through the crown) and the long lever-arm (also referred to as the *effort arm*) of the trunk generates a large *turning moment* (*torque*) that acts to wrench the tree from the ground.

The turning moment is greatest at the point of attachment with the ground (where the lever is longest), so trees are more likely to fail to hold on to the ground than they are to fail at some point higher up their trunk. In fact, catastrophic trunk failure in the absence of decay is quite rare and, when it does occur, is usually associated with high-speed gusts of wind. It is much more likely that trees fail to hold on to the ground, either because their roots have failed in some way (again, this is often associated with decay) or because the soil itself has insufficient strength to resist the forces being transmitted through the tree. Therefore, a tree's ability to be able to anchor itself to the ground is critical for keeping the crown upright (for more detail on how trees achieve this see Chapter 4).

Branches also act as cantilever beams, so the most biomechanically vulnerable parts of a crown are found at the junction between two stems. This is partly because of the branch weight, particularly when fairly horizontal branches are loaded by snow or ice, but mostly because of the drag forces caused by the wind acting on the 'sail-area' of the branch. As with the whole tree, these forces increase their potency as the branch (lever-arm) gets longer. The longer the lever, or the more force applied to the lever, the greater the turning moment at the base of the stem (Niklas and Spatz 2014). The force generated by gravity acting to break a branch at the base may be relatively modest in quite upright branches (where the length of the lever parallel to the ground is small), but it becomes acute in more horizontal branches that produce a longer lever. On the other hand, more upright branches may have to contend with greater forces from the wind: as they grow higher, they are exposed to more wind (more drag force) and have longer lever-arms. In any case, the junction (or union) between the stem and branch (or between co-dominant stems) must allow the transport of sap to and from the leaves, as well as be biomechanically robust.

How, then, do trees ensure their crown is biomechanically safe? In short, they have a range of responses that either reduce drag, or reinforce the key points of biomechanical vulnerability. Critically, these responses require stimulus and so will not occur unless the tree experiences mechanical stress. In much the same way as we can build our muscles through exercise, trees need 'exercise' if they are to reinforce their 'bodies'. For this reason, trees that have grown up in the middle of a sheltered forest that are suddenly exposed to wind loading because surrounding trees have been removed, fail in quite moderate winds. Similarly, high winds from an unusual direction can be more damaging than high winds from the prevailing direction. Branches also require mechanical stimulus to form strong attachments; where this does not happen, weak unions very often form (see Expert Box 3.1).

Trees achieve stability by having thick trunks and scaffold branches that are inherently rigid, but smaller-diameter branches and twigs are much more flexible, mostly because they are thinner rather than any major change in the material properties of the wood. This flexibility means that the outer branches of the crown, which bear the most leaves, can bend in the wind and temporarily change the crown's shape so that it reduces the drag force transmitted to the trunk (Niklas 1992). Tests in wind tunnels have shown that this *crown reconfiguration* in 5 m pine trees exposed to high winds can reduce the drag to under a third of what it would be if the crown was rigid. Angiosperm trees can perform even better than conifers in this respect (Ennos 2016). Pioneer trees that specialise in growing in open and exposed situations are particularly good at making these temporary changes to crown shape, as they often have very flexible branches. Just look at the way species like silver and downy birch *Betula pendula* and *B. pubescens* bend and reshape their crowns in strong winds.

In addition to the general reconfiguration of tree crowns, leaves also temporarily deform in the wind to reduce drag. They deform in at least four distinct ways (Vogel 2009, 2012). Simple, more isolated leaves may curl up into cones, with the underside of the leaf forming the outside of the cone (Figure 3.30a). Clusters of leaves often act together to form a cone with lower leaves pressing themselves against those higher up the stem (Figure 3.30b). Pinnate leaves with a large number of individual leaflets tend to form elongated cylinders, with each leaflet pressing against the next (Figure 3.30c). In the same way, simple leaves along a branch can fold down against one another parallel to the branch, again forming a cylinder-like arrangement (Figure 3.30d). An interesting, although highly complex, avenue of future research would be to see how much the deformation of leaves reduces the wind loading of the whole tree. The fact that leaves from a wide range of families deform to reduce drag suggests that it is beneficial to the whole tree.

Permanent changes to the shape of tree crowns occur whenever a crown experiences consistently strong wind from one direction. Consistent pressure on the windward branches, the desiccating effect of the wind on buds and the impact of wind-borne particles all lead to the crown developing away from the wind in a growth response known as *flagging* (Figure 3.31). This reduces drag whilst maintaining leaf area. Extreme examples of this are found in the almost prostrate forms of trees growing in Arctic regions and close to the treeline on mountainsides. To increase the margin of safety against mechanical failure in windy locations, trees also alter their growth through a process known as *thigmomorphogenesis* (Greek: thigmo – touch; morphê – shape; genesis – creation): literally, the creation of a new shape by touch; in the case of trees this most often relates to the 'touch' of the wind (Telewski 1995). In open-grown trees and trees growing on the windward edge of a group of trees, exposure to wind leads to

(a)

(b)

(c)

(d)

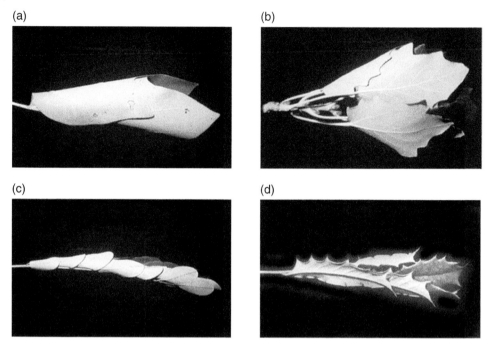

Figure 3.30 Leaves exposed to turbulent wind at 20 m s⁻¹ (45 miles per hour or 72 kilometres per hour): (a) tulip tree *Liriodendron tulipfera*; (b) white poplar *Populus alba*; (c) false acacia *Robinia pseudoacacia*; and (d) American holly *Ilex opaca*. *Source:* Vogel (2009). Reproduced with permission of John Wiley and Sons.

(a)

(b)

Figure 3.31 A wind-altered crown. This reduces drag on the tree as shown in (a) hawthorn *Crateagus monogyna* on a coastal site overlooking Morecambe Bay, UK and (b) beech *Fagus sylvatica* found in an exposed mountain location in Corsica, France.

thicker structural roots, a thicker trunk with increased stem taper and a reduction in tree height, compared to their more sheltered counterparts. This change in stature is particularly apparent in small groups of trees (copses) in exposed locations, where the net result is that the shape of the whole copse takes on a more aerodynamic form.

Thigmomorphogenesis is also responsible for growth that reinforces potential biomechanical weaknesses within the tree. Most obviously, this results in additional wood at the base of the trunk and the production of a trunk flare. If trees do not experience any crown movement, then this adaptive growth does not occur, leaving the tree vulnerable to failure. For this reason, it is important to support newly planted trees at a low point on their stem (less than one-third of the height of the tree) to allow the crown to move and stimulate the reinforcing growth. Trees that have been supported at too high a point have a much reduced stem taper and may not even be fully self-supporting. Furthermore, it is important not to purchase trees that do not have a good stem taper, as it is likely that they have been over-supported in the nursery and will not make structurally sound landscape trees. Adaptive growth is also associated with additional wood around a cavity or region of decay and the reinforcement of weak stem unions. It may be associated with a particular type of wood, known as reaction wood (see later), or may simply be additional growth. For the tree, these responses help maintain the biomechanical integrity of the crown but for us the modifications can be highly instructive when trying to determine the points of vulnerability within a tree's structure. Experienced tree-risk assessors will use these visual symptoms to help to evaluate the risk posed by the tree within the context of where it is growing.

As well as reinforcing the point of attachment with the ground, it stands to reason that trees will also do all they can to make sure that stem unions (junctions) within the crown are as strong as they can be. This has led to modifications within tree junctions that are highly effective from a biomechanical perspective. Evaluation of the strength of bifurcations (forks) in hazel *Corylus avellana* demonstrated that the central fifth of the fork was not only ~20% more dense than adjacent stem wood, but was responsible for about one-third of its strength (Slater and Ennos 2013). Fewer, narrower, shorter and more tortuous vessels, combined with thick-walled fibres and parenchyma, at least partly explains this increase in wood density. In addition, the tortuosity of the vessels at the apex of the junction acts to interlock the tissues going from the parent stem into the arising branch (Figures 3.32 and 3.33); this aligns at least some fibres in their strongest plane (axial) between the adjoining stems, adding extra strength to the attachment (Slater *et al.* 2014). The fact that this whorled grain is found at the top of branch unions in a wide range of species confirms its widespread importance in the biomechanical strength of branch attachment. Where the formation of this interlocking whorled grain is prevented from forming as the result of included bark (bark enclosed inside the junction), the branch attachment loses about 25% of strength, making it more vulnerable to mechanical failure (Slater and Ennos 2015). Duncan Slater discusses the significance of bark-included branch junctions further in Expert Box 3.1.

Modification of the branch junction can only give so much strength. There comes a point when the branch junction with the trunk will not be strong enough to resist the forces imposed upon it. A number of figs *Ficus* spp. have come up with a solution to solve this problem: they produce aerial roots that grow down from branches to form subsidiary trunks (Figure 3.34). These act as natural props and allow extensive sideways growth of the crowns. In urban environments of tropical or sub-tropical regions, it can be quite a challenge to manage these seemingly ever-expanding crowns with their abundant aerial roots, but they do add tremendously to the 'sense of place' in cities such as Hong Kong.

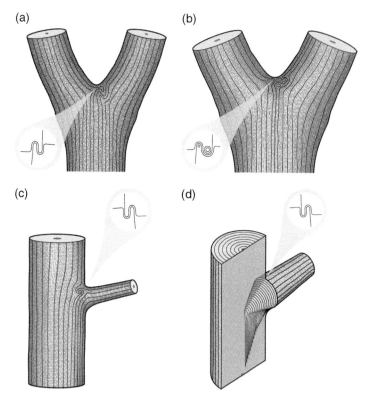

Figure 3.32 Diagrammatic representation of the interlocking grain found at the apex of stem unions in trees: (a,b) show the development of interlocking grain within a fork (see also Figure 3.35); (c,d) represent the interlocking grain being formed at a branch-to-stem attachment. *Source:* Courtesy of Duncan Slater.

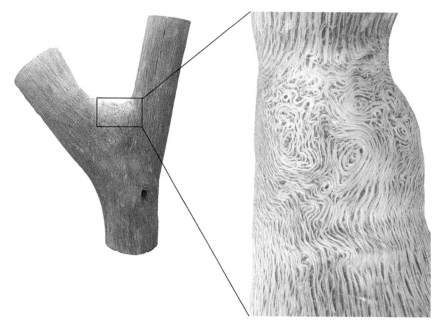

Figure 3.33 Whorled grain at the apex of the branch junction has been shown to have an important biomechanical role in branch attachment. The bark has been removed from this sample so that the grain can easily be seen.

Figure 3.34 Indian banyan *Ficus benghalensis* has aerial roots that grow down from lateral branches and act as natural props, facilitating very wide spreading crowns. This is a relatively small specimen growing in Brisbane City Botanical Gardens, Australia but a champion tree growing in the Royal Botanic Gardens of Calcutta, India, however, covers 1.3 hectares and has 2800 subsidiary trunks formed by these aerial roots.

Expert Box 3.1 Bark-included branch junctions and natural bracing in trees
Duncan R. Slater

Plants adapt their growth and form as a result of a variety of factors. The process of adapting plant shape from mechanical stimulus is known as thigmomorphogenesis (Jaffe 1973; Braam and Davis 1990; Jaffe and Forbes 1993, Telewski 2006).

Providing sufficient mechanical strength within the wood formed at a branch junction is reliant upon the adequate development of at least one of two key anatomical features. We will briefly consider these two anatomical features here, to provide context for the formation of bark-included junctions in woody shrubs and trees.

An anatomical feature of branch junctions that can provide additional mechanical support is the embedding of a smaller branch into the larger branch or stem at the join. When the base of the slower-growing and smaller-diameter branch is fully embedded into the tissues of the other, these embedded tissues are commonly referred to as a 'knot'. An increased level of embeddedness gives significant mechanical support to the branch junction (Kane *et al.* 2008), but this anatomical feature is typically absent where the two branches that form the junction are roughly equal in diameter.

The other anatomical feature that provides support is the specialised wood formed under the branch bark ridge (BBR) that lies within the axil of the branch junction. In this location, denser wood with an interlocking grain pattern is typically formed (Slater *et al.* 2014) and this mixture of wood grain orientations in the axil is usually expressed externally by a visible ruffle on the bark surface that forms the BBR (Figure EB3.1). It is these dense,

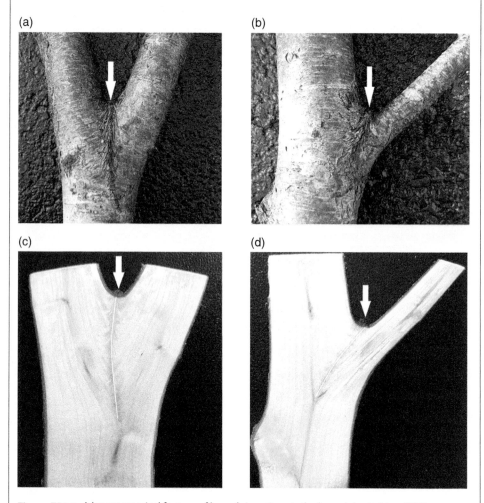

Figure EB3.1 A key anatomical feature of branch junctions is the branch bark ridge (BBR). (a) Co-dominant branch junction in silver birch *Betula pendula*, with a white arrow identifying the apex of the BBR. (b) Branch-to-stem junction formed in the same tree, with a white arrow identifying the presence of a less prominent BBR. (c) Dissection of branch junction A with a white arrow identifying the apex of the BBR and a fine white line running centrally down the denser wood tissues of mixed grain orientation formed centrally under the BBR. (d) Dissection of branch-to-stem junction B with a white arrow identifying the apex of the BBR and a fine white line running centrally down the denser interlocking wood tissues evident in the axil of the branch and stem. This junction also has the base of the smaller branch embedded as a knot into the larger branch, which provides some mechanical support to this junction, in contrast to the co-dominant junction.

interlocking tissues of mixed wood grain orientation that conjoin branches at the apex of a normally formed junction in the aerial parts of a tree. Typically, these interlocking xylem tissues are well developed in co-dominant branch junctions and less developed where a smaller branch conjoins with a larger tree stem, as some support is provided to the latter junction by branch tissue embeddedness. Much as a knot is absent in a co-dominant branch junction, this specialised interlocking wood may be irrelevant or near to absent where a very small diameter twig is conjoined to a very large stem, because of the subsequent knot providing all the necessary mechanical support for this minor twig.

Figure EB3.1 illustrates that the specialised wood formed under the BBR is present in both co-dominant branch junctions and in typical branch-to-stem junctions, whereas it is only the branch-to-stem junctions that will also contain a knot or embedded tissues from the smaller branch into the larger branch or stem.

This anatomical model for branch junctions was initially proposed by Slater and Harbinson (2010) and subsequently validated by a series of experiments (Slater and Ennos 2013, 2015b, 2016; Slater *et al.* 2014).

The growth and development of branch junctions in trees occurs partly in response to both static loading (e.g. by the weight and moment of the branches) and dynamic loading (e.g. by movements in the branches induced by the wind). These mechanical stimuli are critical for the formation of a suitably strong branch junction. This happens as a result of partitioning off the main mechanical role of joining two young branches together to the wood formed under the BBR. This role allocation allows all other sapwood at a juvenile branch junction to be relatively normally formed and capable of conducting sap efficiently (Gartner 1995). A consequence of this partitioning of roles, however, is that when a branch junction is substantially constrained in movement and there is an absence of significant static loading because the two arising branches are relatively upright, the junction will fail to form a sufficient amount of this interlocking xylem under the BBR. Instead, bark will become included within the join, in the absence of these specialised tissues being formed (Figure EB3.2).

Weakening Effect of Bark Inclusions

Where a branch junction forms with bark included into that join, this inevitably weakens the junction mechanically. Arboriculturists are trained that this type of branch junction is more prone to failure and can allow access to diseases and decay as a result of the occlusion of seams of bark as these types of branch junction develop.

Smiley (2003) showed that where bark was included in branch junctions in red maple *Acer rubrum*, this weakened these junctions by an average of 20% when compared with the strength of normal branch junctions, using a static pulling test, but found substantial variability in junction strength. Slater and Ennos (2015a) found a similar reduction in strength in hazel *Corylus avellana* and were able to differentiate stronger bark-included junctions from weaker ones by the relative size and position of the tab of bark formed inside the junction (Figure EB3.3).

However, one needs to ask the more fundamental question: why would this happen? What is it that is holding these branch junctions static and preventing the stimulation of the growth of the important conjoining wood formed under the BBR? I have found a

Figure EB3.2 (a) A bark-included junction formed in southern beech *Lothozonia alpina*. The seam of bark (indicated by the white arrows) lies exactly where the dense wood with interlocking grain patterns would normally form under a BBR. In this case, the BBR and its associated woody tissues are wholly absent at the junction, weakening it greatly. (b) Failure of a bark-included junction in common ash *Fraxinus excelsior*. The presence of bark within a branch junction is a common cause of branch and limb failure in a wide range of tree species.

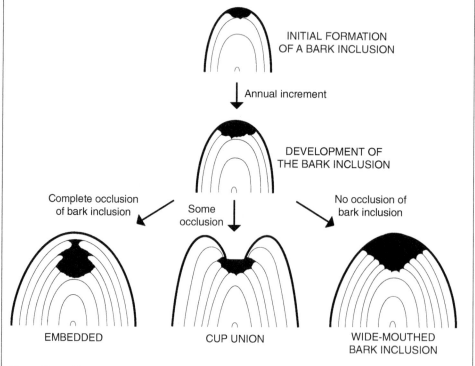

Figure EB3.3 From static pulling tests and analysis of the fracture surfaces of branch junctions in hazel *Corylus avellana*, it was found that weaker junctions are those where a larger area of bark lies at the top of the branch junction (bottom right), and other forms (embedded bark and cup unions) were significantly stronger. *Source:* Slater and Ennos (2015a). Reproduced with permission of International Society of Arboriculture (ISA).

compelling answer to this question – it is the way in which branches come to touch or rest on other branches, stems or other objects in their environment that results in the formation of the majority of these weak bark-included branch junctions.

Natural Bracing in Trees

A 'natural brace' in the crown of a tree is a physical feature that acts to restrict or wholly prevent movement at a branch junction formed lower in the tree's crown. The most common naturally occurring brace is the static touching of two branches higher up in the crown of a tree, with this branch interaction acting to brace the branch junction formed lower down in the tree or shrub (Figure EB3.4).

   If the two or more branches arising from a junction in a young tree are held static in this way, and the branches ascend near to vertical, so that little gravitational loading occurs across the branch junction below, then a bark-included junction is likely to form. As trees mature and growth slows, it takes far more time for a newly developed natural brace to induce the formation of a bark-inclusion in a junction that has developed normally for the past 30 or more years. In most cases, these weak junctions are formed when a tree is young, when it has many small internal lateral branches lower down in its crown, and at least one of these minor branches straddles a junction to brace across it by touching another branch or stem, potentially fusing or entwining with it.

Figure EB3.4 A bark-included junction (lower white arrow) formed in a semi-mature grey alder *Alnus incana* as a result of the natural brace formed higher up in the tree (upper white arrow).

Different Types of 'Natural Brace'

A field study carried out near Lancaster, UK, in 2016 identified that natural bracing was very strongly associated with the presence of bark-included junctions lower down in the tree (an association rate of 93%). The study also identified a wide range of different 'natural braces' that can occur in the crowns of trees, which have been placed into nine categories based on their different forms and likely longevity as braces (Slater 2016).

The most common forms of natural brace found are where branches within the same tree restrict movement: crossing branches, entwining branches and fused branches (Figure EB3.5).

Where a natural brace and associated bark-included branch junction are formed in a young tree by one or more small lateral branches, as the tree develops further and comes to shade out its own lower lateral branches, the death and subsequent decay of these bracing branches can open up a weak bark-included branch junction to movement which it has not experienced for decades. In a similar way, a small lateral branch acting as a brace may fail to be an effective brace as the tree grows, as its growth rate is outstripped by that of the stems it is bracing. This observation explains why the failure of bark-included junctions is particularly frequent at the semi-mature stage of a tree's growth, when its canopy is expanding quickly and slower-growing lateral branches are quickly shaded out – particularly in species in genera such as birch *Betula* spp., maple *Acer* spp. and willows *Salix* spp.

Failure of such a bark-included branch junction that is released to movement is not inevitable. When a natural brace is lost or becomes ineffective, the branch junction will

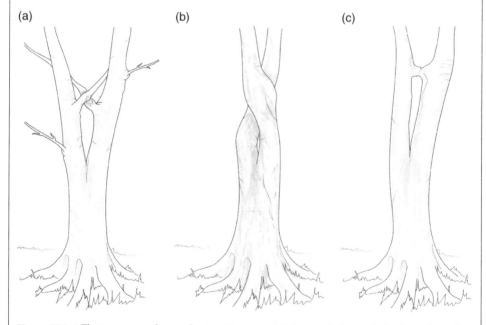

Figure EB3.5 Three common forms of natural brace and their association with the formation of bark-included branch junctions below them: (a) crossing branches; (b) entwining branches; and (c) fused branches.

Figure EB3.6 A bulging bark-included branch junction in a semi-mature beech *Fagus sylvatica* (arrow). The junction is bulged because it has been released to movement after the death of a branch that formed a natural brace above this junction for many years.

modify itself in response to this new movement. During the first growing season after its release to movement, bulges of denser wood will start to form at the base of the seam of included bark, as this is where the greatest strain to the junction will occur and where the vascular cambium straddles the junction and thus can respond to that heightened strain (Figure EB3.6). These bulges develop to restore the branch junction to adequate bending and torsional strength, if failure of the junction does not occur before this repair process is completed.

Implications for Arboricultural Practices

Arboricultural industry guidance has been, for many years, to cut out crossing branches and rubbing branches from the crowns of trees, as they were perceived as defects that needed to be addressed (BSI 2010). Our research highlights that this defect is intrinsically linked with the production of bark-included branch junctions, if these crossing branches act to 'lock-up' a junction so that it experiences little or no mechanical loading. It is better not to think of crossing branches and bark-included junctions as individual defects that need to be treated separately. The very strong association between these two defects found by our research means that the arborist needs to carry out a more comprehensive assessment of the pruning needs of trees under their care, taking into account these branch-to-branch interactions.

(a)

(b)

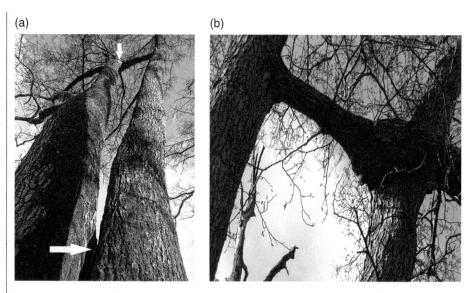

Figure EB3.7 (a) A mature common alder *Alnus glutinosa* where a lateral branch straddles across the junction between the two main stems of the tree. The lower white arrow identifies the bark-included junction with some minor bulging present, and the upper white arrow identifies the natural brace. (b) The swelling of the branch rubbing against the other stem identifies that this natural brace has been *in situ* for many years. It would be very foolish to cut away this rubbing branch, despite current industry guidance that encourages this action, for that would make it much more likely that this bark-included junction would fail and half of the tree would fall down in the next strong wind event.

In young trees it is clear that early intervention by formative pruning of the tree can prevent the production of bark-included junctions if crossing branches or entwining stems are also prevented (Gilman 2012). When a natural brace has been in place for many years, however, and it is associated with a bark-included junction in a large tree, the arborist needs to take this association into account and not to remove the natural brace in an unplanned way – unwittingly releasing a bark-included junction to dynamic movement it has not experienced for many years (Figure EB3.7).

Reaction Wood

As the crown of a tree develops, branches can adjust their direction of growth so that they move towards the light (heliotrophy) to maximise photosynthesis, or restore vertical growth to reduce biomechanical strain (gravitrophy). As lateral branches develop, they become bigger and heavier and sag, taking the leaves down into the shade of surrounding foliage. Therefore, adjustment of branches within a crown is needed to maintain leaves in an efficient position for photosynthesis. Without this capacity for adjustment, trees would be vulnerable to a wide variety of disturbances and would have limited capability to respond to dynamic environments. Therefore, as well as producing new cells, the vascular cambium is able to perceive environmental cues and change the structure of cells in addition to the amount of wood it produces at locations across the tree that experience high mechanical stress. This *reaction wood* has a specialised wood structure to resist compressive or tensile forces, and is only produced where the tree

Figure 3.35 Reaction wood. Here, compression wood in a gymnosperm is seen as the darker portion on the lower side of this stem. If this was an angiosperm stem, the tension wood would be in the upper part of the stem. The inset illustrates how it is sometimes necessary to reorientate the crown to achieve vertical growth. *Source:* Schweingruber *et al.* (2006). Reproduced with permission of Springer.

needs to adjust the direction of growth or provide reinforcement to biomechanically vulnerable areas. Stems with reaction wood are often asymmetrical because more wood is laid down on one side. This has the advantage of producing elliptical trunks and branches so that the geometry of the stem also helps resist bending in a particular direction. As differences are most pronounced on the compression side of gymnosperms and the tension side in angiosperms, reaction wood is termed *compression wood* and *tension wood*, respectively. As an example, where strong prevailing winds bend a tree away from vertical growth, a gymnosperm tree would seek to buttress the leeward (compression) side of the stem (Figure 3.35) whilst an angiosperm tree would reinforce the windward (tension) side of the stem. In the same way, gymnosperms tend to produce compression wood on the underside of a leaning stem and angiosperms tend to produce tension wood on the upperside.

Whilst it is tempting to contrast reaction wood to 'normal' wood, this implies that reaction wood is in some way not normal. This is certainly not the case as reaction wood is vital to tree development and is found in every mature tree. Therefore, when the two types of wood need to be compared, it is better to refer to use the terms *non-reaction wood* and *reaction wood*.

Compression wood is usually darker, often taking on a brown or reddish-brown appearance. Growth rings within the compression wood are usually wider than in the non-reaction wood. This results in the pith (anatomical centre of the stem) being found off-centre and closer to the side opposite the compression wood development, as seen in Figure 3.35. Compression wood is almost always denser than non-reaction wood, because its tracheids have thicker cell walls (with around 40% more lignin) and are slightly rounder than non-reaction wood.

In tension wood, the wider growth rings occur on the tension wood side but a change in colour is much less apparent. Vessels are generally less frequent and have smaller lumen, and fibres make up a higher proportion of the wood. In many angiosperm species, modifications in the fibre cell wall structure can be seen: for example, the addition of a cellulose rich gelatinous layer (G-layer) (such as in *Castanea*, *Fagus*, *Populus* and *Quercus* spp.) or a polylaminate structure. However, these modifications, if present, are not consistent across broad taxonomic groups (Ruelle 2014).

The use of reaction wood clearly results in a system that can fine-tune the orientation of stems by buttressing the stem (in the case of compression wood) or acting like a guy rope (in tension wood). This increases the overall stability of the crown, helping it to react in an efficient way as the environment around it changes or it experiences the environment differently.

Although reaction wood is clearly vital for the biomechanical integrity of the living tree, once a tree is harvested for wood it has a number of undesirable characteristics. Shrinking, warping, brittleness and weakness have all been reported as more pronounced in reaction than in non-reaction wood of the same species. Therefore, timber producers frequently seek to minimise the production of reaction wood in the trunk.

Branch Shedding as a Natural Process

As the crown expands, it is inevitable that some branches are less productive and come to the end of their useful life. This is usually when they are excessively shaded by new growth above or from neighbouring trees, and so are less productive and potentially represent a net importer rather than a net exporter of carbon. Most trees have the ability to shed or *abscise* not just unproductive leaves, but whole unproductive branches. As well as removing a carbon liability, abscission of branches (and other plant parts) allows decomposers to degrade the constituent parts and recycle nutrients. Moreover, in riverside (riparian) trees such as *Populus* and *Salix*, the loss of branches may be an important part of their vegetative reproduction strategy, as detached shoots are capable of rooting if they land in soil adjacent to the parent tree or in a riverbank further downstream (Braatne *et al.* 1996).

The process of active shedding or *abscission* of small shoots is covered by two terms that are sometimes used interchangeably in the literature. Technically, the loss of a branch with at least partially green leaves is termed *cladoptosis*; when without leaves it is referred to as *natural pruning*. A problem arises because 'natural pruning' (also referred to as 'self-pruning') includes both the active or deliberate shedding of branches, and also the passive dying and rotting away of branches; some also use it to mean the careful pruning of trees by humans to create a natural appearance.

Cladoptosis, the shedding of branches with green leaves, is particularly important in a number of conifers that shed short shoots instead of needles (e.g. *Araucaria*, *Calocedrus*, *Chamaecyparis*, *Glyptostrobus*, *Juniperus*, *Libocedrus*, *Metasequoia*, *Pinus*,

(a)

(b)

Figure 3.36 Cladoptosis in pedunculate oak *Quercus robur*. (a) Small twigs with green leaves can be seen littered around the base of this veteran tree. (b) Evidence of cladoptosis is confirmed by the clean abscission zone at the original point of attachment.

Sequoia, Taxodium and *Thuja*), but it also occurs in some angiosperm trees (e.g. *Populus, Quercus* and *Salix*) as a means of removing unproductive shoots (Figure 3.36). As well as being part of natural crown development, cladoptosis tends to increase during periods of environmental stress and so can be an important indicator of poor tree health when it occurs at levels atypical for the species concerned. Natural pruning, as a deliberate process, is used to shed larger twigs and small branches up to about 5 cm (exceptionally up to 10 cm) in diameter.

The process of shedding small-diameter branches by either cladoptosis or natural pruning involves the same mechanism. Within the branch base at the point of attachment to the parent stem, there is an abscission zone made up mainly of parenchyma cells, just as there is in leaves before they are shed. In trees that typically have a pronounced branch collar (swelling at the base of the branch), branch shedding is helped by a separation layer within the abscission zone that secretes enzymes capable of digesting the pectin-rich middle lamella and portions of the cell walls. This enables the separation of the branch from the stem, leaving a relatively smooth branch scar. In species without an obvious branch collar, a separation layer is generally not apparent but cells in the abscission layer become sufficiently weakened for separation to occur (Addicott 1991). The outcome of this branch abscission is a clean wound at the point of branch attachment that can be rapidly covered (occluded) by subsequent stem growth. Shedding of small branches in this way is characteristic of a wide range of trees, such as *Acer, Fraxinus, Juglans, Populus, Quercus* and *Salix* (Pallardy 2008).

As branches increase in size, an abscission zone has less of a role in branch shedding. Nevertheless, the lower branches of trees in dense forest stands and internal branches of larger crowns can still be shed. In these cases, natural pruning is preceded by relatively slow but progressive senescence of the branch to the point of death, presumably as a result of carbon starvation. During this process, the branch base is often impregnated with kinos, resins, gums and tyloses, depending on the species concerned (Fink 1999). This, in combination with the anatomical modifications to branch junctions described earlier (higher wood density, greater tortuosity of vessels and more vessel endings), results in a *branch protection zone* that inhibits the colonisation of the stem by all but the most specialised decay fungi. These dead branches are gradually rotted by saprotrophic fungi up to the branch protection zone and eventually fall off when they can no longer support their own weight or they are knocked or blown off. In such circumstances, a stub remains on the tree and can delay the covering or occlusion of the wound. To counter this, some species, including a number of *Eucalyptus*, can push out or extrude the branch stub over a few years by the force of the new wood growing adjacent to the wound, so that its closure is much quicker (Addicott 1982). Arborists who regularly work with *Eucalyptus* species will recognise their capacity to expel these dead stubs: it is rarely a good idea to stand on them when climbing!

A great deal of variation occurs across species in their ability to abscise branches, to form branch protection zones and naturally shed branches. Different species make different 'choices' in how energy is allocated to a range of processes. Some trees invest in more defence, and therefore longevity, while others prioritise rapid growth. Nonetheless, the understanding of these natural processes of branch abscission and shedding has contributed much towards our approach to pruning trees.

Tree Pruning

Nature has evolved highly efficient tree crowns that, although somewhat constrained by their genetics, have the capacity to respond to constantly changing environmental conditions. Tensions between the allure of more light and the investment needed to safely intercept it have been resolved; compromises made. Environmental niches have been sought and fought for; innovations have been tried and tested to secure advantage, no matter how marginal, over neighbours. Architecture, anatomy, physiology and phenology have been honed over vast expanses of time so that each generation can have the opportunity to pass its genes to the next. Therefore, it may seem wrong to suggest intervening in the design of tree crowns. Yet, tree pruning is an important operation for the management of tree crowns, especially in managed landscapes.

It is critical to realise that pruning trees has a series of biological consequences (listed in Figure 3.37) which should be considered carefully prior to the pruning operation itself; most of these act to the disadvantage of tree health and have a cascade of effects that are harmful to the tree. Perhaps the most obvious consequence of pruning will be a reduction in the leaf area of the crown. This almost certainly leads to a reduction in carbon fixation and other products associated with photosynthesis. At the same time, the wounding created increases the carbon allocated to recovery and defence, just at the time when photosynthesis is reduced. Cutting off stems, particularly if they are under water deficit, will rapidly lead to cavitation in the xylem; this, in turn, may lead to further hydraulic stress by restricting hydraulic pathways between the roots and crown. Exposure of sapwood and heartwood can increase the likelihood of infection by pathogens, and the physiological stress imposed by cutting will often further increase the

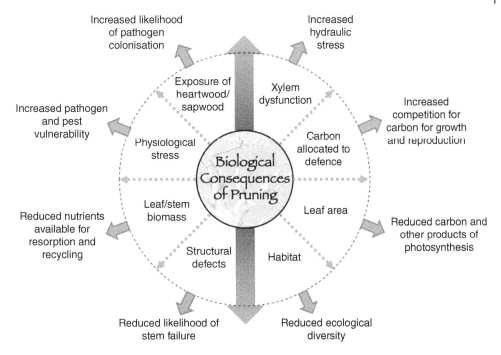

Figure 3.37 The biological consequences of pruning. The inner circle represents some important main outcomes whilst the outer circle gives likely secondary outcomes. The lower portion of the figure gives characteristics that are likely to be decreased as a consequence of pruning; the upper portion of the figure gives characteristics that are likely to be increased.

tree's vulnerability to pests and diseases. Loss of stem and leaf biomass that is removed from the site reduces the amount of nutrients available for recycling and, consequently, uptake by the tree. Pruning, especially of dead wood, may also result in loss of habitat for other species; this will have implications for broader ecosystem processes. Given the fact that trees are capable of shedding branches that are surplus to requirements without our help and pruning has a number of detrimental consequences, not pruning trees must be considered a valid management option.

Having noted all this, however, it is important to say that the removal of structural defects can help reduce subsequent stem failure, helping to preserve the integrity of the tree crown for years to come. For this reason, formative or structural pruning at an early stage in crown development should be a central component of any crown management strategy.

In urban environments and other managed landscapes, there are numerous, perfectly legitimate reasons to prune trees. As defects within the tree crown can lead to partial tree failure, the most compelling reason to prune a tree is to produce a structure that will reduce future biomechanical defects and help promote tree longevity. However, reducing tree risk, providing clearance for infrastructure, reducing the tree crown's impact on light and the improvement of aesthetics are all valid reasons for pruning urban trees. The key objectives and the advantages and disadvantages of alternative pruning practices used on amenity trees are outlined in Table 3.2. National standards, codes of practice and specifications should be used to create robust prescriptions for all

Table 3.2 Summary of important pruning practices in amenity trees.

Pruning practice	Key objectives	Advantages	Disadvantages
Safety/risk management	• Removal of deadwood to a defined diameter • Removal of branches with cracks and splits • Removal or subordination (reduction of a branch to aid the natural dominance of an associated branch) of branches with weak branch unions	• Reduces the risk of branch failure damaging persons or property • Can often be achieved with minimal impact on leaf area	• Removes potential wildlife habitat
Formative pruning (not including formative pruning for specialised objectives, such as pollarding or pleaching)	• To create the conditions that will lead to the mature tree being free from major biomechanical weakness • Removal or subordination of branches with weak branch unions • Removal or subordination of branches with undesired patterns of growth • Removal of rubbing branches • Establishment of the desired crown height	• Provides the conditions for stable crown development • Reduces the future pruning requirement • Removes biomechanical defects	• Pruning wounds offer potential infection courts for microorganisms • Loss of leaf area may temporarily reduce photosynthesis
Crown thinning	• To remove a defined percentage (<30%) of leaf area while creating an even density of foliage throughout the crown	• Reduces the shade cast by the tree crown • Reduces wind-load on retained branches • Pruning wounds tend to be small so will rapidly occlude	• Advantages tend to be temporary as new growth rapidly fills voids in the remaining crown • Pruning wounds offer potential infection courts for microorganisms
Crown lifting	• Removal of branches to achieve a defined vertical clearance from a given surface while maintaining >85% of the live crown	• Provides clearance from infrastructure, traffic (pedestrian and vehicular) and sight-lines	• Creates wounding on the main stem, unless the lift can be achieved through the removal of higher-order lateral branches • Pruning wounds offer potential infection courts for microorganisms

Table 3.2 (Continued)

Pruning practice	Key objectives	Advantages	Disadvantages
Crown reduction (whole crown or selective portions)	● Reduce the length of branches and stems by a defined amount that accumulatively results in the removal of <30% of the live crown ● Reduce the volume of space occupied by the tree crown in proportion to its original shape	● Alleviates biomechanical stress placed on the tree by reducing leaf (sail) area and leverage ● Allows the retention of a tree in a confined space ● Provides desired clearance from a structure without impacting the entire crown (selective reduction only)	● In tree species that bear the majority of leaves on the outer portion of their crown, a modest reduction in branch length can result in very high proportions of leaf-bearing structures being removed ● Thin-barked species may be vulnerable to sun-scorch
Crown cleaning	● Removal of dead, diseased and broken branches throughout the crown ● Selective removal or reduction of sprouts ● Removal of foreign objects and parasitic plants (e.g. mistletoe) from the crown	● Creates a safe, clean crown ● Only removes a small amount of photosynthetic (leaf) area ● Typically results in smaller pruning wounds	● Pruning wounds offer potential infection courts for microorganisms ● Removes potential wildlife habitat

N.B. Pruning practices may be combined if required by tree management objectives.

arboricultural operations (for examples see Table 3.3), and all relevant legislation and best practice must also be followed. This is likely to include legislation relating to risk assessment, equipment fitness, personnel training, tree protection and habitat (or wildlife) conservation.

Pruning Practices

Throughout much of the twentieth century, it was commonplace for trees to be pruned so that the final pruning cut was both very close to and parallel to the parent stem. These *flush cuts* (Figure 3.38) invariably caused substantial damage to the remaining stem and removed any branch protection zones. By using a series of stem dissections, Alex Shigo, of the US Forest Service, was able to demonstrate that flush cuts increased the wood discoloration and decay associated with the wound. This contrasted with wood that was associated with naturally shed branches that had less discoloration and decay. The outcome of this highly instructive study was the recommendation of *natural target pruning* (Shigo 1989), a technique that requires the finishing cut to mimic the natural pruning caused by branch senescence (Figure 3.39). Crucially, this cut respects

(a)

(b)

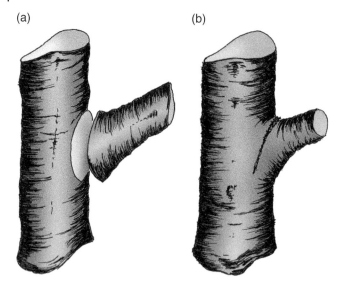

Figure 3.38 Poor pruning practices, such as flush cuts and stub cuts, should be avoided. (a) Flush cuts remove the branch collar and branch bark ridge to leave a large wound relative to the size of branch being removed. This results in considerable damage to the stem that is being kept. (b) Stub cuts leave an excessive branch stub that is easily colonised by microorganisms. They are also likely to grow epicormic sprouts that could lead to future management problems. *Source:* Adapted from Gilman (2012), drawn by Keith Sacre.

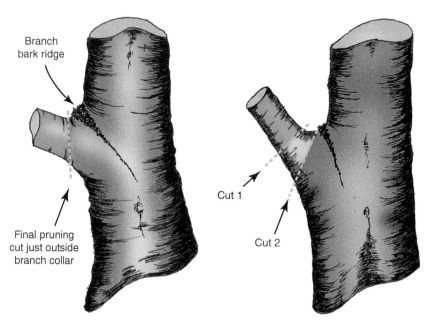

Branch bark ridge

Final pruning cut just outside branch collar

Cut 1

Cut 2

Figure 3.39 Locations of correct removal cuts for branches with and without branch collars. Where a branch collar is present (left), the final pruning cut should be completed just outside of the collar. If there is no visible branch collar and only a branch bark ridge is present (right), the finishing cut should be between cuts 1 and 2. Cut 1 has the advantage of minimising the surface area of the cut and reducing subsequent decay, but cut 2 reduces the likelihood of some stub dieback. No cut should be made closer to the main stem than cut 2. *Source:* Adapted from Gilman (2012), drawn by Keith Sacre.

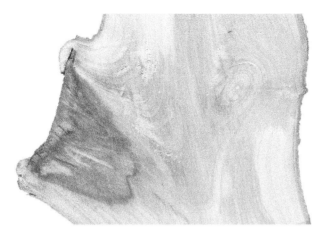

Figure 3.40 Natural target pruning can help restrict the development of wood discoloration and decay associated with the wound. Here, a correctly pruned apple *Malus* sp. branch shows only a very discrete area of discoloration 3 years after the original pruning cut was made. Note, also, that the growth of new wood to cover over (occlude) the wound has begun around the wound margins.

the base of the branch and does not induce any substantive wounding on the remaining stem tissue. Consequently, it affords the best opportunity for the tree to heal itself (Figure 3.40).

A further recommendation of the natural target pruning technique is the removal of branch ends that have been left too long (stubs) which not only hinder the occlusion of wounds, but may also harbour decay fungi at a potential gateway into the remaining stem. The final pruning cut should therefore remove as much of the branch tissue as possible while protecting any stem tissue that is to remain. Alex Shigo's advocacy for this revision in tree pruning technique transformed best practice in pruning during the latter part of the twentieth century and the concept of natural target pruning is now firmly embedded into several national standards on tree pruning (Table 3.3).

As the removal of stems, even if performed correctly, is injurious to trees, pruning should be carefully considered and follow a number of general principles:

- The amount of plant material removed should be the minimum required to achieve the pruning objectives.
- Wherever practicable, pruning cuts should be made on small-diameter parts of the tree so that no heartwood (or non-conducting xylem) is exposed within the wound. This allows active responses within the xylem to occur and reduces the time period for wound occlusion. Small-diameter pruning cuts are to be preferred over fewer large-diameter pruning cuts which expose larger areas of secondary xylem, often including heartwood.
- Formative pruning of young trees is always preferable to remedial pruning of mature trees because the relatively small-diameter pruning wounds can be quickly occluded by rapid growth. Such formative pruning also reduces future pruning requirements by ensuring a good branch structure is established.

Table 3.3 Important examples of national standards or work codes for arboricultural practice (adapted from Johnston and Hirons 2014).

Country	Standard
Australia	AS 4373 – Pruning of Amenity Trees
	AS 4970 – Protection of Trees on Development Sites
Britain	BS3998 – Tree Work Recommendations
	BS5837 – Trees in Relation to Design, Demolition and Construction – Recommendations
	BS8545 – Trees: From nursery to independence in the landscape – Recommendations
Czech Republic	SPPK A02 002 – Řez stromů – Tree pruning
Germany	DIN 18320 – German construction contract procedures (VOB) – Part C: General technical specifications in construction contracts (ATV) – Landscape works
	DIN 18919 – Vegetation engineering/management in landscaping; development and maintenance of green spaces
	DIN 18920 – Vegetation engineering/management in landscaping; protection of trees, plant populations and vegetation areas during construction
	ZTV – Baumpflege – Additional technical contractual terms and guidelines for tree care
New Zealand	Approved Code of Practice for Safety and Health in Arboriculture
USA	ANSI A300 – for Tree Care Operations – Tree, shrub and other woody plant management
	ANSI Z133 – Safety Requirements for Arboricultural Operations

Source: Dixon and Aldous (2014). Reproduced with permission of Springer.

- Pruning should be avoided, where possible, during periods of known or suspected tree water deficit, otherwise extensive xylem dysfunction caused by high levels of vessel embolism is likely.
- The final removal cut should be as small as possible and in accordance to the natural target pruning technique (Figure 3.39) outlined in national pruning standards (Table 3.3), or in established pruning texts (Gilman 2012; Brown and Kirkham 2017). Stub cuts outside the correct pruning position and flush cuts that damage the adjacent stem tissue must be avoided (Figure 3.41).
- Where it is necessary to shorten a stem, the cut should be made just above a branch junction (union) that is greater than one-third the diameter of the removed portion at the cut surface (Figure 3.41).
- Flush cuts that injure the stem tissue being retained and expand the area of exposed xylem should be avoided, unless they are part of a specialised pruning operation to accelerate fungal colonisation for the purposes of veteranisation – the creation of trees with the characteristics of veteran trees.
- Internodal cuts (cutting between branch junctions), resulting in stubs, should not be used unless this is desired for a particular pruning objective, such as the formation of a pollard. Stub and internodal cuts can often stimulate epicormic branches that are poorly attached when compared to 'normal' lateral branch formation.

(a)   (b)

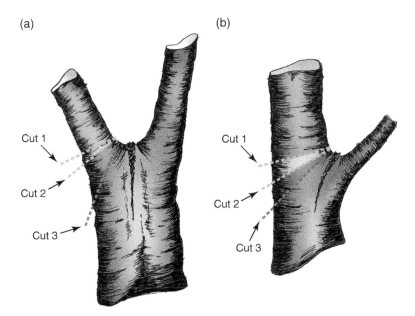

Cut 1   Cut 1

Cut 2   Cut 2

Cut 3   Cut 3

Figure 3.41 When removing a branch that is (a) the same size (co-dominance) or (b) bigger that the branch being left (reduction cut), cut 2 is preferred. Making a cut nearly parallel to the branch bark ridge (cut 3) leaves the union weak because supporting wood is too thin and the surface area of the cut larger than needed. Often, less decay occurs in response to cut 1 if the cambium does not die back under the cut, so it provides a safe alternative cut to cut 2. Retained lateral branches should be at least one-third the diameter of the pruning cut. *Source:* Adapted from Gilman (2012), drawn by Keith Sacre.

- The removal of branches with their points of attachment growing close together on the parent stem should be avoided to reduce potential coalescence of xylem dysfunction and decay within stems.

Tree Crown Support

Trees may develop in such a way that biomechanical defects threaten the structural integrity of the tree; for example, included bark within branch unions or minor splits where branches join. Where biomechanical defects exist and tree failure has the potential to cause damage to persons or property, expert evaluation of the tree should be sought. Indeed, it may be a legal obligation of the tree owner to obtain a professional risk assessment from a well-qualified and experienced arboriculturist. Supplementary crown support is normally considered an option where the risk of damage by branch (union) failure cannot be mitigated by pruning or target management. It may also be appropriate where preserving the whole crown is desirable. Under these circumstances, crown support can be offered through *flexible bracing, ridged bracing, propping* or *guying.*

Flexible bracing uses a system of cables, ropes or belts constructed within the crown of the tree to reduce the likelihood of structural failure. This technique has been used to stabilise tree crowns for around 100 years (Le Sueur 1934), although the systems and

(a)

(b)

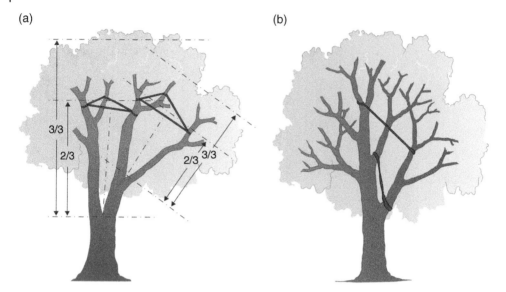

3/3

2/3

2/3 3/3

Figure 3.42 Bracing in trees. (a) Non-invasive, flexible bracing should be installed at approximately two-thirds of the branch length from the defect to branch tip: these may be static, as shown, or include an energy absorber to allow movement of stems within the bracing system. (b) Further support can be installed at the branch base using a tethering system. If this type of support system is installed, it must be carried out by a professional arboriculturist. *Source:* Adapted from FLL (2007).

materials have evolved since the early 1900s. Historically, eye-bolts have been used invasively to attach steel cables to the branches. However, recognition that invasive systems can lead to decay (Stobbe *et al.* 2000) has led to the widespread adoption of non-invasive bracing of tree branches. These systems use some form of belt that holds a synthetic rope or steel cable in place. Where ropes are used, these may be coupled with an energy absorber to allow movement that helps stimulate adaptive growth at the branch union and reduce peak loads on the system. In general, the point of attachment should be approximately two-thirds of the distance out from the weak branch union to its far (distal) end (Figure 3.42) and on stems strong enough to withstand the forces on the bracing. The exact configuration of the supporting braces will depend on the nature of the defect(s), existing crown architecture and potential loads on the system (Figure 3.43). Often, arboricultural standards, such as BS3998 and ZTV – Baumpflege (Table 3.3), provide additional guidance on the installation of crown bracing.

Rigid bracing is sometimes appropriate on weak structures where preventing independent movement of adjacent stems (or branches) is desired. This usually involves the invasive installation of a steel rod that is held in place with washers and nuts.

Alternatives to bracing include supporting stems or branches from the ground using props or, in cases of root instability, guys from the main stem. Some trees of great cultural value have been successfully supported with elaborate support systems engineered around the tree (Figure 3.44). Deciding on the most appropriate materials and configuration is often complex, and should always be carried out by an expert arboricultural practitioner.

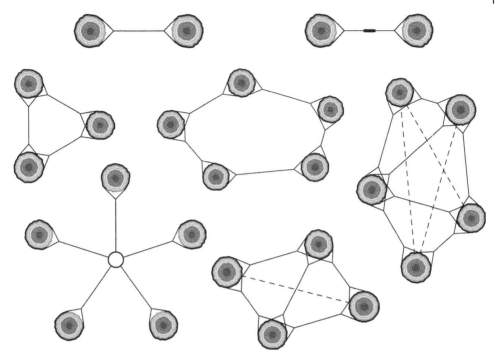

Figure 3.43 Potential configurations of non-invasive bracing in trees. Braces may have an integrated energy absorber, as indicated by the top right image. *Source:* Adapted from FLL (2007).

Figure 3.44 Bespoke support system used for a 400-year-old Chinese banyan *Ficus microcarpa* in Kowloon Park, Hong Kong. Cables fixed to vertical steel pillars support the crown. In turn, a ridged steel framework mounted on piles prevents the heavy structure from damaging the tree roots. Such intensive (and expensive) approaches are only really viable for trees of high cultural or historical importance.

References

Addicott, F.T. (1982) *Abscission*. University of California Press, Berkeley, USA.
Addicott, F.T. (1991) Abscission: Shedding of parts. In: Raghavendra, R.S. (ed.) *Physiology of Trees.* John Wiley and Sons, New York, USA, pp. 273–300.

Aerts, R. (1996) Nutrient resorption from senescing leaves of perennials: Are there general patterns? *Journal of Ecology*, 84: 597–608.

Alexander, R.M. (1997) Leaning trees on a sloping ground. *Nature*, 386: 327–329.

Augspurger, C.K. and Bartlett, E.A. (2003) Differences in leaf phenology between juvenile and adult trees in a temperate deciduous forest. *Tree Physiology*, 23: 517–525.

Basler, D. and Körner, C. (2012) Photoperiod sensitivity of bud burst in 14 temperate forest tree species. *Agricultural and Forest Meteorology*, 165: 73–81.

Braam, J. and Davis, R.W. (1990) Rain-, wind- and touch-induced expression of calmodulin and calmodulin-related genes in *Arabidopsis*. *Cell*, 60: 357–364.

Braatne, J.H., Rood, S.B. and Heilman, P.E. (1996) Life history, ecology, and conservation of riparian cottonwoods in North America. In: Stettler, R.F., Bradshaw, H.D. Jr., Heilman, P.E. and Hinckley, T.M. (eds) *Biology of Populus and its Implications for Management and Conservation*. NRC Research Press, Ottawa, Canada, pp. 57–85.

Brown, G.E. and Kirkham, T. (2017) *Essential Pruning Techniques: Trees, Shrubs and Conifers*. Timber Press, Portland, Oregon, USA. p. 408.

BSI (2010) *BS3998: Tree work – Recommendations*. British Standards Institution, London, UK.

Cescatti, A. and Niinemets, U. (2004) Leaf to landscape. In: Smith, W.K., Vogelmann, T.C. and Critchley, C. (eds) *Photosynthetic Adaptation; Chloroplast to Landscape*. Springer, Berlin, Germany, pp. 42–85.

Chabot, B.F. and Hicks, D.J. (1982) The ecology of leaf life spans. *Annual Review Ecology and Systematics*, 13: 229–259.

Craine, J.M. (2009) *Resource Strategies of Wild Plants*. Princeton University Press. Princeton, USA.

Dawson, J. and Lucas, R. (2011) *New Zealand's Native Trees*. Craig Potton Publishing, Nelson, New Zealand.

Debreczy, Z. and Rácz, I. (2011) *Conifers Around the World; Conifers of Temperate Zones and Adjacent Regions*, Volume 1. DendroPress, Budapest, Hungary.

Dixon, G. & Aldous, D. (2014) *Horticulture: Plants for People and Places*, Volume 2. Springer, Berlin, Germany, pp. 693–711.

Eckenwalder, J.E. (1980) Foliar heteromorphism in *Populus* (Salicaceae), a source of confusion in the taxonomy of Tertiary leaf remains. *Systematic Botany*, 5: 366–383.

Ennos, A.R. (2016) *Trees: A Complete Guide to Their Biology and Structure*. Comstock Publishing, New York, USA.

Fink, S. (1999) *Pathological and Regenerative Plant Anatomy*. Gebruder Borntraeger Verlagsbuchhandlung, Stuttgart, Germany.

FLL (Forschungsgesellschaft Landschaftsentwicklung Landschaftsbau e. V.) (2007) *ZTV Baumpflege: Additional Technical Contractual Terms and Guidelines for Tree Care*. FLL, Bonn, Germany.

Gartner, B.L. (1995) Patterns of xylem variation within a tree and their hydraulic and mechanical consequences. In: Gartner, B.L. (ed.) *Plant Stems; Physiological and Functional Morphology*. Academic Press, New York, USA, pp. 125–149.

Ghelardini, L., Santini, A., Black-Samuelsson, S., Myking, T. and Falusi, M. (2010) Bud dormancy release in elm (*Ulmus* spp.) clones: A case study of photoperiod and temperature responses. *Tree Physiology*, 30: 264–274.

Gilman, E.F. (2012) *An Illustrated Guide to Pruning*, 3rd edition. Delmar Cengage Learning, New York, USA.

Givnish, T.J. (1978) On the adaptive significance of compound leaves, with special reference to tropical trees. In: Tomlison, P.B. and Zimmermann, M.H. (eds) *Tropical Trees as Living Systems. Proceedings of the Fourth Cabot Symposium.* Cambridge University Press, Cambridge, UK, pp. 351–380.

Givnish, T.J. (1995) Plant stems: Biomechanical adaptation for energy capture and influence of species distributions. In: Gartner, B.L. (ed.) *Plant Stems: Physiology and Functional Morphology.* Academic Press, San Diego, USA, pp. 3–49.

Givnish, T.J. (2002) Adaptive significance of evergreen vs. deciduous leaves: Solving the triple paradox. *Silva Fennica*, 36: 703–743.

Hallé, F., Oldeman, R.A.A. and Tomlinson, P.B. (1978) *Tropical Trees and Forests: An architectural Analysis.* Springer, Berlin, Germany.

Horn, H.S. (1971) *The Adaptive Geometry of Trees.* Princeton University Press, Princeton, USA.

Ishii, R. and Higashi, M. (1997) Tree coexistence on a slope: an adaptive significance of trunk inclination. *Proceedings of the Royal Society of London Series B: Biological Sciences*, 264: 133–139.

Jaffe, M.J. (1973) Thigmomorphogenesis: The response of plant growth and development to mechanical stimulation. *Planta*, 114: 143–157.

Jaffe, M.J. and Forbes, S. (1993) Thigmomorphogenesis: The effect of mechanical perturbation on plants. *Plant Growth Regulation*, 12: 313–324.

James, K.R., Haritos, N. and Ades, P.K. (2006) Mechanical stability of trees under dynamic loads. *American Journal of Botany*, 93: 1522–1530.

Johnson, E.A. (1987) The relative importance of snow avalanche disturbance and thinning on canopy plant populations. *Ecology*, 68: 43–53.

Johnston, M. and Hirons, A.D. (2014) Urban trees. In: Dixon, G. and Aldous, D. (eds) *Horticulture: Plants for People and Places*, Volume 2. Springer, Berlin, Germany, pp. 693–711.

Jones, H.G. (2014) *Plants and Microclimate: A Quantitative Approach to Environmental Plant Physiology*, 3rd edition. Cambridge University Press, Cambridge, UK.

Kane, B., Farrell, R., Zedaker, S.M., Loferski, J.R. and Smith, D.W. (2008) Failure mode and prediction of the strength of branch attachments. *Arboriculture and Urban Forestry*, 34: 308–316.

Keskitalo, J., Bergquist, G., Gardeström, P. and Jansson, S. (2005) A cellular timetable of autumn senescence. *Plant Physiology*, 139: 1635–1648.

Kikuzawa, K. and Lechowicz, M.J. (2011) *Ecology of Leaf Longevity.* Springer, Berlin, Germany.

King, D.A. (1990) The adaptive significance of tree height. *The American Naturalist*, 135: 809–829.

King, D.A., Davies, S.J., Tan, S. and Noor, N.S.M. (2009) Trees approach gravitational limits to height in tall lowland forests of Malaysia. *Functional Ecology*, 23: 284–291.

Kramer, P.J. and Kozlowski, T.T. (1979) *Physiology of Woody Plants.* Academic Press, New York, USA.

Kuuluvainen, T. (1992) Tree architectures adapted to different light utilization: Is there a basis for latitudinal gradients? *Oikos*, 65: 275–284.

Kwit, M.C., Rigg, L.S. and Goldblum, D. (2010) Sugar maple seedling carbon assimilation at the northern limit of its range: The importance of seasonal light. *Canadian Journal of Forest Research*, 40: 385–393.

Lambers, H., Chaplin, F.S. III and Pons, T.L. (2008) *Plant Physiological Ecology*, 2nd edition. Springer, Berlin, Germany.

Larcher, W. (2003) *Physiological Plant Ecology: Ecophysiology and Stress Physiology of Functional Groups*, 4th edition. Springer, Berlin, Germany.

Laube, J., Sparks, T.H., Estrella, N., Höfler, J., Ankers, D.P. and Menzel, A. (2014) Chilling outweighs photoperiod in preventing precocious spring development. *Global Change Biology*, 20: 170–182.

Le Sueur, A.D.C. (1934) *The Care and Repair of Ornamental Trees in Garden, Park and Street*. Country Life, London, UK.

Lechowicz, M.J. (1984) Why do temperate deciduous trees leaf out at different times? Adaptation and ecology of forest communities. *The American Naturalist*, 124: 821–842.

Margalef, R. (1997) *Our Biosphere*. Excellence in Ecology, Book 10. Ecology Institute, Oldendorf/Luhe, Germany.

Matsuzaki, J. Masumori, M. and Tange, T. (2006) Stem phototropism of trees: A possible significant factor in determining stem inclination on forest slopes. *Annals of Botany*, 98: 573–581.

McMahon, T. (1975) The mechanical design of trees. *Scientific American*, 233: 92–102.

Meyer-Berthaud, B., Scheckler, S.E. and Wendt, J. (1999) *Archaeopteris* is the earliest known modern tree. *Nature*, 398: 700–701.

Niklas, K.J. (1992) *Plant Biomechanics: An Engineering Approach to Plant Form and Function*. University of Chicago Press, Chicago, USA.

Niklas, K.J. and Spatz, H.-C. (2014) *Plant Physics*. Chicago University Press, Chicago, USA.

Pallardy, S.G. (2008) *Physiology of Woody Plants*, 3rd edition. Academic Press, London, UK.

Park, S., Keathley, D.E. and Han, K.-H. (2008) Transcriptional profiles of the annual growth cycle in *Populus deltoides*. *Tree Physiology*, 28: 321–329.

Polgar, C.A. and Primack, R.B. (2011) Leaf-out phenology of temperate woody plants: From trees to ecosystems. *New Phytologist*, 191: 926–941.

Polgar, C.A. and Primack, R.B. (2013) Leaf-out phenology of temperate forests. *Biodiversity Science*, 21: 111–116.

Reich, P.B., Walters, M.B. and Ellsworth, D.S. (1997) From tropics to tundra: Global convergence in plant functioning. *Proceedings of the National Academy of Science*, 94: 13730–13734.

Ruelle, J. (2014) Morphology, anatomy and ultrastructure of reaction wood. In: Gardiner, B., Barnett, J., Saranpää, P. and Gril, J. (eds) *The Biology of Reaction Wood*. Springer, Berlin, Germany, pp. 13–35.

Scheffer, M., Vergnon, R., Cornelissen, J.H.C., Hantson, S., Holmgren, M., van Nes, E.H., *et al.* (2014) Why trees and shrubs but rarely trubs? *Trends in Ecology and Evolution*, 29: 433–434.

Schweingruber, F.H., Börner, A. and Schulze, E.D. (2006) *Atlas of Woody Plant Stems: Evolution, Structure, and Environmental Modifications*. Springer Science & Business Media, Berlin, Germany.

Shigo, A.L. (1989) *Tree Pruning*. Shigo and Trees Associates, Durham, USA.

Slater, D. (2016) *Assessment of Tree Forks: Assessment of Junctions for Risk Management*. Arboricultural Association, Stroud, UK.

Slater, D. and Ennos, A.R. (2013) Determining the mechanical properties of hazel forks by testing their component parts. *Trees*, 27: 1515–1524.

Slater, D. and Ennos, A.R. (2015a) The level of occlusion of included bark affects the strength of bifurcations in hazel (*Corylus avellana* L.). *Journal of Arboriculture and Urban Forestry*, 41: 194–207.

Slater, D. and Ennos, A.R. (2015b) Interlocking wood grain patterns provide improved wood strength properties in forks of hazel (*Corylus avellana* L.). *Arboricultural Journal*, 37: 21–32.

Slater, D. and Ennos, A.R. (2016) An assessment of the ability of bifurcations of hazel (*Corylus avellana* L.) to remodel in response to bracing, drilling and splitting. *Journal of Arboriculture and Urban Forestry*, 42: 355–370.

Slater, D. and Harbinson, C.J. (2010) Towards a new model for branch attachment. *Arboricultural Journal*, 33: 95–105.

Slater, D., Bradley, R.S., Withers, P.J. and Ennos, A.R. (2014) The anatomy and grain pattern in forks of hazel (*Corylus avellana* L.) and other tree species. *Trees*, 28: 1437–1448.

Slater. D and Ennos, A.R. (2015) The level of occlusion of included bark affects the strength of bifurcations in hazel (*Corylus avellana* L.). *Arboriculture and Urban Forestry*, 41: 194–207.

Smiley, E.T. (2003) Does included bark reduce the strength of co-dominant stems? *Journal of Arboriculture*, 29: 104–106.

Smith, W.K. and Carter, G.A. (1988) Shoot structural effects on needle temperatures and photosynthesis in conifers. *American Journal of Botany*, 75: 496–500.

Springmann, S., Rogers, R. and Spiecker, H. (2011) Impact of artificial pruning on growth and secondary shoot development of wild cherry (*Prunus avium* L.). *Forest Ecology and Management*, 261: 764–769.

Stobbe, H., Dujesiefken, D. and Schröder, K. (2000) Tree crown stabilization with the double belt system Osnabrück. *Journal of Arboriculture*, 26: 270–274.

Sumida, A., Terazawa, I., Togashi, A. and Komiyama, A. (2002) Spatial arrangement of branches in relation to slope and neighbourhood competition. *Annals of Botany*, 89: 301–310.

Taiz, L. and Zeiger, E. (2010) *Plant Physiology*, 5th edition. Sinauer Associates, Sunderland, USA.

Telewski, F.W. (1995) Wind-induced physiological and developmental responses in trees. In: Coutts, M.P. and Grace, J. (eds) *Wind and Trees*. Cambridge University Press, Cambridge, UK, pp. 237–263.

Telewski, F.W. (2006) A unified hypothesis of mechanoperception in plants. *American Journal of Botany*, 93: 1466–1476.

Terborgh, J. (1985) The vertical component of plant species diversity in temperate and tropical forests. *American Naturalist*, 126: 760–776.

Terbough, J. (1992) *Diversity and the Tropical Forest*. Scientific American Library, New York, USA.

Thomas, P.A. (2014) *Trees: Their Natural History*, 2nd edition. Cambridge University Press, Cambridge, UK.

Valladares, F. and Niinemets, U. (2007) The architecture of plant crowns. From design rules to light capture performance. In: Pugnaire, F.I. and Valladares, F. (eds) *Functional Plant Ecology*, 2nd edition. CRC Press, Boca Raton, USA, pp. 101–149.

Vogel, S. (2009) Leaves in the lowest and highest winds: Temperature, force, shape. *New Phytologist*, 183: 13–26.

Vogel, S. (2012) *The Life of a Leaf*. University of Chicago Press, Chicago, USA.

Walker, D.J. and Kenkel, N.C. (2000) The adaptive geometry of boreal conifers. *Community Ecology*, 1: 13–23.

White, J. (1984) Plant metamerism. In: Dirzo, R. and Sarukhán, J. (eds) *Perspectives on Plant Population Ecology*. Sinauer, Sunderland, USA, pp. 176–185.

Wrigley, J. and Fagg, M. (2010) *Eucalypts: A Celebration*. Allen and Unwin, Sydney, Australia.

4

Tree Roots

Roots, the 'hidden branches' of a tree, are vital for the development and persistence of trees. They are the primary means of water and nutrient uptake, and the tree's stability is dependent on the roots' capacity to provide anchorage. Roots grow in response to internal genetic controls but also respond to their environment, allowing them to forage for resource-rich 'patches' within soil. Critically, this interaction between the soil environment and the roots results in highly diverse spatial configurations (architecture), both within and between species. Significant variation exists in the precise anatomy of roots (this can be useful for root identification purposes), but the basic structure of roots is the same; this is briefly described in Box 4.1.

Whilst the uptake of resources and mechanical support are the main functions of roots, they also provide a number of other services. Roots are vital conduits for water, nutrients, carbohydrates and hormones travelling basipetally (from the roots tips toward the shoots), acropetally (from the shoots towards the root tips) and from one portion of the root system to another. Roots act as storage organs for surplus carbohydrates in times of plenty when photosynthetic production of carbon exceeds growth and maintenance requirements. Compounds that help regulate tree development are produced in the roots and transported through the tree to act on the shoots. Roots also produce chemicals that are capable of modifying soil pH and other aspects of the rooting environment, either to their advantage or to the detriment of competitive vegetation. In addition, most root systems also act as a habitat for fungi and a suite of other allied microorganisms within what has been termed the *rhizosphere zoo* ('rhiza' is Greek for root) (Buée *et al.* 2009).

In spite of roots being mostly hidden from sight beneath the soil surface, those seeking to manage trees must have an understanding of how tree roots develop, and their role in tree development. Many symptoms of ill health in trees can be attributed to an impoverished or damaged root system. Conversely, interventions to improve root function and health usually lead to an improved condition of the tree crown.

Root Growth and Development

All below-ground organs originate from the *root apical meristem (RAM)* which is first developed in the *radicle* (first root) that emerges from the seed. This RAM is organised around the *quiescent centre* (Figure 4.1) which produces radial lines of cells in two directions. Those cells on the inside develop into the main root tissues,

Applied Tree Biology, First Edition. Andrew D. Hirons and Peter A. Thomas.
© 2018 John Wiley & Sons Ltd. Published 2018 by John Wiley & Sons Ltd.

> **Box 4.1 The Anatomy of a Root**
>
> Roots all have the same basic structure. The skin or *epidermis* of the root lies over the *cortex*. The cortex is made up of layers of loosely connected cells, except for the innermost layer in which the cells fit tightly together to form an 'inner skin', the *endodermis*. This inner skin is made more impenetrable by being impregnated with corky substances (suberin) where the cells touch, to form the Casparian strip. Water can pass between the cells of the cortex but, to get into the centre of the root, water must go through the cells of the endodermis; in this way the tree can control what gets into the root.
>
> Just underneath the endodermis is another layer of cells called the *pericycle*, which is responsible for growing new side (lateral) roots out through the endodermis. The centre of the root contains the plumbing of the roots, the *vascular tissue* (also sometimes called the vascular cylinder or the stele, composed of the water-conducting xylem sitting at the centre) and the sugar-conducting phloem sitting around the outside. Figures 4.1 and 6.5 also help to describe root anatomy.

including the vascular tissue, cortex and epidermis, while cells produced to the outside form the *root cap*. This cap protects the root from abrasive particles as it pushes through the soil, although on some short lateral fine roots a distinct root cap may be absent. Lateral roots originate from pericycle cells someway back along the root; exactly where is likely to be dependent on the tree's genetic makeup and its interaction with the soil environment. Several consequent divisions of the cells generate the *lateral root primordium* which grows through the outer cell layers within the cortex. Further differentiation produces a new RAM that either remains dormant (analogous to pre-ventitious epicormic buds of the shoot system) or emerges as a new lateral root and continues to grow. To ensure that a large volume of soil is explored, lateral roots are initiated in a spiral sequence around the parent root, so that they emerge in numerous directions. This is similar to the way in which shoots often grow in spiral patterns (phyllotaxy) to intercept light efficiently; however, roots tend to have a less systematic approach to the production of new lateral roots.

In addition to the normal lateral root development, branching also occurs in response to root death, injury or deflection (Figure 4.2). New roots may arise just behind the injury or from callus tissue associated with wounds. These new roots tend to copy the juvenile pattern of root development in a process known as *immediate reiteration*. As can be seen in Figure 4.2, the new roots tend to carry on growing in the same direction as the original roots (a process known as *exotrophy*), thus ensuring roots continue to expand into new soil volumes. Exotrophy also has relevance to root growth in nurseries, as it allows compact root systems to be developed by root pruning. However, in container-grown or containerised trees, it can lead to pot-bound roots continuing to grow in circles even after transplantation. Containers (pots) that prevent (or at least postpone) roots from circling around the internal container wall should be favoured for this reason. It is also important that such root defects are ameliorated at planting.

An inevitable consequence of root growth is that the most active portion of the root system (close to the outermost reaches) shifts outwards, away from the stem like a slow-moving ripple. Mortality of the early fine roots potentially leaves fertile soil near the trunk only playing host to older coarse roots, which have limited ability to absorb water and nutrients. To get round this, the basal region of the trunk (the hypocotyl)

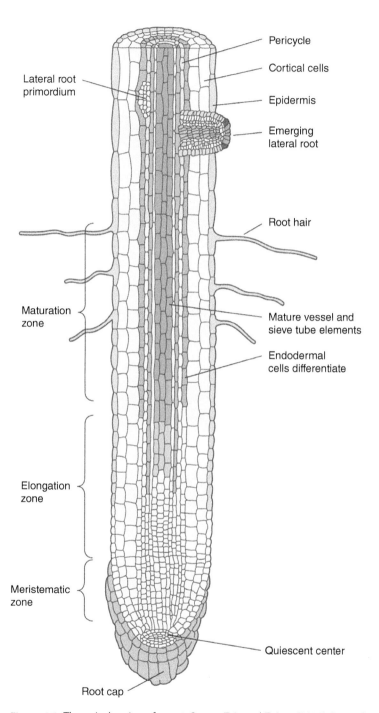

Pericycle

Cortical cells

Epidermis

Emerging
lateral root

Lateral root
primordium

Root hair

Maturation
zone

Mature vessel and
sieve tube elements

Endodermal
cells differentiate

Elongation
zone

Meristematic
zone

Quiescent center

Root cap

Figure 4.1 The apical region of a root. *Source:* Taiz and Zeiger (2010). Reproduced with permission of Oxford University Press.

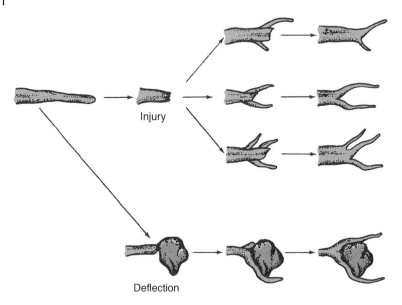

Injury

Deflection

Figure 4.2 Root development after injury or deflection around an object. *Source:* Wilson (1984). Reprinted from *The Growing Tree.* Copyright © 1984 by the University of Massachusetts Press.

and older coarse roots use a process known as *delayed reiteration* to produce new roots that infill regions of the soil close to the trunk where the originally established root system has few remaining actively absorbing roots. A detailed study of red maple *Acer rubrum* identified three phases of root system development that characterise this type of growth (Figure 4.3):

1) Development of the seedling root system;
2) Expansion of the coarse root system with highly branched 'root fans' (predominantly fine roots) in their furthermost portions; and
3) Slowing in growth of the original coarse root system, and development of adventitious roots near the stem (Lyford and Wilson 1964).

It is likely that many temperate tree root systems follow a similar pattern of development.

In red maple *Acer rubrum*, large diameter root tips (containing apical meristems or RAM) of ~2 mm in diameter produced coarse, woody roots, while root tips ≤1 mm in diameter produced fine, non-woody roots (Lyford and Wilson 1964). Similarly, in red oak *Quercus rubra* large diameter root tips ~1–4 mm in diameter developed into coarse roots while root tips <1 mm developed into fine roots. Remarkably, in a small terminal portion taken from a root 0.2 mm in diameter, 1230 root tips were found, most of which also had mycorrhizal associations (Lyford 1980).

As young roots develop, the tissues age and mature over time, leading to three distinctive zones in the apical region referred to as white, condensed tannin (CT) and cork zones (Kumar *et al.* 2007). The very youngest portions of the root appear white and contain a living cortex, an endodermis with abundant non-corky cells (sometimes referred to as 'passage' cells because they allow the passage of water) and a central stele that contains the vascular tissues. As the root matures, condensed tannin (these are non-hydrolysable tannins: a group of polymers formed by the condensation of flavans)

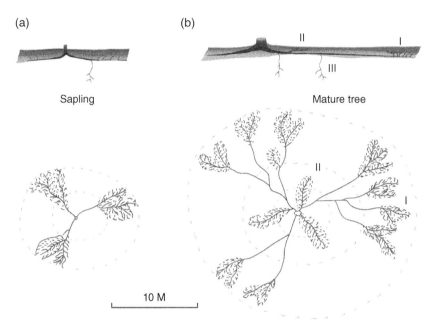

(a)

Sapling

(b)

Mature tree

II   I

III

II

I

10 M

Figure 4.3 Tree root development in red maple *Acer rubrum*. (a) Root fans growing from the younger portions of the developing coarse root system. (b) Reoccupation of a soil area near the base of the tree: I, root fans growing from the younger portions of the woody roots have extended a distance of several metres from the tree; II, root fans on adventitious roots have only recently emerged from the zone of rapid taper or root collar, now occupying the area near the base of the tree; III, sinker roots penetrating down into the soil. *Source:* Adapted from Lyford and Wilson (1964). Reproduced with permission of Harvard Forest, Harvard University.

is deposited in the walls of all cells outside of the central stele, giving the root a brown colour that is easily seen. Cells within the cortex are dead (or dying) at this point and there is a decrease in passage cells; less water is therefore taken up along this part of the root. The cork zone indicates the onset of secondary growth with secondary vascular tissues and a complete ring of mature cork cells. Although the delineation between the CT and cork zones is clear in the internal anatomy, it is not easily seen from the outside, compared with the contrast between white and CT zones. For this reason, earlier studies often categorised young roots into two zones: non-suberized or suberized; or white or brown. Regardless of the precise definition of these root zones, the white zone, closest to the apex of the root, appears to be most effective at water and nutrient absorption, although some resources may also be taken up through the CT and cork zones (Danjon *et al.* 2013).

Roots can also arise from stems, or other plant parts, in which case they are referred to as *adventitious roots*. All trees that are vegetatively propagated rely on the capacity of stem tissues to produce adventitious roots. In flood-plain trees, such as many poplars *Populus* and willows *Salix*, the rooting of detached shoots in riverbanks downstream is an important strategy for colonising new ground. The capacity to produce adventitious roots is also essential to cope with sudden changes in soil level caused by post-flood silt deposition. Further, in palm trees and other arborescent monocotyledons, adventitious root development from the base of stems is the main way of producing a root system.

Root Systems

Tree root systems differ substantially from those of annual plants, as the perennial nature of trees needs roots that persist and expand with the increasing demands of the crown. This is achieved in trees through a framework of long-lived, woody coarse roots (*syn.* structural roots) that support a labyrinth of short-lived *fine roots*. Although somewhat arbitrary, the division between the coarse and fine root classes is widely taken as 2 mm in diameter, however, some studies categorise fine roots anywhere between <0.5 mm and <5 mm diameter. Fine roots can vary quite considerably between species (Figure 4.4). In some species, fine roots show slow, limited extension, are heavily branched and do not get fatter by secondary growth. In other species, fine roots have the ability for secondary growth and rapid extension should this be required. While there is some overlap in function between coarse and fine roots, their dominant roles differ. The finer, lower order roots have a very high *specific root length* (SRL: m g^{-1}), that is, have a high surface area per unit weight, so have a dominant role in taking up water and nutrients. In an assessment of fine roots in 12 temperate tree species, SRL was found to vary from ~9 m g^{-1} in the tulip tree *Liriodendron tulipifera*, to ~90 m g^{-1} in pignut hickory *Carya glabra* (McCormack *et al.* 2012). Higher order fine roots

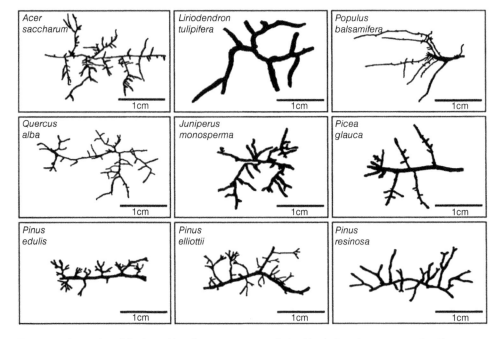

Figure 4.4 Examples of the branching fine root systems of nine North American tree species. These are intact segments of the fine root systems that were washed free of soil, their images digitised on a flatbed scanner and converted to black and white. The species are: sugar maple *Acer saccharum*, tulip tree *Liriodendron tulipifera*, balsam poplar *Populus balsamifera*, white oak *Quercus alba*, one-seed juniper *Juniperus monosperma*, white spruce *Picea glauca*, pinyon pine *Pinus edulis*, slash pine *P. elliottii* and red pine *P. resinosa*. *Source:* Pregitzer *et al.* (2002). Reproduced with permission of John Wiley and Sons.

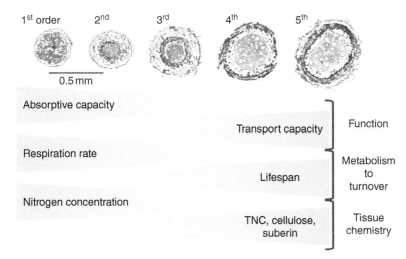

Figure 4.5 Root cross-sections of Norway maple *Acer platanoides*, showing a typical pattern of increasing root diameter and secondary (woody) development with increasing root order. Notice that first- and second-order roots have little or no secondary development, and first- to third-order roots still possess intact root cortical cells; fourth- and fifth-order roots have lost all cortex and instead have secondary xylem. Triangles depict simplified patterns of root function (absorptive and transport capacity) and root traits (respiration rate per gram of root; lifespan; total non-structural carbohydrates (TNC); and other aspects of tissue chemistry) with increasing root order. Root function may not change linearly, depending upon the trait and species. It is also worth noting that despite their recognised importance to root function, many aspects of tissue chemistry (including cellulose, suberin and phenolic content) are not well studied, and patterns of root function with root order may vary across species. *Source:* McCormack *et al.* (2015). Reproduced with permission of John Wiley and Sons.

(typically fourth and fifth orders) are predominantly involved with conduction (transport), anchorage and storage of carbohydrates, but will also be able to take up some water (Figure 4.5).

The contrasting roles of fine and coarse roots lead to different life expectancies of individual roots within a single root system. Coarse roots usually increase in diameter (via secondary growth) to improve their ability to conduct water, their biomechanical performance and their resistance to decay. As a result, they may persist for years, decades and even centuries. Conversely, the highly dynamic fine root system, particularly the finest lower order roots, rapidly proliferates within resource-rich patches until they are depleted, at which point the roots rapidly die off, whilst, elsewhere in the root system, other fine roots will be growing. This *fine root turnover* can result in an individual root lifespan of just a few days. In temperate trees, the median fine root lifespan has been found to vary, between 95 days in aspen *Populus tremuloides* to 336 days in white oak *Quercus alba*. This results in quite different fine root survivorship profiles (Figure 4.6). For example, in aspen, just 30% of the fine roots survived longer than 200 days, whereas in white oak 80% of fine roots survived longer than 200 days (McCormack *et al.* 2012). It is also clear that some of the higher-order fine roots may persist for several years, particularly in regions where there are less marked seasonal differences in water availability and soil temperature.

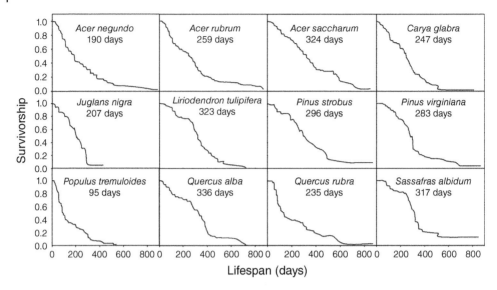

Figure 4.6 Survivorship curves of fine roots of 12 temperate tree species grown together in central Pennsylvania, USA. Survivorship as a proportion of all the fine roots is shown over 800 days (over 2 years). The median root lifespan (in days) is given for each species. The species are: manitoba maple *Acer negundo*, red maple *A. rubrum*, sugar maple *A. saccharum*, pignut hickory *Carya glabra*, black walnut *Juglans nigra*, tulip tree *Liriodendron tulipifera*, eastern white pine *Pinus strobus*, Virginia pine *P. virginiana*, quaking aspen *Populus tremuloides*, white oak, *Quercus alba*, red oak *Q. rubra* and sassafras *Sassafras albidum*. Source: McCormack *et al.* (2012). Reproduced with permission of John Wiley and Sons.

Secondary Root Growth

Coarse roots and higher-order fine roots achieve greater stability, and are better able to conduct water, by undergoing secondary growth. The process is very similar to secondary growth in stems (see Chapter 2), and starts when the vascular cambium begins to form as a continuous wavy line between the primary xylem and phloem tissue (Figure 4.7). Gradually, the vascular cambium attains a more circular form when viewed in cross-section, although this may become eccentric over time as the root matures. The newly formed vascular cambium is capable of forming a continuous layer of secondary xylem to its inside and secondary phloem to its outside. Shortly after the vascular cambium has formed, some of the cells in the pericycle differentiate into the cork cambium, which produces layers of corky cells (phellum) on the outside and a few cells inside the cambium called the phelloderm (Figure 4.7). These three layers (cork cambium, phellum and phelloderm) together are technically termed the periderm, and mirror bark formation in the shoot system. After both cambia are formed and cork cells begin to form, the cortex with its endodermis is increasingly stretched, broken and, finally, shed. Subsequent secondary growth within the roots proceeds in a similar way to that of the stem but the wood anatomy may not be the same as in the stem. For example, a ring porous stem will not necessarily have a ring porous vessel distribution in its root.

In seasonal climates, the secondary growth of roots starts in the coarse roots at the base of the stem, before migrating towards the outer (more distal) roots. As growth goes

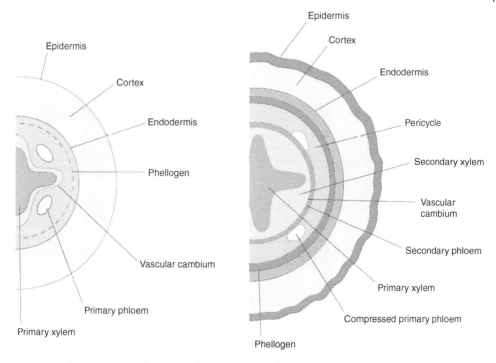

Figure 4.7 Secondary growth of a woody root, showing development of vascular cambium and production of secondary xylem and phloem. *Source:* Adapted from Beck (2010). Reproduced with permission of Cambridge University Press.

on for longer around the base of the trunk than the more distant roots, a *zone of rapid taper* (ZRT) develops around the base of the trunk. This gradual addition of woody material just beneath the soil surface can result in uneven surfaces and/or a raised region at the base of the stem. Despite being a normal feature of tree development, this process can cause problems with urban infrastructure. Where roots have insufficient soil volume and hard landscaping is close to the stem, this problem is often compounded. To reduce the risk of such potential conflicts occurring, those responsible for the installation of trees into hard landscapes must anticipate tree development and take opportunities to increase soil volumes wherever possible.

Secondary growth in roots results in distinct growth rings (seen in cross-section), just as it does in the stem. However, these rings may have little or no correlation to the pattern of growth rings shown by the stem of the same tree. This makes it extremely difficult, if not impossible, to estimate the age of a tree using woody roots. In roots, some rings may appear to be missing (or 'discontinuous') because of the variation in cambial activity occurring around a root. Younger coarse roots are usually circular in section but, as they age, xylem deposition becomes increasingly uneven, resulting in discontinuous growth rings and eccentric form, often leading to in irregular elliptical transverse sections (Figure 4.8). This is particularly characteristic of horizontal roots in the zone of rapid taper, as the greater increase in diameter tends to be at the top of the root where there is little to impede growth, rather than underneath where the soil is more likely to restrict root growth physically. This modified growth also has the advantage of

(a) (b)

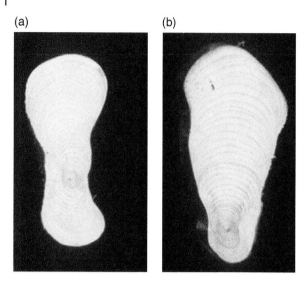

Figure 4.8 Cross-sections of structural roots of Sitka spruce *Picea sitchensis*. (a) 'I-beam' shape that was well developed on the prevailing wind side of the tree; (b) 'T-beam shape' that was more characteristic on the leeward side of the tree. *Source:* Coutts *et al.* (1999). Reproduced with permission of Springer.

maximising the resistance to bending for a given investment in woody material. For example, in Sitka spruce *Picea sitchensis*, a 'T-beam' development is common in roots on the leeward side of the prevailing wind as this shape is highly resistant to compressive forces. On the windward side, an 'I-beam' shape is frequently formed as this shape is more effective at resisting vertical flexing (Figure 4.8) (Coutts *et al.* 1999). I-beam and T-beam configurations, as well as intermediate forms, can be observed in a number of species: this type of modification appears to be widespread.

Root Architecture

Five key processes determine root architecture: the production of new growing points; branching; elongation of roots; radial (secondary) root growth; and root mortality.

At germination, the juvenile root (the *radicle*) breaks through the seed coat and develops into the first proper root (see Chapter 5). Typically, the radicle develops into a *taproot* that then branches to produce multiple lateral roots; approximately 8–12 of these persist and subsequently develop into the woody framework. These elongate, expand and branch, often crossing over each other and grafting together to produce the mature root system. Roots on the upper portion of the taproot tend to develop into *lateral* and *oblique* roots (sometimes termed horizontal roots, heart roots or slope runners; Figure 4.9). Roots on the lower part of the taproot tend to be positively geotrophic and descend vertically, forming a *side taproot*. Similarly, downward growing *sinker* roots can develop the lateral roots. In many species, the initial tap root dies back as the tree matures, leaving an extensively branched lateral root system. Further branching, elongation and development of several orders (up to ~7) of lateral roots result in the exploration of large soil volumes, even in young trees. For example, roots of a 3-year-old plum tree were found to occupy more than 50 m^3 of soil (Vercambre *et al.* 2003). Although not every form of root will be found within every root system, Figure 4.9 shows typical features of a mature tree root system.

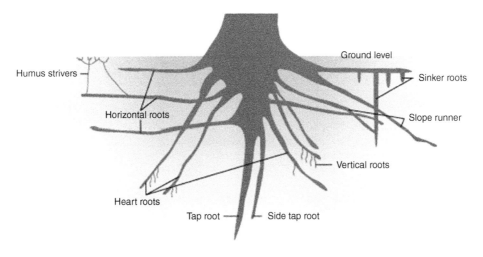

Figure 4.9 Cross-section of a tree root system with names that are used in several European schools of classification. Note that not all features will be present in every root system. *Source:* Persson (2002). Reproduced with permission of Taylor and Francis Group LLC Books.

Soil is a complex medium, varying spatially and over time, consequently, the three-dimensional architecture of tree roots, even within the same species, can differ widely between individuals. For example, physical obstructions (e.g. from stones, woody debris or underground urban infrastructure), soil drainage, the existence of impermeable layers (soil pans), areas of toxicity, variability in soil compaction, accessibility to mineral resources and rhizosphere competition all influence the final architecture of the tree root system. In addition, most tree production systems substantially modify root architecture (at least initially), and planting technique can also influence subsequent root development.

Underlying these variations caused by the environment, however, different species do tend to have a characteristic architecture. Evidence based on a root excavations of European tree species (Kostler *et al.* 1968; Kutschera and Lichtenegger 2002) suggests four basic types of root architecture in mature temperate trees (Figure 4.10):

a) *Plate root systems* are dominated by lateral roots growing at a relatively consistent depth below the soil surface, as found in fir *Abies* spp., spruce *Picea* spp. and beech *Fagus* spp.

b) *Sinker root systems* have a plate of lateral roots that develop vertical sinker roots: the tap root may or may not persist. These are found in some oaks *Quercus* spp., many southern beeches *Fuscospora* and *Lophozonia* spp. and kauri *Agathis* spp. (Figure 4.11).

c) *Heart root systems* display a central complex of sinker roots, as well as oblique and shallow lateral roots giving the more compact rooting structure found in birch *Betula* spp., larch *Larix* spp., oaks *Quercus* spp. and lime *Tilia* spp.

d) *Tap root systems* that maintain a persistent positively geotrophic root from the radicle in addition to developing a plate of shallow lateral roots. Examples include many oaks *Quercus* spp. and pines *Pinus* spp. from Mediterranean climates.

(a)

(b)

(c)

(d)

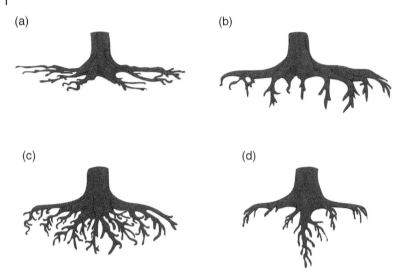

Figure 4.10 Four principle tree root types found in temperate trees: (a) plate root system; (b) sinker root system; (c) heart root system; (d) tap root system. Details of each are given in the text.

Figure 4.11 Kauri *Agathis australis* has an example of a sinker root system, the remains of which can be seen in this old kauri root plate on display outside the Kauri Museum in Matakohe, Northland, New Zealand.

Tree Anchorage

The different root systems found in trees produce a number of variations in the way that trees are anchored in the soil. To some extent at least, all trees rely on their weight to keep them in the soil and, in mature trees, the width of their trunk to give them stability (stiffness). The wider and stiffer the base, the more stable the tree, regardless of how well it can attach itself to the ground. Additionally, roots frequently graft together to form ridged, interlocked systems, both within a single tree's root system and potentially between neighbouring trees. In trees with plate root systems, the interlocked roots create a stiff *root plate*. Anchorage of trees with a root plate relies on the weight of the tree pushing the root plate into the soil, as well as the tensile strength and stiffness of the lateral roots. As the bending stresses are felt most acutely in the roots on the leeward side of the tree, the root stiffness is particularly important here and is helped by the eccentric root shapes that are very resistant to bending (Figure 4.8). The stiffness is created by shape rather than additional wood, so it is a very efficient way to increase tree stability. In many ways, these root plates are analogous to the base of a wine glass. For the tree to topple, the tree has to be lifted to pivot around the edge of the stiff root plate, just as a wine glass only falls when its weight is lifted over the edge (pivot) of the flat base. This has to involve the failure of some lateral roots in tension; a force sufficient to overcome the weight of the tree; and usually a failure within the root plate that allows the tree to hinge over, close to the trunk on the leeward side. This effectively reduces the force required to pivot the tree in the same way that a wine glass with a narrower base would be much easier to push over.

Trees that have sinker roots have further advantage in that the *root cage* produced holds a large amount of soil, which adds to the weight of the tree that needs to be lifted for it to fall. Sinker roots also strongly resist being pulled upwards out of the soil (Ennos 2016). As a consequence, the spread of the root cage can be reduced whilst still providing the tree with good anchorage. The value of this is seen especially in many tropical trees, where lateral roots have been cut through within a metre or less of the trunk (as happens, for example, with some road building schemes) without greatly reducing the stability of the tree (Ghani *et al.* 2009).

The heart root system (Figure 4.10b) is shaped more like a root ball; when the tree is pushed by the wind, this tends to rotate in the soil like a ball and socket joint. In this case, it is the weight of the root ball and the strength of the soil, rather than the strength of the roots, that holds the tree up. As the soil strength reduces significantly when it is wet, trees with this type of root system tend to fail most spectacularly in wet soils.

Trees with extensive taproots stand up much like a fence post. As the crown sways in the wind, the bottom of the trunk provides a pivot point and the taproot is pushed in the opposite direction to the sway of the crown. It is effectively the stiffness of the root and the resistance of the soil that keeps the tree standing. This type of root system is common in young trees before they develop an extensive system of lateral roots but it is also found in some mature trees such as pine *Pinus* spp.

Some tropical trees, particularly the emergent species that hold their crowns above the forest canopy, have developed a further strategy to stay upright. *Buttress roots* extend downward and outward from the lower portion of the trunk (Figure 4.12a). Their growth appears to be stimulated by the mechanical stresses (caused by the wind) being concentrated on the top of the lateral roots (Ennos 1995). To reinforce the areas that are most highly stressed, the tree puts on additional woody material that, over time, leads to structures that greatly increase the mechanical security of the lower trunk, where wind

(a)   (b)

Figure 4.12 (a) Large buttress roots formed on an emergent species in Khao Sok National Park, Thailand. (b) Special stilt roots formed on the lower stem of a young *Uapaca* sp. growing in the littoral forest near Sainte Luce, Madagascar.

stresses are focused most acutely. However, unlike the architectural buttresses used to resist compressive loads from tall buildings, the thin diameter of root buttresses means that they would easily buckle under compression. They must therefore act mainly as guy ropes that transmit the forces from the crown out to the root plate, thus reducing the stress at the stem–root junction (Vogel 2012). Although buttress roots offer a very efficient way of improving anchorage without increasing the dimensions of the trunk, it appears that some species have managed to achieve the same qualities with even less investment in woody material (Figure 4.12b). By growing aerial roots out from the base of the trunk rather than developing buttresses that grow upwards from the lateral roots, *Uapaca* spp. have developed woody guy ropes that reinforce the lower portion of the trunk without creating solid buttresses. It should be noted that not all tropical trees, and not even all emergent tropical trees, form buttress roots. Those trees that do not form buttresses will rely on the other forms of anchorage described above.

Extent of Root Systems

The vertical and lateral extent of root systems is of interest to those managing trees as it is helps to determine their influence on other plants and the built environment, as well as their capacity to take up large quantities of water and avoid drought by accessing deep ground water. This, in turn, can have a big influence on which trees can be grown where, and their potential to cope with a changing climate.

Most tree roots (whether measured by weight or length) occur in the upper portions of the soil. Oxygen and mineral nutrients are more readily available closer to the soil surface and the interception of rainfall favours a relatively shallow root system. For most trees, the top 50 cm of soil hosts 80–90% of root biomass and 90–99% is in the top 1 m (Figure 4.13). Maximum rooting depths do vary, however. In temperate trees, this tends to range 2–8 m in conifers and 2–4 m in deciduous species. At a global scale, particularly in dry environments, roots have been found exceptionally deep in the soil (Figure 4.14). Mediterranean trees, particularly oaks *Quercus* spp. and eucalypts *Eucalyptus* spp., are often thought of as having particularly deep roots, presumably because they often experience long periods without rain and require access to deep-soil water to survive. However, the Natal fig *Ficus natalensis* holds the current record: its roots were found at an astonish-

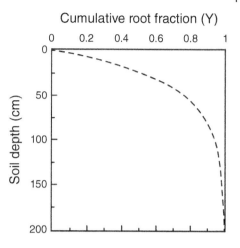

Figure 4.13 The cumulative proportion of root biomass with depth into the soil. The data is an average across trees from temperate deciduous, temperate coniferous, tropical deciduous, tropical evergreen and tropical savannah biomes. *Source:* Adapted from Jackson *et al.* (1996). Reproduced with permission from Springer.

ing depth of 120 m in the Echo Caves, Mpumalanga, South Africa. Other species noted for their particularly deep roots are the shepherd's tree *Boscia albitrunca*, which was found with roots located 68 m under the Kalahari desert, and two other species, *Vachellia erioloba* (previously called *Acacia erioloba*) from Africa and mesquite *Prosopis juliflora*

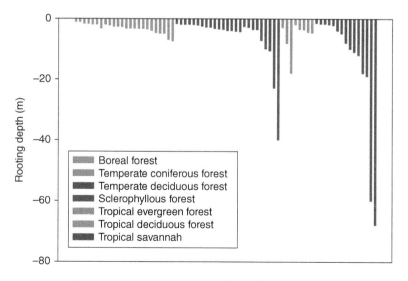

Figure 4.14 Maximum rooting depth of trees from different types of forest. *Source:* Data from Canadell *et al.* (1996).

from North America, have both been found with roots below 50 m. Most roots do not go this deep, simply because the soil or rock is too dense for roots to grow through and there is not enough oxygen. Nevertheless, given the challenges in locating very deep roots, it is likely that under the right soil conditions (and with enough time), a number of other species will also be capable of reaching these depths.

The spread of lateral tree roots out from the trunk can be readily seen by carefully excavating the roots with hand or air tools. As a general rule, however, the roots of most trees spread out to 2–2.5 times the width of the canopy. In tall, relatively columnar trees, this also equates to roughly the height of the tree. To put this into perspective, in mature (~60 years) red maple *Acer rubrum* and red oak *Quercus rubra* growing together, lateral roots extended around 20 m from the stem (Figure 4.15). In general, the maximum distance of tree root spread would appear to be in the region of 20–30 m, but some genera (e.g. *Acacia*, *Adansonia*, *Juglans*, *Metrosideros*, *Populus*, *Quercus*, *Sequoiadendron* and *Ulmus*) are apparently capable of growing lateral roots >30 m long (Stone and Kalisz 1991). In fact, Grandidier's baobab *Adansonia grandidieri* (see Figure 10.12) has been found to have lateral tree roots 50 m from the trunk. This helps intercept as much rainfall as possible, an important strategy for survival in arid environments (Petignat and Jasper 2015). While the ultimate distance roots grow is likely to be genetically controlled, soil conditions are also likely to be very important in determining how far roots will spread.

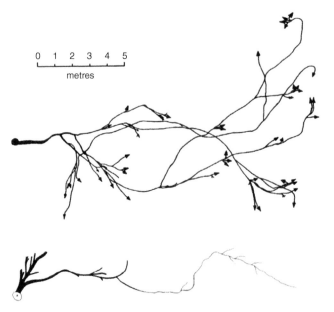

Figure 4.15 Scale diagram of an excavated lateral root of red maple *Acer rubrum*, at the top, and red oak *Quercus rubra*, at the bottom growing in the same area of Harvard Forest, Massachusetts. Both trees were around 60 years of age. *Acer rubrum* roots were found in the top 10 cm of soil, while *Quercus rubra* roots were found at a depth range of 5–50 cm with an average depth of 30 cm, thus demonstrating a species preference for rooting depth that will limit competition between these species. Arrows indicate that the root tips were not found and therefore these roots continue somewhat further than is shown. (Top) *Source:* Adapted from Lyford (1964). Reproduced with permission of Harvard Forest, Harvard University. (Bottom) *Source:* Adapted from Lyford and Wilson (1980). Reproduced with permission of Harvard Forest, Harvard University.

When Do Roots Grow?

As early as 300 BC, Theophrastos of Lesbos observed that roots often start growing before shoots in spring. Subsequent investigations have supported this claim for most species in seasonal climates, as new fine roots need to be established to acquire the water and nutrients necessary for crown development (Figure 4.16). Nevertheless, measuring root growth in trees is fraught with difficulty. Different segments of the root system may start and stop growth at different times; different trees, even within the same species, show variation in when their roots grow. There is also variation in the timing of root growth between different sites and different years. Despite these complexities, it has been possible to distil several types of root production. Where the availability of resources is strongly seasonal, a *concentrated* pattern of root production occurs in early to mid summer, before the soil becomes too dry. If several periods of favourable environmental conditions occur, a *bimodal* pattern of root production is

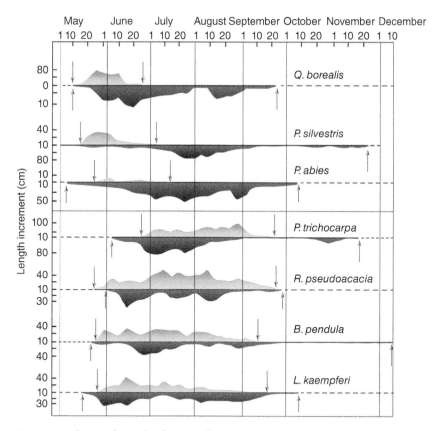

Figure 4.16 Seasonal root development (brown) tends to start in advance of shoot development (green) in a range of temperate tree species. Here, variations in seasonal shoot and root growth are shown for seven species of forest tree. Seasonal starting and stopping of growth are indicated by arrows. The species are: northern red oak *Quercus borealis*, Scots pine *Pinus sylvestris*, Norway spruce *Picea abies*, black cottonwood *Populus trichocarpa*, false acacia *Robinia pseudoacacia*, silver birch *Betula pendula* and Japanese larch *Larix kaempferi*. *Source:* Adapted from Lyr and Hoffmann (1967). Reproduced with permission of Elsevier.

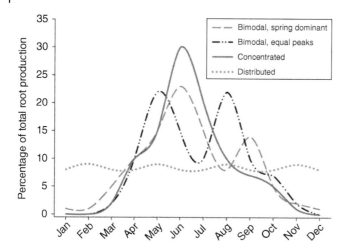

Figure 4.17 Patterns of root growth in temperate regions of the northern hemisphere. *Source:* McCormack *et al.* (2014). Reproduced with permission of John Wiley and Sons.

typical. In regions with fairly uniform environmental conditions, root growth may be more evenly distributed or arranged in multiple pulses (Figure 4.17). McCormack *et al.* (2014) found that a concentrated pattern of root production in early to mid summer was most frequent in a study of 12 temperate tree species over 3 years (Figure 4.18).

Patterns of root growth may also be somewhat dependent on the development of the shoot system. If not enough fine roots survive winter, new roots will be required prior to leaf flush. This results in a key period of root development after the soil warms to ~5 °C but before leaves emerge (Figure 4.19). After this, root production may be reduced during periods of rapid shoot growth (especially those with determinate shoot growth) as the shoots can compete strongly for water, nutrients and carbohydrates.

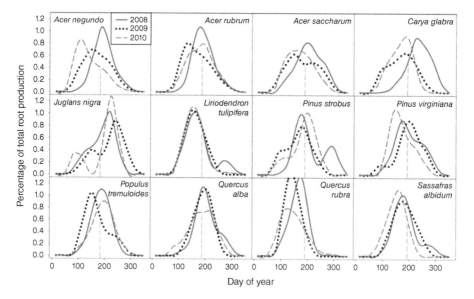

Figure 4.18 Patterns of root growth during 2008–2010 for 12 temperate tree species in central Pennsylvania, USA. Vertical grey bars are included for reference to show 1 July. Common names are given in Figure 4.6. *Source:* McCormack *et al.* (2014). Reproduced with permission of John Wiley and Sons.

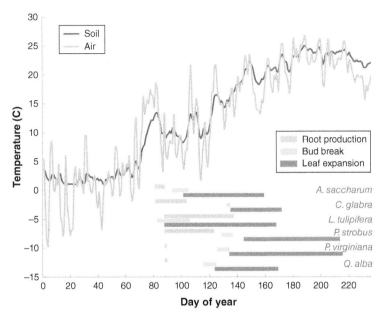

Figure 4.19 Soil temperature at 20 cm depth (dark grey line) and air temperature (light grey line) recorded at the Russell E. Larson Agricultural Research Center in central Pennsylvania, USA, with corresponding above- and below-ground growth periods in 2012. Growth periods are reported for six temperate tree species and include: the period from first root production to peak root production (brown bars); timing of bud break (light green bars); and duration of active leaf expansion (dark green bars). Full scientific names are given in Figure 4.18 and common names in Figure 4.6. *Source:* McCormack *et al.* (2015). Reproduced with permission of Springer.

Seasonal soil water deficits frequently slow or halt root growth during summer, but root growth often resumes when soil water is recharged, the shoots are not growing much and the soil remains warm. Root elongation rates vary substantially across species, from a few millimetres to ~80 mm per day.

To a large extent, the timing of root development is decoupled from shoot development and can occur at almost any time of year, if key factors are met. Internal signalling from the shoot system (by food supply and hormones) has an important role but the soil environment is also critical. In temperate trees, root growth starts at a soil temperature of >5 °C (Figure 4.19). Above 5 °C, root elongation generally increases, up to a soil temperature of around 20 °C, providing there is adequate soil moisture.

Well-aerated soil is also critical to root development. Roots and many other soil organisms, such as mycorrhizae, which contribute to healthy roots, all require a good supply of oxygen. However, oxygen levels need to be quite low for root growth to cease entirely. Roots often continue to grow below 10% oxygen (compared to the 21% in air) and only stop when oxygen reaches <3%.

Waterlogging, soil compaction and hard surfaces can all substantially reduce the movement of gases (gas flux) between the soil and the atmosphere. A useful way of describing the ability of soils to transport gases by diffusion is the relative gas-diffusion coefficient (Gliński and Stępniewski 1985). On a 0–1 scale, it quantifies the ease of gas flow through the soil compared to the atmosphere. For example, a soil gas-diffusion coefficient of 0.2 describes a soil in which the gas flux is equal to 20% of the gas flux within the atmosphere.

In a study on urban trees in Kassel, Germany, Weltecke and Gaertig (2012) were able to demonstrate the relevance of surface characteristics to soil aeration. Sealed surfaces reduced soil gas-diffusion coefficients to less than 0.05, whilst more 'natural' surfaces often showed soil gas diffusion coefficients above 0.20, comparable to forest environments. This measure of soil aeration was shown to influence the fine root development, particularly in the upper soil layer studied (0–15 cm) as shown in Figure 4.20 (Weltecke and Gaertig 2012). This supports the observation that healthy root development is dependent on well-aerated soil, capable of freely exchanging gases with the atmosphere. In hard landscapes, design features that improve soil aeration will always enhance

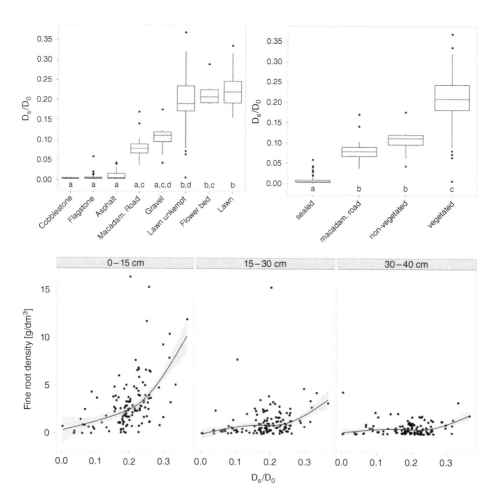

Figure 4.20 (Top) Gas-diffusion coefficient (shown as Ds/Do) of several typical urban soil covers. On the right side, the soil cover types are grouped into four classes. Means with the same letters are not significantly different, with 95% probability. The bottom and top of the box are the 25th and 75th percentiles, respectively, and the band near the middle of the box is the 50th median. The ends of the whiskers represent 1.5 times the interquartile range from the box, and outliers are indicated by points. (Bottom) The relationship between fine root density and relative gas diffusivity at three different depths in the soil. The black lines and shaded areas show key trends in the data. *Source:* Weltecke and Gaertig (2012). Reproduced with permission of Elsevier.

rooting environments. Therefore, opportunities to improve soil aeration through design and specification should always be taken where possible. This involves a number of strategies, including the use of porous surfaces and the integration of technology to prevent soil compaction. However, it is important to remember that good rooting environments are about balance. The most highly aerated soils are also those that drain most freely and are prone to rapid drying. Consequently, it is important not to exacerbate water deficits in pursuit of the best possible gas exchange between soil and atmosphere.

Soil Compaction

A number of variables that either directly or indirectly affect root growth occur as a consequence of soil compaction or, more accurately, over-compaction. Soil compaction is widespread in forest plantations, arboreta, urban parks and streets. Most sources of compaction can be attributed to humans, and are often related to vehicle and/or pedestrian traffic. However, even under natural conditions, where no or minimal human activity occurs above the root zone, animals may still exert sufficient surface pressure to compress soil aggregates into a higher bulk density (for definitions see Box 4.2). Resilience of a soil to compaction is determined by a host of chemical (e.g. pH, cation exchange capacity) and physical (e.g. texture, organic matter content, water content) characteristics that influence the cohesive forces between soil particles and their aggregates. Once external forces overcome these intrinsic cohesive forces, a change in bulk density occurs.

A range of methods can be used to assess the extent of compaction. Visual indicators, such as localised waterlogging (at the surface or just sub-surface), or a smooth soil surface, associated with a visible reduction in soil porosity, can be used to rapidly screen areas that may warrant more detailed inspection. On larger field sites, examination of a vertical soil profile exposed in a trench, or series of trenches, provides an effective way of identifying soil compaction. Very dense regions of soil that show few, if any, macropores (pores or gaps between soil particles larger than 75 μm) are likely to be compacted: smooth surfaces indicate this. Looking at adjacent areas that are thought not to have experienced high surface loads can be useful for comparison.

Under heavy traffic loads, root-limiting soil compaction may be present to around 1 m in depth but generally the highest soil compaction is found in the upper 30 cm of soil, a region critical for root development. Quantitative data collection using penetrometers or collecting soil samples for bulk density testing can be valuable, but they may also provide a misleading picture as they only assess a very discrete volume of soil, and soil compaction can vary considerably across short distances, both sideways and vertically. Many samples across a large area are required to give a meaningful assessment of the compaction and, even then, the vertical variation in soil compaction may be poorly evaluated, as deeper soil is difficult to access. However, quantification of compaction using bulk density or soil strength data can help show the magnitude of the problem and help instruct future management. The collection of data across an area relevant for a single mature tree is clearly much more feasible than the collection of similar data for a group of trees or a woodland.

Soil compaction can very rapidly become limiting to tree root development. In a simple assessment of soil bulk density measured after a series of passes by a vehicle over

Box 4.2 Describing Soil

Soil provides a vital medium for tree growth and development by providing water and nutrients, and by acting as a substrate for plant anchorage (Kozlowski *et al*. 1991). Soil is a mix of rock or *mineral soil* particles, such as clay or sand (typically 45% of the soil volume) and organic matter (typically 5%, but varies tremendously). These form *soil particles* that can clump together into *aggregates*. The spaces between soil particles (the *pores*) are occupied by water and gases, together typically making up 50% of the soil volume, in varying ratios (Brady and Weil 2016).

The relative amount of mineral particles and organic matter will affect the bulk density (how much a set volume weighs), usually measured in units of grams per cubic centimetre of soil ($g \, cm^{-3}$) or kilograms per cubic metre ($kg \, m^{-3}$), if strict scientific convention is being followed. Organic matter is very light and an increase in it lowers the bulk density. The bulk density can be increased by compaction that squeezes the soil particles closer together. This comes at the expense of its ability to hold water and gases, as well as making it physically more difficult for roots to grow through the soil.

Soil texture describes the look and feel of a soil in terms of the particles it is made from. A sandy soil has mostly large particles between 0.2 and 2 mm diameter, while a clay soil is composed of very small particles less than 0.002 mm diameter.

The acidity of the soil is measured as pH, and this is often very much misunderstood. The pH is important because nutrients vary in their availability depending upon pH, and, for a variety of physiological reasons, most plants have a range of acidity they can cope with: it is well known that acidic-loving rhododendrons do badly on alkaline soil. The pH of a soil is affected by the mineral components (e.g. chalk and limestone are very alkaline), but also by the amount and type of organic matter. As it decomposes, the organic matter releases organic acids into the soil; litter from some trees species produce more acids than others. Soils also tend to be acidic because rainwater has carbon dioxide dissolved in it, producing carbonic acid. A chemistry textbook will talk about the pH scale running from 0 (very acidic) to 14 (very alkaline), with pH 7 being chemically neutral. However, different professional communities tend to use soil pH terms in slightly different ways. For ecologists, a *neutral soil* tends to be on the acidic side of chemical neutrality at pH 5.0–6.5. Only below pH 5.0 can a soil be described as *acidic*, and above pH 6.5 as *alkaline*. For agriculturists, a soil tends to be considered neutral if pH 6.5–7.5, acid if below pH 6.5 and alkaline if above pH 7.5. Therefore, caution must be applied when discussing soil acidity and wherever possible terms such as acid, alkaline and neutral should be accompanied by the actual measured pH.

Another important part of soils are the nutrients they hold. Many of the nutrients needed by plants are *cations*, positively charged elements such as calcium, magnesium, potassium and sodium. The ability of the soil to hold these nutrients and to be able to exchange them with plant roots is referred to as the *cation exchange capacity*. This is usually formed by clay particles in the soil. A high cation exchange capacity means that the soil can readily hold these nutrients and relatively easily give them up to the tree roots.

Soils scientists classify soils according to a range of textural, hydrological, structural, chemical and biological criteria. Historically, each country would have had a slightly different way of classifying soils but efforts are now underway to reach a consensus on soil classification and comparable soil maps are available for large regions (e.g. see Jones *et al*. 2005).

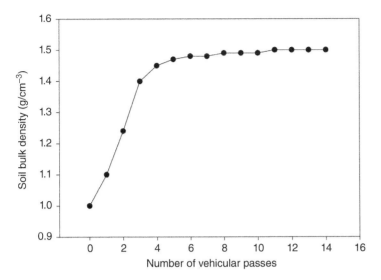

Figure 4.21 Increase in soil bulk density after a series of passes by a pick-up truck. Most of the increase in bulk density occurs as a result of the first four passes, after which the increases become more modest until a maximum soil bulk density is reached. Based on data courtesy of Glynn Percival.

the same area, the first four passes yielded the greatest change in bulk density; subsequent passes had a reduced impact: the damage was already done (Figure 4.21). This is a typical pattern of soil compaction development. Surface pressure from tyres, equipment, construction materials, pedestrians, hooves and so on, that exceeds the soil's inherent resistance to compaction immediately increases soil bulk density. For this reason, the prevention of soil compaction is central to the maintenance of a soil environment suitable for tree root development.

Soil Resistance to Root Development

Tree roots encounter mechanical resistance to their growth as a result of the force required to push past soil particles. This can greatly affect the spatial distribution of roots and the soil volume that is available for the uptake of water and minerals. Two interdependent variables are important for describing the resistance of soil to root penetration: soil strength and soil bulk density. Soil strength describes the mechanical impedance to root elongation and is the force required to penetrate the soil. Typically, this is assessed by a penetrometer which measures the force required for a steel cone to be pushed into the soil (Figure 4.22). As a result of various biological characteristics of roots, such as the exudation of lubricating mucilage from the root tip, the soil strength recorded by penetrometers tends to be 2–8 times higher than that experienced by a root in the same soil (Whalley and Bengough 2013). However, as this is still proportional to the soil strength experienced by a root, it remains an important variable to assess. It is widely accepted that soils recording penetrometer readings in excess of 2 MPa (2 megapascals = 20 bars) will seriously restrict root elongation. This value should not be treated as a threshold though, as root elongation rates have been shown to reduce in a linear fashion in soils less than 2 MPa (Figure 4.22a).

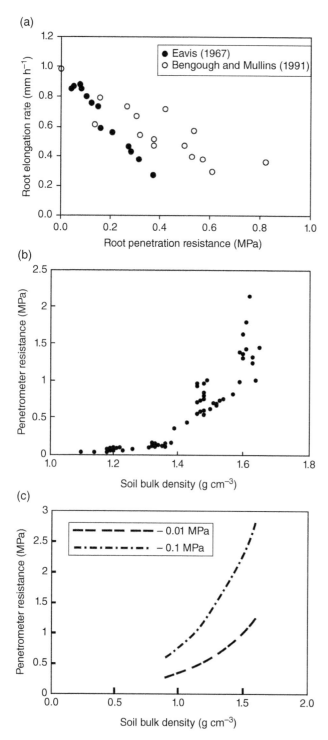

Figure 4.22 (a) Elongation rate of roots is reduced as root penetration resistance increases in peas (black circles) and maize (open circles). It should be noted that penetrometer resistance is 2–8 times greater than the root penetration resistance. (b) Penetrometer resistance plotted against soil bulk density. Penetrometer resistance becomes increasingly sensitive to bulk density at higher values of bulk density (>1.4 g cm^{-3} in this example). (c) The penetrometer resistance of soil at two matric potentials (moisture levels) of –0.01 and –0.1 MPa, plotted against soil bulk density. *Source:* (a) Adapted from Eavis (1967). Reproduced with permission of Springer. (b) Adapted from Whalley (2005). Reproduced with permission of Elsevier. (c) Adapted from Whalley (2013). Reproduced with permission of Taylor and Francis Group LLC Books.

Soil compaction measured in terms of bulk density (g cm^{-3}) typically displays a curvilinear relationship with soil strength. At relatively low bulk densities ($\sim < 1.4$ g cm^{-3}), only modest increases in soil strength are apparent with an increase in bulk density. However, as bulk density increases, soil strength becomes increasingly sensitive to soil compaction, and small changes in bulk density can have a profound effect on soil strength (Figure 4.22b) and, consequently, on root elongation. The exact nature of this relationship will vary with soil type and how much water the soil contains (Figure 4.22c). Soil strength in a drier soil is more readily affected by changes in soil bulk density than in wetter conditions. Consequently, accurate assessment of soil strength is highly complex, as soil bulk density and water content can vary substantially across the root spread of even one mature tree.

Although different species vary in their ability to grow through compacted soils, as general guidance, most roots are unable to penetrate moist soils of a bulk density greater than 1.4–1.6 g cm^{-3} in fine textured soils (high clay content) and 1.7 g cm^{-3} in more coarsely textured soils (high sand content). These values will be lower in drier soils.

In addition to directly impeding the growth of roots, soil compaction acts on a range of other variables that can also limit tree performance. As soil aggregates break down, pore space (the volume for liquid and air in the soil) is diminished. This is harmful to roots and soil organisms, primarily because it slows down gas diffusion and thus, aeration within the soil. Soil oxygen is less readily replaced by diffusion from above ground, allowing hypoxic (low oxygen) conditions to develop. As mineral uptake is an active process involving the expenditure of energy, hypoxic conditions reduce nutrient uptake. This, together with the disruption to (or loss of) mycorrhizal fungi and other soil organisms, can cause an acute shortage of nutrients, exacerbating any tree nutritional deficiencies. Further, resource-rich patches of soil may be unavailable to the tree if soil compaction physically limits root growth in these regions. Modification of soil structure also changes hydraulic properties significantly, reducing water movement through the soil that can lead to both water deficits and waterlogging. Increased surface run-off caused as a result of poor infiltration into a compacted surface will also limit soil water recharge, while associated erosion removes valuable soil and accelerates nutrient leaching. Therefore, reduced tree performance may result from direct mechanical impedance of roots, or indirectly via a number of biologically relevant variables (Figure 4.23). Similarly, amelioration of soil compaction may enhance tree performance via multiple mechanisms. It is therefore important to remember that the most obvious symptoms of a sick tree (such as nutrient deficiency) may not point to the ultimate cause. In most cases where soil compaction, or other problems within the root zone are limiting the ability of the roots to supply resources to the crown, a decline in the condition of the crown is apparent (Figure 4.24). Therefore, where a decline in crown condition can be seen, it is very important to evaluate the condition of the root zone before deciding on the best management strategy.

Management of Soil Compaction

Where tree roots are close to urban infrastructure or hard landscaping, the provision of low bulk density soil, protected from compaction, may be essential for the establishment of new trees or the survival of existing trees. By far the cheapest and most effective

(a)

(b)

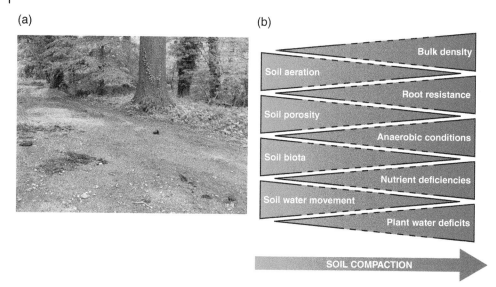

Bulk density

Soil aeration

Root resistance

Soil porosity

Anaerobic conditions

Soil biota

Nutrient deficiencies

Soil water movement

Plant water deficits

SOIL COMPACTION

Figure 4.23 (a) Soil compaction caused by construction traffic running alongside a woodland. (b) Soil characteristics modified by soil compaction. Increasing or decreasing band width indicates the impact of soil compaction on the named soil characteristic. Dashed lines indicate that trends are likely to be non-linear. Hirons & Percival (2012). Licensed Open Government Licence v3.0, http://www.nationalarchives.gov.uk/doc/open-government-licence/version/3/.

(a)

(b)

Figure 4.24 The impact of soil compaction on the crown of a copper beech *Fagus sylvatica* 'Purpurea'. Both trees are on the same site: (a) was protected from soil compaction during the expansion of a car park, whilst (b) was subjected to significant soil compaction and has suffered serious decline within the crown as a result of damage to the root system. *Source:* Courtesy of Duncan Slater.

Figure 4.25 The use of boardwalks in Waipoua Forest, New Zealand prevents soil compaction whilst allowing people to experience the forest.

form of soil compaction management is prevention. This may mean fencing off areas, providing physical barriers (such as bollards) to limit vehicular access to a site or re-orientating pathways away from vulnerable rooting zones. In parks and gardens where compaction is often caused by foot traffic, keeping the grass long under mature trees, or (even better) mulching, can subtly influence people's behaviour and discourage them from walking over the root zone. However, it may not be appropriate or desirable to exclude people from walking over tree root zones. For example, in sensitive forest sites the use of boardwalks can still allow visitors to enjoy the forest whilst reducing the impact on the root environment (Figure 4.25).

Where soil compaction is already a problem, there are two options: prevent it getting any worse and let it recover naturally, or physically ameliorate the soil. Clearly, the extent and depth of compaction are important considerations, but natural rehabilitation of the soil is likely to require many years, if not decades. Freeze–thaw cycles in winter can have a substantial role in breaking up compacted layers. However, these will obviously not occur at all in warm climates that never experience ground frosts. Plant roots as well as worms and other soil macro-fauna will gradually break up compacted soil, reducing bulk density and improving water infiltration. Therefore, *naturalisation* of garden or parkland sites by restricting access and allowing natural recovery is useful in some circumstances but tree decline may occur more rapidly than the recovery of the soil structure: more proactive treatments should be considered to help valuable trees.

In natural forests, low bulk density, well-aerated, biologically active soil, rich in organic matter, is the perfect host for the fine roots of trees. This is what we should aim to mimic as closely as possible in any tree root management plan.

Mulching

A simple, low-cost treatment that can be used to help ameliorate soil compaction around the base of the tree is mulching. This has the dual effect of encouraging biological activity and discouraging people from walking in the vicinity of tree roots (more detailed information on mulching can be found in Chapter 5).

Decompaction

Where there is evidence that the actual or potential rooting environment is compacted, reducing the soil strength and bulk density should be a fundamental objective of any tree health care programme. On some sites that are yet to be planted, or where tree roots are yet to develop, cultivation using hand tools or agricultural equipment can reduce the bulk density of large volumes of soil. However, where compaction occurs within an existing root zone, mechanical cultivation becomes much riskier because it is likely to wound roots, leaving them vulnerable to pathogens whilst reducing their ability to take up water and minerals.

The best methods to use, where roots are present, break up the soil without significantly damaging the root system. The use of *air tools*, including air-injection systems and compressed air soil-excavation tools, are particularly valuable for soil decompaction. Air-injection systems (e.g. VOGT® Geo-Injector) can be used to fracture compacted soil and, thus, improve soil aeration. Some tools can also inject granules or liquids into the newly made fissures (Figure 4.26a). Using this approach, it is possible to improve the soil structure (and fertility) of quite large areas relatively quickly. Extensive changes in the structure of compacted soil can be brought about by *air cultivation* using air tools, such as the Arb-Ex, Air Lance, Air Spade®, Soil Pick© or Supersonic Air Knife. Such tools are capable of breaking up soil to around 30 cm depth, with minimal disturbance to existing coarse roots. They blast air at high velocity (around 500 m s^{-1}) into the compacted soil to break it up (Figure 4.26b). Any non-porous objects, such as coarse roots, are left intact by the air-stream; the outcome is a much lower soil bulk density, normally around 1.2 g cm^{-3}, around a preserved root system.

Soil-excavation air tools can be used in several ways. *Radial mulching* uses the air tool to generate a series of trenches (usually around 20 cm wide and 30 cm deep) that radiate from the base of the stem like spokes on a wheel (Figure 4.27a). Once the soil is broken up, it can be removed (using hand tools or a soil vacuum), mixed with mulch or organic compost, before being backfilled into the same trenches. A top-dressing of mulch (5–10 cm in depth) can then be applied out to the dripline, where possible. The back-filled trenches provide enriched, decompacted channels that facilitate rapid fine root development from nearby coarse roots. This mimics forest conditions as tree roots often grow along the low bulk density tunnels, rich in organic matter, left by the decay of old coarse roots.

A disadvantage of radial mulching is the effort needed to extract, mix and backfill the soil from trenches, as this can be rather laborious and potentially expensive.

(a)

(b)

Figure 4.26 (a) Soil aeration can be improved by using specialist equipment, such as the VOGT® GeoInjector, which fractures compacted soil by injecting compressed air into the soil, potentially in combination with granular material or a liquid. (b) Air cultivation around the root-zone using an air tool, such as the Air Spade® (shown), can break up compacted soil and integrate organic matter into the root-zone. *Source:* (b) Courtesy of Glynn Percival.

There is also potential for roots exposed in the trenches to desiccate and die, although laying moist hessian over exposed roots reduces this risk. As a result, for badly compacted tree root-zones, *air cultivation* is favoured where possible (Figure 4.27b). Initially, the area of soil around the base of the tree and large segments within the dripline are broken up with an air tool. Mulch and any necessary nutritional supplements (fertilisers) are then added to the surface of the newly cultivated soil; this is then

(a)　　　　　　　　　　　　　　(b)

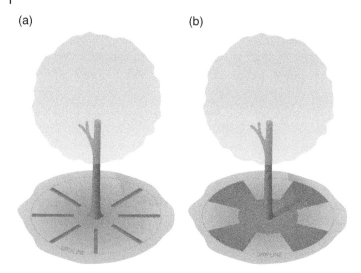

Figure 4.27 (a) Radial mulching where a series of trenches are dug with air tools and backfilled with soil mixed with mulch or other organic matter. (b) Air cultivation decompacts a larger area, typically a region around the whole stem with several segments that extend to the dripline.

mixed and integrated with the soil, using the air tool. Finally, a top-dressing of mulch is applied as described above. Rather than a series of radial trenches, air cultivation provides a larger connected volume of enriched, low bulk density soil that provides an ideal substrate for fine root development.

Both radial mulching and air cultivation are most easily achieved if the soil surface is free from other vegetation. Some preparatory work of the site may need to be carried out to achieve this, providing it does not conflict with other management aims of the site. If the soil is dry, it is also worth damping-down the area with water so that the air stream does not blow soil away so readily.

Given the high velocity of the air and the particles of soil that are inevitably blasted around, it is unsurprising that some damage to the roots occurs whilst using air tools (Kosola *et al.* 2007). Intuitively, the delicate fine roots will not be able to withstand such forces and this must be seen as a drawback of the technique. However, natural turnover of the fine roots is normally rapid (see Root Systems earlier in this chapter) so their loss can be viewed as just accelerating fine root mortality rather than irreversibly damaging the root system. As the vast majority of the root system remains intact and functional in the above methods, new fine roots will quickly emerge and proliferate within the amended soil. Therefore, while it should be acknowledged that fine root damage occurs, the marked increase in low bulk density rooting volumes vastly outweighs the modest damage caused by these techniques. Damage to fine roots can be minimised by using air tools in late autumn, when fine roots are likely to be dying off anyway and the demand for water from the crown is low. The use of air tools should also be kept to a minimum, using them just long enough to reduce soil bulk density; no longer. Excessive use of air tools in one area of the root zone may unnecessarily damage roots, without additional gains made in soil decompaction.

Estimating Appropriate Soil Volumes for Tree Roots

Before looking at options to increase the soil (rooting) volume for tree roots, it is important to try to assess how much soil a tree will require to reach maturity and thrive. While the capacity of the soil to provide anchorage and nutrition for the tree is vital, it is the soil's ability to supply sufficient water to the tree that is most frequently the limiting factor. Therefore, the volume of soil required to meet the tree's demand for water during the course of the growing season drives the soil volume required by the tree. Typically, if this volume is met then the other services provided by the soil in terms of nutrition and anchorage will also be met; exceptions to this may be in very sandy soils and/or heavily leached soils.

The difficulty in working out how much soil is needed is the variation in the amount of water that is required by individual trees of different sizes and species growing in different climates. Substantial variation also exists in the ability of different soils to supply water: their *water release characteristics* (see Chapter 6).

There has been much interest in water use by trees growing in natural forests because of their role in regulating water supply to rivers, influencing climate and the threat climate change has on the natural distribution of trees. A large-scale experiment in a North American temperate deciduous forest found that a mature tree would use between 40 and 340 litres of water per day. The three species that used the most water were a 45-cm diameter tulip tree *Liriodendron tulipifera* (340 L per day), a 38-cm diameter hickory *Carya* sp. (250 L per day) and a 48-cm diameter white oak *Quercus alba* (90 L per day) (Wullschleger and Hanson 2003). A wider survey of 67 species from 35 genera found that in 90% of cases, maximum daily water use was between 10 and 200 L per day for mature trees averaging around 20 m in height (Wullschleger *et al.* 1998). It is therefore clear that very large trees are capable of using hundreds of litres of water a day, providing the root system and soil is capable of its supply.

How much water, then, does a young landscape tree use? A highly instructive study was conducted by Richard Beeson in Florida using a series of instruments, known as lysimeters, which measure tree water by assessing the weight change in a container during the course of a day (1 g = 1 mL water) (see Beeson 2011a,b and Expert Box 4.1). The study determined that red maple *Acer rubrum* (12–14 cm diameter at 1 m) used up to 100 L per day (~30 US gallons) during the growing season, but this varied substantially depending upon leaf area and local weather conditions (Figure 4.28a). Other broadleaved trees, such as southern magnolia *Magnolia grandiflora* and Chinese elm *Ulmus parvifolia*, used comparable quantities of water at similar stages in development, but the conifer slash pine *Pinus elliottii* used about half as much water per day (Figure 28b–d). These data reveal some important characteristics of tree water use. First, leaf area strongly regulates whole tree water use. Both the deciduous species shown in Figure 4.28 used about 10 L per day without leaves; after leaf flush, daily water use increased to around 100 L per day within a few weeks. Secondly, pronounced variation in tree water use can occur as a function of changing microclimate and species. Thirdly, seasonal trends in tree water use include a rapid increase in spring to an early summer maxima, followed by a decline in late summer as leaves begin to age and senesce (Beeson 2011a).

These data also show why it is so important to have a root system that is well coupled to the bulk soil as quickly as possible after planting. If the root system is prevented from

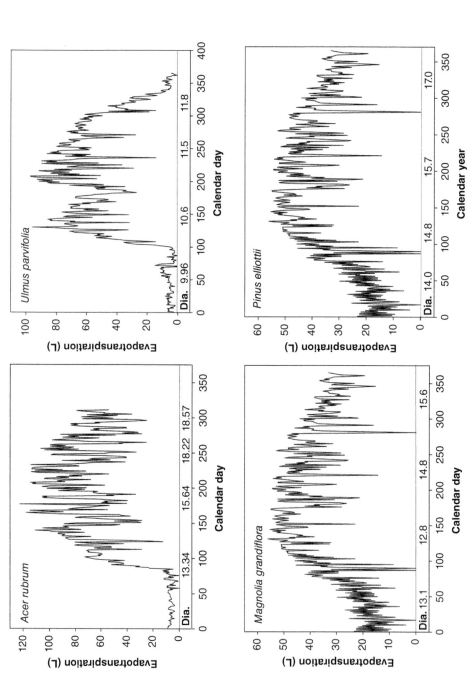

Figure 4.28 Water use (in litres) per day of red maple *Acer rubrum*, Chinese elm *Ulmus parvifolia*, southern magnolia *Magnolia grandiflora* and slash pine *Pinus elliottii* grown in tree lysimeters (see text for an explanation) as part of a project on water use of landscape plants, based at the University of Florida. Note the different scales on the axes. *Dia.* indicates the diameter at ground level in centimetres. See Expert Box 4.1 for more details on this research. *Source:* Courtesy of Richard Beeson.

establishing, through problems with the roots or soil, water deficits may quickly appear as the supply for water fails to meet transpirational demand. This is often the biggest causal factor in transplant shock and tree mortality in recently transplanted trees. Therefore, every attempt should be made to procure high quality plant material and ensure that the rooting environment does not limit the availability of water (see Chapter 6).

Accurate estimation of future tree water use is highly complex, because the volume of water demanded by a tree will vary depending on species, development stage (seasonal and maturity), developmental history, climate, microclimate (light, humidity, wind speed, temperature) and the interaction between the individual tree and soil drying cycles. This makes general estimations of required soil volumes somewhat speculative. Nevertheless, an estimation of suitable soil volumes for landscape trees is valuable to those tasked with establishing trees or designing the infrastructure required for trees to grow within.

Expert Box 4.1 Water Consumption of Landscape Trees
Richard C. Beeson Jr.

To reach their potential for shade, rainfall management, noise reduction and carbon sequestration, trees must have sufficient water available for uptake within the rooted soil volume for growth and for transpiration. In open areas where rooting volume is minimally impacted, tree roots often extend three times the width of the crown; in upright (fastigiate) cultivars, the ratio of root plate diameter to crown diameter can be even greater. In these locations, supplementary water is rarely needed. However, in urban and suburban areas, where restricted soil volumes inhibit natural root development, supplementary irrigation is often necessary to ensure high crown quality.

Researchers have measured tree water use in numerous ways, from measuring sap flow by various methods to weighing trees on scales. Mostly, these have been short snapshots of tree water use, ranging from a few hours to a few weeks. Most often, tree transpiration has been reported in volumes per day, with little background information such as solar radiation, vapor pressure deficits or tree size that would allow comparisons between sites or species. Thus, there has been insufficient information to develop robust models that could predict tree water consumption. In response to regional government agencies' restrictions on water allocations to tree farms, in 2001, I set out to quantify tree water use in trees from seedlings or cuttings to trees with a minimum trunk diameter of 13 cm above the soil (or 6 years of growth, whichever was greater). By the end of 2016, I had quantified tree water use for nine urban forest species, using large weighing lysimeters.

Lysimeters are weighing devices use to quantify changes in water content in soil-plant systems. For the research conducted here, I started with tripod suspension lysimeters using 23 kg load cells and 27 L containers for accuracy. The second year select trees were transplanted individually into 95 L containers and placed in nine large suspension lysimeters (Figure EB4.1). Each triangular basket weighed 136 kg, and was suspended at each apex from a 341 kg load cell. With a working capacity of 150%, each lysimeter could measure up to 1534 kg with an accuracy of ±250 g. Each species was replicated three times.

The combined results verified that tree water use of all nine species, while in leaf, could easily and accurately be estimated using a simple linear equation [4.1]:

$$\text{Daily tree } ET_A = \text{area} \times ET_o \times \text{a species specific coefficient} \qquad [4.1]$$

ET_A is the actual evapotranspiration of a lysimeter. It consists of water evaporation from the soil surface, and transpiration from leaves and lenticels in the bark. Area is the planar cross-sectional area of a tree trunk. Three area measurements (based on specific locations of 15 or 30 cm above ground, and below the swelling of the first major branch or DBH) can be used to scale ET_A to the size of juvenile trees. A fourth measurement can also be used; it is based on the horizontal projected area of a tree crown as a rectangle, calculated from measurements of the width in two perpendicular directions. For moderate and large trees, especially ring porous species, projected crown area is likely to be more accurate than trunk area if the tree crown is full. ET_o is reference evapotranspiration, calculated daily from the previous day's weather data using the UN-FAO 56 model. This is the measure of forces intercepted by leaves and soil that cause evapotranspiration. Results from these species served as a basis for the American Society of Agriculture and Biological Engineers (ASABE) S623 Standard for Landscape Irrigation, accepted in 2015. The aim of this standard is to reduce water volume applied to landscapes to a level required for healthy growth.

Inspection of the graphs provided elsewhere (Figure 4.28) demonstrates obvious differences between deciduous and evergreen species. In spring, increases in ET_A occur very rapidly with bud break for deciduous species (maple and elm). If available soil water is not increased comparatively, shoot growth and leaf area will be impeded. Even during the winter months, there was measurable transpiration if the soil was not frozen, most likely from lenticels. Flowering of maples, which occurs before leaf-out, doubled the daily water loss compared to barren branches. In fall, ET_A declined slowly in contrast to the rapid rise in the spring. Even though leaves were changing colour, stomata were still functioning and trees transpiring water until the last leaf fell. In contrast, ET_A of the two evergreen species, pine and magnolia, increased much more gradually in the spring and ended the year at a higher level of ET_A than in the beginning of spring, as a result of increases in overall leaf area.

Wide swings in daily ET_A are caused mainly by daily variations in solar radiation. Solar radiation generally accounts for nearly 90% of the energy that drives water in leaves to evaporate and pass through the stomatal pores into the atmosphere. Much of this is because of steep gradients of high water vapour in leaves compared to low concentrations of water vapour in air surrounding leaves. These gradients are more pronounced in spring and fall, when concentrations of water vapour in the air are low because of cooler night air temperatures, relative to generally warm days. The other factor driving transpiration is the rate of air movement across leaves. Higher air flows reduce the stagnant air boundary layer around leaves which, in turn, reduces the distance water molecules must traverse to reach the moving air around the leaf.

For evergreen trees, the relationship between transpiration, ETo and tree size is generally constant throughout the year, deviating towards high values during shoot flush. However, for deciduous trees the relationship falls apart shortly after leaf senescence begins, and is not re-established until the first leaves to flush in the spring have fully expanded. This can be seen in the graphs of both the red maple and Chinese elm (Figure 4.28). Caretakers of urban trees should be cognisant of the rapid increase in ET_A in spring, especially if it is a dry spring or if air temperatures are high and humilities are low.

While this research was conducted in central Florida, the thermodynamic forces that drive ET_A are universal. Shorter or longer days than the 10.5–14-hour days that occurred during this research would extend or shorten the periods of ET_A. Differences in energy levels driving ET_A, such as sun angle and the vapour pressure of water in the air would also moderate ET_A, but these are accounted for by ETo. The main factor is the water vapour pressure deficit (VPD) of air. Some tree species are anisohydric and are mostly unresponsive to VPD. These species exert little control over stomata until water potentials in the xylem become very negative. For *Quercus virginiana*, stomata do not fully close until after −4.5 MPa. However, most temperate species are isohydric; stomata of these species are sensitive to moderate decreases in VPD. Most have stomata that are closed, or nearly so, at −1.5 MPa. Thus, under high VPDs (more negative), ET_A would be less than predicted by the equation.

Although somewhat unsatisfactory, by making some rather gross assumptions (necessary to simplify the calculations) it is possible to come up with some plausible estimates of required soil volumes for landscape trees. Information needed includes the percentage of available soil water, the approximate daily water requirement of the

Table 4.1 Soil volumes required to deliver a determined quantity of water based, on an available soil water content ranging from 10% to 20%. This assumes no rain for 14 days and that roots occupied the entire soil volume.

Available soil water (%)	Soil volume (m³) required to deliver a given number of litres per day to a tree (maximum soil water recharge period = 14 days)					
	50 L	100 L	200 L	300 L	400 L	500 L
10	7.0	14.0	28.0	42.0	56.0	70.0
11	6.4	12.7	25.5	38.2	50.9	63.6
12	5.8	11.7	23.3	35.0	46.7	58.3
13	5.4	10.8	21.5	32.3	43.1	53.8
14	5.0	10.0	20.0	30.0	40.0	50.0
15	4.7	9.3	18.7	28.0	37.3	46.7
16	4.4	8.8	17.5	26.3	35.0	43.8
17	4.1	8.2	16.5	24.7	32.9	41.2
18	3.9	7.8	15.6	23.3	31.1	38.9
19	3.7	7.4	14.7	22.1	29.5	36.8
20	3.5	7.0	14.0	21.0	28.0	35.0

mature tree(s) in question and the frequency of soil water recharge. Available soil water is determined by the difference between the soil water content at field capacity (when the soil is holding as much water as it can when freely drained) and the point at which the tree can no longer extract water from the soil (the permanent wilting point or turgor loss point). This will depend on the relationship between the soil water release characteristics and the tree species, but will typically be equivalent to 10–20% of the soil volume (see Chapter 6). The daily water requirement will depend on the leaf area of the tree, species and how much water the tree is losing through evapotranspiration. Mature landscape trees that have full crowns not excessively shaded or restricted by competing trees usually have a larger demand for water than similar trees in woodland. Assuming that soil water will not be recharged by rain for 14 days during summer, a tree requiring 200 L per day over the 14 days would require 20 m³ of soil if 14% of the soil water was available to the tree. However, the same tree would require 28 m³ of soil if only 10% of the soil water was available, or 14 m³ of soil if 20% of the soil water was available (Table 4.1). This assumes that the root system is developed throughout the entire volume of soil, and that it continues to transpire water at the same daily rate for all 14 days (both of these factors are unlikely in reality). If soil water recharging events are more frequent, the soil volumes may be decreased; if they are less frequent, soil volumes should be increased.

It can be more straightforward to use an estimate of mature crown projection (area under the dripline of the tree; Figure 4.29) to specify soil volumes. As a general estimate

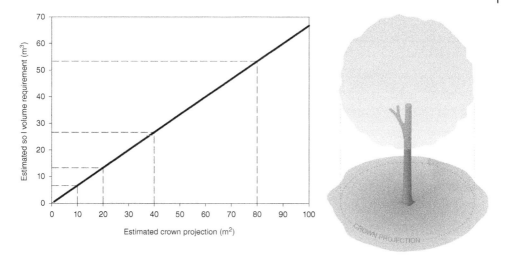

Figure 4.29 The estimated soil volume needed for trees, based on the potential mature tree crown projection – the area inside the edge of the crown (or the dripline). This should be used as a guide only, and used with particular caution if estimating soil volume requirements for more upright, fastigiate cultivars.

in temperate climates, 0.6 m^3 of soil is recommended for every 1 m^2 of crown projection (Lindsey and Bassuk 1991).

While there are many variables that will influence the volume of soil required by a tree, it is clear that the 'standard' 1 m^3 of soil often specified for urban planting pits is totally inadequate for any large tree intended to make a substantial contribution to the landscape. The contrast between adequate soil volumes for the supply of water and the reality of many rooting environments is a major reason for the reduced performance of urban trees, and highlights why tree response to water deficit can have such an important role in tree establishment (see Chapter 6). The amount of soil available to a tree does not just affect survival, but will also affect growth. In an overview of many pot-based trials, Poorter *et al.* (2012) found that, on average, for every doubling in pot size, plant weight increased by 43%, even without any other change in environment or growth conditions. In most cases, reduced growth in smaller pots was caused by a reduction in photosynthesis. Scaling these findings up a little, it is clear that the rooting volume of landscape trees can also have a profound effect on tree size. For example, Figure 4.30 shows the same species grown across a car park. Trees around the edge of the tarmac with access to larger volumes of soil are larger than those with more con-stricted soil volumes in the middle of the car park: trees need adequate soil volume if they are to thrive, rather than just survive.

Trees in natural environments, particularly on mountain slopes or very rocky condi-tions, regularly have to contend with restricted soil volumes. While many trees cannot adjust to such impoverished ciircumstances, and so soon die, some trees possess a remarkable capacity to survive on seemingly miniscule soil volumes (Figure 4.31). Other things being equal, these, surely, would do well in urban environments.

Figure 4.30 The impact of rooting environment on tree growth in a car park in Gelsenkirchen, Germany. Trees planted in central areas have minimal soil volume; trees around the edge of the car park share a more expansive soil volume and so have grown larger. *Source:* Courtesy of Johan Östberg.

(a)　　　　　　　　　　　　　　　　(b)

Figure 4.31 Trees growing in very small amounts of soil: (a) Corsican pine *Pinus nigra* subsp. *laricio* growing on the slopes of mountains in Corsica, France; (b) *Eucalyptus* spp. growing in incredibly tough conditions in the Watarrka National Park, Northern Territory, Australia.

Improving Soil Volumes in Urban Environments

In urban environments, opportunities to maximise soil volumes should be sought and taken wherever possible. A number of approaches can be used, through careful design of the below-ground environment and the use of specialist products. These products are broadly classified as raft systems, structural growing media or crate systems (TDAG 2014).

Raft systems are designed to sit on top of existing root systems and dissipate vertical load forces so that the underlying soil is not compacted. They consist either of deep (~150 mm) plastic tiles that can be filled with supplementary soil or a honeycomb-shaped cellular mattress (also known as *cellular confinement systems* or *geocells*) that is filled with a washed granular material. These 'rafts' are then covered with a geotextile membrane before surface construction takes place. Their principle value is in restricting further compaction to existing rooting environments, rather than expanding soil volumes, as they do not have the void space or structural qualities found in other systems.

Structural growing media are manufactured substrates that have been designed to be load-bearing, resistant to compaction and therefore able to provide an environment that does not physically limit root development. The media can be made of various substrates but they tend to fall into three main categories: sand-based substrates (tree soils); medium-sized aggregate substrates; and large-stone skeleton substrates (also known as the Stockholm system; Embrén *et al.* 2009). In practice, these are rather loose categories as different sized load-bearing aggregates are frequently used in the same planting beds. *Sand-based substrates* contain (by weight) around 90% sand (0.2–2 mm sized particles), 4–5% organic matter and 3–5% clay (<2 μm sized particles). The sand component provides a load-bearing matrix that holds the organic matter and clay for water and nutrient retention. Where relatively low surface loads are expected, such as under footpaths and bicycle tracks, these 'tree soils' can provide a substrate that makes root development compatible with modest surface loading. *Medium-sized aggregate substrates* use an angular matrix of stones from 25–100 mm in size, which holds organic matter and clay 'soil' components. *Large stone skeleton substrates* use larger stones 100–150 mm in diameter as the matrix. A combination of good quality soil and/or biochar is then flushed with water into the large voids within the stone matrix. A bearing layer of smaller washed granite (60–90 mm) is added on top of the larger stone base followed by a geotextile layer upon which the street surface is constructed. These layers are frequently punctuated with vertical aeration channels that ensure gas exchange throughout the substrate (Figure 4.32).

Although structural growing media are valuable for root development in that they resist soil compaction and provide good aeration, they do have a number of potential limitations. Unless correctly specified, sand-based and medium-sized aggregate substrates are often supplied with a wide range of particle sizes. This means that the finer material clogs soil pores, reducing aeration and drainage. It is therefore essential to use a very specific and narrow range of particle sizes to prevent this. For medium and large stone-based substrates, angular aggregates are vital to ensure good void space within the load-bearing matrix. High proportions of sand and stone, necessary to resist surface loading, make these substrates vulnerable to rapid drying and leaching of nutrients.

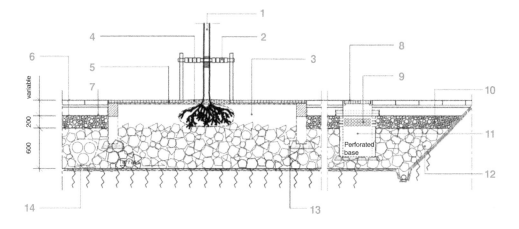

1. New tree size 20–25 cm	9. Air hole placed at level of aerated bearing layer
2. Tying in tree support	10. Aerated bearing layer
3. Planting soil.	11. Air and water supply
4. Crushed rock at grid 4–8 mm thick c. 50 mm	12. Crushed rock structural soil with planting soil
5. Surface grid 1400 × 2800 mm	13. Fertiliser at each structural soil level
6. Surfacing superstructure	14. Pipes in structural soil protected with geotextile and gravel surround.
7. Geotextile	
8. Stormwater cover, dished for laying by gutter	

Figure 4.32 Section of large stone skeleton substrate installation for a new planting. *Source:* Adapted from Embrén *et al.* (2009). Reproduced with permission of Trees and Design Action Trust, 2014.

Trees may then be subject to water and nutrient deficits, particularly as they mature and their crowns begin to demand greater volumes of water. In order to increase the range of species suitable for these sites, it is necessary to supplement the water supply by integrating irrigation and rainwater harvesting systems into the planting beds. In addition, nutritional supplements are likely to be required as soil nutritional status may suffer in the mid to long-term because of the small amounts of clay and organic matter used in the structural growing media, despite fine root turnover contributing some organic matter to the root zone. To counter the effect of nutrient leaching, biochar is being used to aid nutrient retention in these manufactured growing substrates. Nevertheless, despite these potential limitations to structural growing media, there is good evidence that at least some species can cope well with these conditions. For example, Grabosky and Bassuk (2016) found that the performance of swamp white oak *Quercus bicolor* and willow oak *Quercus phellos* growing in a 'structural soil' was comparable to trees planted in an adjacent lawn area over a 17-year period. Furthermore, in Stockholm, large stone skeleton substrates, in combination with diligent tree management, have been used to achieve excellent rates of tree establishment (Embrén *et al.* 2009).

A further method to prevent soil compaction is to use *crate-based systems* such as *RootSpace*^TM, *StrataCell*^TM, *Silva Cell, Tree Box High-Performance* or *TreeBunker*^TM.

(a)

(b)

Figure 4.33 A crate system, in this case a RootSpace™ system, designed to provide the trees with the maximum amount of uncompacted soil to root into. The crate itself bears the load so the soil inside remains uncompacted. (a) An artist's impression of the crates being used underneath a paved surface; (b) the RootSpace™ crate. *Source:* Reproduced with permission of GreenBlue Urban.

These rely on modular 'structural cells' made from plastic or concrete to provide the load-bearing capacity that prevents soil compaction (Figure 4.33). In comparison to structural growing media, crate systems hold a much larger quantity of good quality soil: around 90% of the crate volume can be filled with high-quality soil. This means that much more substantial soil volumes can be designed into urban landscapes, and that trees can be provided with soil volumes capable of supporting their growth in the longer term. As below-ground constraints are frequently the greatest limitation to tree performance in the urban environment, this type of root-zone technology represents a major advance in our ability to combine urban infrastructure with trees. Such systems have also been used in *sustainable urban drainage systems* (SUDS) to reduce flooding. Careful design will also make these systems compatible with other below-ground infrastructure, such as utility pipes and cables.

Figure 4.34 shows an experiment on tree performance in a paved site laid on either compacted soil, structural growing media or a suspended pavement analogous to a crate system. The suspended pavement clearly provided the best environment for tree growth (Smiley *et al.* 2006). This is likely to have resulted from the greater water and nutrient availability created by the higher proportion of quality soil within the crate system. In comparison, by 8 years after planting, several of the trees in the structural growing media needed replacing.

Each of the above systems has its advantages and limitations; each appropriate for a specific set of circumstances that may be unsuitable in other situations. Therefore, before utilising raft systems, structural growing media or crate systems for a planting scheme it is essential that expert advice be sought. Their installation will require professional oversight and substantial planning if they are to deliver high quality rooting environments that meet the engineering requirements of the site and minimise any potential conflicts with other stakeholders of below-ground space.

Figure 4.34 The Urban Tree Plaza experiment based at the Bartlett Tree Experts research laboratories in Charlotte, North Carolina, USA. Trees growing in the suspended pavement system that mimics the crate-based systems discussed in the text, out-performed those growing in structural soils (Stalite and Gravel) or compacted soil (Compacted). In lower image, 'New' represents replacement trees where the first planted tree has previously died. *Source:* Courtesy of Thomas Smiley.

References

Beck, C.B. (2010) *An Introduction to Plant Structure and Development: Plant Anatomy for the Twenty-first Century*, 2nd edition. Cambridge University Press, Cambridge, UK.

Beeson, R.C. (2011a) Evapotranspiration of woody landscape plants. In: Gerosa, G. (ed.) *Evapotranspiration: From Measurements to Agricultural and Environmental Applications*. InTech, Rijeka, Croatia, pp. 347–370.

Beeson, R.C. (2011b) Weighing lysimeter systems for quantifying water use and studies of controlled water stress for crops grown in low bulk density substrates. *Agricultural Water Management*, 98: 967–976.

Bengough, A.G. and Mullins, C.E. (1991) Penetrometer resistance, root penetration resistance and root elongation rate in two sandy loam soils. *Plant and Soil*, 131: 59–66.

Brady, N.C. and Weil, R.R. (2016) *The Nature and Properties of Soils*, 15th edition. Pearson Education, Prentice Hall, New York, USA.

Buée, M. Boer, W.D. and Martin, F. (2009) The rhizosphere zoo: An overview of plant-associated communities of microorganisms, including phages, bacteria, archaea, and fungi, and of some of their structuring factors. *Plant and Soil*, 321: 189–212.

Canadell, J., Jackson, R.B., Ehleringer, J.R., Mooney, H.A., Sala, O.E. and Schulze, E.-D. (1996) Maximum rooting depth of vegetation types at the global scale. *Oecologia*, 108: 583–595.

Coutts, M.P., Nielsen, C.C.N. and Nicoll, B.C. (1999) The development of symmetry, rigidity and anchorage in the structural root system of conifers. *Plant and Soil*, 217: 1–15.

Danjon, F., Stokes, A. and Bakker, M.R. (2013) Root systems of woody plants. In: Eshel, A. and Beeckman, T. (eds) *Plant Roots: The Hidden Half*, 4th edition. CRC Press, Boca Raton, USA, pp. 29–26.

Eavis, B.W. (1967) *Mechanical Impedance to Root Growth*. Agricultural Engineering Symposium, Paper 4/F/ 39, 1–11, Silsoe, UK.

Embrén, B., Alvem, M.B., Stål, O. and Orvesten, A. (2009) *Planting Beds in the City of Stockholm: A Handbook*. City of Stockholm, Sweden.

Ennos, A.R. (1995) Development of buttresses in rainforest trees: The influence of mechanical stress. In: Coutts, M.P. and Grace, J. (eds) *Wind and Trees*. Cambridge University Press, Cambridge, UK, pp. 293–301.

Ennos, A.R. (2016) *Trees: A Complete Guide to Their Biology and Structure*. Cornell University Press, Ithaca, USA.

Ghani, M.A., Stokes, A. and Fourcaud, T. (2009) The effect of root architecture and root loss through trenching on the anchorage of tropical urban trees (*Eugenia grandis* Wight). *Trees*, 23: 197–209.

Gliński, J. and Stępniewski, W. (1985) *Soil Aeration and Its Role for Plants*. CRC Press, Boca Raton, USA.

Grabosky, J. and Bassuk, N. (2016) Seventeen years' growth of street trees in structural soil compared with a tree lawn in New York City. *Urban Forestry and Urban Greening*, 16: 103–109.

Hirons, A.D. and Percival, G.C. (2012) Fundamentals of tree establishment. In: Johnston, M. and Percival, G.C. (eds) *Trees People and the Built Environment: Proceedings of the Urban Trees Research Conference, 13–14 April 2011*. Forestry Commission, Edinburgh, UK, pp. 51–62.

Jackson, R.B., Canadell, J., Ehleringer, J.R., Mooney, H.A., Sala, O.E. and Schulze, E.D. (1996) A global analysis of root distributions for terrestrial biomes. *Oecologia*, 108: 389–411.

Jones, A., Montanarella, L. and Jones, R. (2005) *Soil Atlas of Europe*. European Commission, Brussels, Belgium.

Kosola, S., Workmaster, B.A.A., Busse, J.S. and Gilman, J.H. (2007) Sampling damage to tree fine roots: Comparing air excavation and hydropneumatic elutriation. *Hortscience*, 42: 728–731.

Köstler, J., Brückner, E. and Bibelriether, H. (1968) *Die Wurzeln der Waldbäume. Untersuchungen zur Morphologie der Waldbäume in Mitteleuropa*. Paul Parey, Hamburg, Germany.

Kozlowski, T.T., Kramer, P.J. and Pallardy, S.G. (1991) *The Physiological Ecology of Woody Plants*, Academic Press, San Diego, USA.

Kumar, P., Hallgren, S.W., Enstone, D.E. and Peterson, C.A. (2007) Root anatomy of *Pinus taeda L.*: Seasonal and environmental effects on development in seedlings. *Trees*, 21: 693–706.

Kutschera, L. and Lichtenegger, E. (2002) Wurzelatlas mitteleuropäischer Waldbäume und Sträucher. Leopold Stocker Verlag. Graz, Austria.

Lindsey, P. and Bassuk, N. (1991) Specifying soil volumes to meet the water needs of mature urban street trees and trees in containers. *Journal of Arboriculture*, 17: 141–148.

Lyford, W.H. (1980) *Development of the root system of northern red oak (*Quercus rubra *L.)* Harvard Forest Paper No. 21. Harvard University, Petersham, USA.

Lyford, W.H. and Wilson, B.F. (1964) *Development of the root system of* Acer rubrum *L.* Harvard Forest Paper No. 10. Harvard University, Petersham, USA.

Lyr, H. and Hoffmann, G. (1967) Growth rates and growth periodicity of tree roots. *International Review of Forestry Research*, 2: 181–236.

McCormack, L.M., Adams, T.S., Smithwick, E.A. and Eissenstat, D.M. (2012) Predicting fine root lifespan from plant functional traits in temperate trees. *New Phytologist*, 195: 823–831.

McCormack, M.L., Adams, T.S., Smithwick, E.A. and Eissenstat, D.M. (2014) Variability in root production, phenology, and turnover rate among 12 temperate tree species. *Ecology*, 95: 2224–2235.

McCormack, M.L., Dickie, I.A., Eissenstat, D.M., Fahey, T.J., Fernandez, C.W., Guo, D., *et al.* (2015) Redefining fine roots improves understanding of below-ground contributions to terrestrial biosphere processes. *New Phytologist*, 207: 505–518.

Persson, H.A. (2002) Root systems of arboreal plants. In: Waisel, Y., Eshel, A. and Kafkafi, U. *Plant Roots: The Hidden Half*, 3rd edition. Marcel Dekker, New York, USA, pp. 287–313.

Petignat, A. and Jasper, L. (2015) *Baobabs of the World: The Upside-down Trees of Madagascar, Africa and Australia*. Struik Nature, Penguin Random House South Africa, Cape Town, South Africa.

Poorter, H., Bühler, J., van Dusschoten, D., Climent, J. and Postma, J.A. (2012) Pot size matters: A meta-analysis of the effects of rooting volume on plant growth. *Functional Plant Biology*, 39: 839–850.

Pregitzer, K.S., DeForest, J.L., Burton, A.J., Allen, M.F., Ruess, R.W. and Hendrick, R.L. (2002) Fine root architecture of nine North American trees. *Ecological Monographs*, 72: 293–309.

Smiley, E.T., Calfee, L., Fraedrich, B.R. and Smiley, E.J. (2006) Comparison of structural and non-compacted soils for trees surrounded by pavement. *Arboriculture and Urban Forestry*, 32: 164–169.

Stone, E.L. and Kalisz, P.J. (1991) On the maximum extent of tree roots. *Forest Ecology and Management*, 46: 59–102.

Taiz, L. and Zeiger, E. (2010) *Plant Physiology*, 5th edition. Sinauer Associates, Sunderland, MA, USA.

TDAG (2014) *Trees in Hard Landscapes: A Guide for Delivery*. Trees and Design Action Group.

Vercambre, G., Pagès, L., Doussan, C. and Habib, R. (2003) Architectural analysis and synthesis of the plum tree root system in an orchard using a quantitative modelling approach. *Plant and Soil*, 251: 1–11.

Vogel, S. (2012) *The Life of a Leaf*. University of Chicago Press, Chicago, USA.

Weltecke, K. and Gaertig, T. (2012) Influence of soil aeration on rooting and growth of the Beuys-trees in Kassel, Germany. *Urban Forestry and Urban Greening*, 11: 329–338.

Whalley, W.R. and Bengough, A.G. (2013) Soil mechanical resistance and root growth and function. In: Eshel, A. and Beeckman, T. (eds) *Plant Roots: The Hidden Half*, 4th edition. CRC Press, Boca Raton, USA, pp. 29–26.

Whalley, W.R., Leeds-Harrison, P.B., Clark, L.J. and Gowing, D.J.G. (2005) Use of effective stress to predict the penetrometer resistance of unsaturated soils. *Soil Tillage Research*, 84: 18–27.

Wilson, B.F. (1984) *The Growing Tree*. University of Massachusetts Press. Massachusetts, USA.

Wullschleger, S.D. and Hanson, P.J. (2003) Sensitivity of sapling and mature-tree water use to altered precipitation regimes. In: Hanson, P.J. and Wullschleger, S.D. (eds) *North American Temperate Deciduous Forest Responses to Changing Precipitation Regimes*. Springer, Berlin, Germany, pp. 121–139.

Wullschleger, S.D., Meinzer, F.C. and Vertessy, R.A. (1998) A review of whole-plant water use studies in tree. *Tree Physiology*, 18: 499–512.

5

The Next Generation of Trees: From Seeds to Planting

The Next Generation of Trees: From Seeds to Planting

Flowers, Seeds and Fruits

All angiosperm trees (the 'flowering plants') are, by definition, capable of producing flowers. Gymnosperms (Figure 5.1) do not have flowers: they have male and female cones (officially called *strobili*). In both cases, the flowers and strobili are capable of producing seeds that are necessary for creating the next generation.

Once the male and female cones are produced in gymnosperms, seed production starts with pollination: pollen from male cones is delivered to the female cones, usually by wind. Once there, the pollen either slides down between the scales of the female cone or it lands in a sticky pollination drop that is reabsorbed at the base of the cone scales, pulling the pollen deeper into the cone. This delivers the pollen close to the female egg (or ovule). The pollen then germinates, grows towards and penetrates the egg, thus fertilising it. Once fertilised, the egg grows into a seed. In conifers, there are normally two seeds on each scale of the female cone.

Seed production in angiosperms follows a similar but slightly more complex process. During pollination, pollen from the anthers lands on the top part of the female carpel, called the stigma (Figure 5.1). The stigma normally has microscopic undulations within which the pollen from the same species fits snugly. The biochemical signals that the pollen receives from this close contact with the stigma causes the pollen to germinate and grow down through the style to reach the ovule, and fertilise it. This is very similar to what happens in gymnosperms, but the pollen growth happens inside the carpel rather than in the open air between two cone scales. Another difference is that, in angiosperms, the seeds become encased within a fruit, made up of the carpel tissue from the mother, while in gymnosperms the seeds are naked.

Once the seeds are released and reach a suitable niche that gives them the right amount of water and heat (and sometimes light), the seed will germinate and perhaps survive to grow into a new tree. This, in turn, will flower, producing new seeds, ensuring new generations of trees and perpetuation of the species.

As well as producing new trees, seeds also have several other important roles. Adult trees are fixed in one place, so seeds are the main mechanisms that allow trees to invade new areas and spread across the landscape. By this means, as new habitats open up or gaps appear in woodlands, well-placed seeds can take advantage of the opportunities. Seeds also have another role: producing genetic variation within a population. In most

Applied Tree Biology, First Edition. Andrew D. Hirons and Peter A. Thomas.
© 2018 John Wiley & Sons Ltd. Published 2018 by John Wiley & Sons Ltd.

(a) (b)

Figure 5.1 (a) Typical angiosperm flower of the deciduous camellia *Stewartia pseudocamellia*, composed of white petals (collectively called the corolla) surrounding a ring of male stamens (each made of the pollen producing anther borne on a filament). In the centre is the female carpel, composed of the ovary at the bottom (containing the ovules which will become the seeds) from which grows the style surmounted by the stigma, the part of the carpel that receives the pollen. (b) Female cones of western hemlock *Tsuga heterophylla*. Gymnosperms do not have flowers but produce seeds using male and female cones (called strobili).

cases (see below for the exceptions), seeds are the product of two parents and include genetic material from both of them, creating new combinations of genes.

Normal plant cells contain chromosomes each made up of DNA twisted together into the famous double helix. Each cell normally contains two sets of *chromosomes* (referred to as 2n or diploid, di meaning two). When a plant's reproductive cells are forming, the chromosomes separate so that each pollen grain and ovule contain one set of chromosomes (haploid, 1n); upon fertilisation these come together to produce the normal two sets of chromosomes. However, other processes happen in the formation of pollen and ovules that allow the swapping over of some parts of the chromosomes (called *recombination* or chromosomal crossover). The result is that although offspring of two parents all have one chromosome from each parent, they may be very different from either parent. For this reason, when a batch of tree seeds germinates, even if all are from the same mother tree, there can be variation in such things as leaf shape, speed of growth or colour of the leaves. There can also be differences not visible to the naked eye, such as variation in drought resistance, ability to extract nutrients from the soil or sensitivity to pollution. For people managing trees, this can be a double-edged sword. If you want uniformity in new trees it is better to take cuttings or use a similar method of vegetative propagation to produce a set of trees that are genetically the same, and so are more likely to grow in a similar way. As generations of plant breeders have found, however, the variation thrown up in seedlings can produce a huge range of new varieties. For the tree species, this variation is important because it is the basis of evolution. As habitats and climate constantly change, it may well be that a genetically new individual is better adapted to the new conditions and will prosper, producing many seeds that contain the new adaptations. This *natural selection* of the individuals that best suit the current environment leads to progressive change within a species and, eventually, the production of new species. The truth of this selection of new adaptations can be seen within one species growing in different areas. For example, beech *Fagus sylvatica*, in colder parts of Europe (such as Poland), have evolved to be more frost tolerant than they are in western Europe (Kreyling *et al.* 2014). We can take advantage of this when breeding new varieties. A good example is the copper beech *Fagus sylvatica* 'Atropunicea' or 'Purpurea'. A very small proportion of seedlings naturally have some purple coloration in the leaves; by growing those seedlings until they produced seeds, and selecting the most purple of the resulting seedlings, we have artificially 'selected' a more coppery beech tree than is found in the wild.

The difference between angiosperms and gymnosperms is that the former produce fruits that fully enclose their seeds, while the conifers produce 'naked seeds' in cones – you can see the seeds by bending the scales of the mature female cone apart, the seeds are not completely enclosed. These cones hold the seeds until they are mature, when the scales open to release the seeds. Some gymnosperms produce cones that are fruit-like, such as those that look like berries on junipers *Juniperus* spp., but these are just fleshy cones that still contain naked seeds.

This raises the question: why have fruits in the angiosperms? The answer is twofold. First of all, a fruit protects the seeds from being eaten while they are developing. This can be physical protection, such as is provided by the tough woody skin of a nut (in the gymnosperms, the cone does the same job), or it can protect the seed by being unpleasant to eat, as in hard, sour, green apples. The second purpose of the fruit is often to help the seeds be dispersed: the hard, sour, green apple that protects the developing seeds will become softer and sweet to encourage animals to take the fruit away to eat, in the process dropping the seeds some way away from the parent trees. The colour of the apple changes from green to something more attractive to signal to the animal dispersers that it is now palatable. However, the fruit does not always help seed dispersal; in the case of nuts, the hard woody shell does not help the seed be moved, it simply protects it.

Variation in Flowers and Pollination

Tree flowers come in many different shapes and sizes. The most basic division in flower type is based on pollination method. In trees pollinated by animals, and this means primarily insects, the flowers are just like any other plant in their contents (Figure 5.1). The petals of the flowers advertise where the flowers are to the insects and, in some cases, are carefully crafted to make sure the insect lands in the correct place on the flower to deliver and pick up pollen. The reward for the insect comes in the form of nectar – a sugary solution that often has other things added, such as amino acids, to supplement the diet of the pollinator. This is not altruism on the part of the plant: looking after your pollinators ensures they stay alive to carry on pollinating. Some pollinators, such as bees and some butterflies, also feed on the pollen. Although the loss of pollen may seem counterproductive for the trees, the amounts being taken are fairly small in comparison to what a plant produces, so it is a small price to pay.

Individual flowers can be very attractive to pollinating insects, but the problem for a large tree is often attracting enough pollinating insects from over a large area to ensure good pollination. This problem is solved in many trees by flowering before the leaves are produced, and by either putting many small flowers together to make an inflorescence (Figure 5.2a) or by using the whole tree as one large visual signal (Figure 5.2b).

(a)

(b)

Figure 5.2 (a) The individual small flowers of the staghorn sumac *Rhus typhina* are fairly insignificant, but when grouped into a tight inflorescence (flower head) they are more visually attractive to pollinating insects from a distance. (b) Pink trumpet tree *Tabebuia heptaphylla*, native to tropical South and Central America and used as a street tree in sub-tropical areas, flowers before the leaves are opened to increase the visibility of the flowers: the large number of flowers on a single tree act as a very prominent visual signal to draw pollinating bees from a very wide area.

(a)

(b)

Figure 5.3 (a) Wind pollinated male flowers of European ash *Fraxinus excelsior*. The petals are reduced to a lobed fringe around the base of the flower, and the visible parts of the flowers are masses of unopened male anthers. (b) Male catkins of hop hornbeam *Ostrya carpinifolia*. Each little scale in the catkin is really a short branch ending in a brown bract, behind which are sheltered a group of male flowers consisting of nothing more than stamens. A female catkin looks similar, but the stigmas and styles from the female flowers poke out around the bracts in order to catch the pollen.

By contrast, many temperate trees have fairly insignificant flowers because they are wind pollinated (Figure 5.3a). In this case, petals would just get in the way so they have been lost, and the flowers consist of just the naked male and female parts. Some trees, such as birches *Betula* spp. and willows *Salix* spp., have separate male and female flowers that are concentrated together in catkins (literally 'a little cat' because they resemble a kitten's tail) (Figure 5.3b).

We are fairly used to animals being either male or female, but plants, of course, take little notice of this convention. A basic flower contains both male and female parts – the pollen-producing stamens and the female carpel containing the ovary. Such a flower is referred to as being a *hermaphrodite* or *perfect* flower. In many trees, particularly those that are large and wind pollinated, the sexes are separated out into male and female unisex (or *imperfect*) flowers. In the case of a male flower, the central carpel may be entirely absent or remain as a shrunken, non-functional vestige; in female flowers, the stamens are likewise greatly reduced or absent. That gives us three basic flower types (hermaphrodite, male and female), but things get more complicated when looking at whole trees because these three flower types can occur in many different configurations. Some trees are fairly straightforward, such as flowering cherries *Prunus* that have just simple hermaphrodite flowers covering the whole tree. Many trees, especially those that are wind pollinated, have separate male and female flowers, often with the males congregating at the bottom of the tree and the females nearer the top. Segregating males and females through the crown has the dual advantage that it helps prevent pollination by one's own flowers, because pollen very rarely rises vertically up a tree (even when moved by insects), and it also means that the seeds are produced at the top of the trees, making them more accessible to birds or other animals, or to the wind if they spread by being blown away. It is possible, however, to get almost any combination of the three flower types within a tree. The European ash *Fraxinus excelsior* is a good example: some trees are all males while others can be all female, or they may be mostly one gender but with branches of the other gender, and these unisex flowers may be liberally mixed with hermaphrodite flowers (Tal 2011). Moreover, as gender in plants is not as genetically fixed as it is in many animals, trees can change from year to year: a male tree can become female or contain flowers from both sexes.

Some trees are a little more straightforward and remain either solely female or male. These *dioecious* trees (as opposed to those that have both male and female flowers on the same tree – *monoecious*) may be angiosperm or gymnosperm and include about 15% of maples *Acer* spp., most *Ilex, Juniperus, Podocarpus, Populus, Salix* and *Taxus*, as well as many tropical trees. Knowing the gender of a tree is of practical use as you can preferentially select male trees for planting schemes if you do not want a lot of fruit litter. For example, the female maidenhair tree *Ginkgo biloba* has fruits that smell unpleasantly of rancid butter or vomit, and so the male trees tend to be preferred. However, care is needed when buying trees that are normally only one gender because they can change over time. For example, all of the upright Irish yews *Taxus baccata* 'Fastigiata' descend from cuttings from one female yew tree and are therefore all female, except a few which have now switched to become male. Similarly the normally male Italian poplar clones *Populus* x *euroamericana* 'Serotina' (produced in the same way) now have female individuals.

Seeds and fruits are usually more expensive to produce than pollen, so the reproductive costs of being female are usually higher than for the male; although in some wind-pollinated male trees that produce huge quantities of nitrogen-rich pollen, the costs can be more equal. This helps explain why completely male trees tend to grow faster than females, reach a bigger final size and have a lower death rate (Barrett and Hough 2013), something else to bear in mind when planning a planting scheme.

Not All Seeds Require Pollination

Vegetative reproduction produces individuals that are genetically identical to the parent, while sexual reproduction using seeds produces new individuals that are genetically mixed. However, it is possible to produce seeds that are all genetically identical if seeds form without using pollen. In this process of *apomixis*, all the genes (i.e. both sets of chromosomes) come from the mother and the offspring will, consequently, be genetically identical to her. Many members of the rose family (Rosaceae) are apomictic including species of *Malus*, *Sorbus*, *Crataegus* and *Amelanchier*. Strictly speaking, all but the apples *Malus* spp. physically need the pollen to be present to stimulate seed production, but there is no fertilisation by the pollen so all the genetic material still comes from the mother. The net result is that in the wild, groups of individuals will look remarkably alike if they all come from the same mother, but slightly different from a group of the same species further away that comes from a different mother. Whether you think these different *clones* are just variants on a theme or should be separated into *microspecies* depends on your viewpoint. From a landscape design perspective, such diversity in clones means that there are often tree characteristics that can be chosen to improve the desirability of the tree in some way; for example, particularly profuse flowering, vibrant autumn colours or an upright branching habit. Whether these clones are commercially available is another matter.

Some plants go a step further and are *parthenocarpic*; they produce fruits without pollination, but in this case there are no seeds either. A wide range of temperate trees including birch *Betula* spp., maples *Acer* spp., elms *Ulmus* spp., hollies *Ilex* spp. and conifers such as firs *Abies* spp. and junipers *Juniperus* spp., will produce fruits with no seed. These fruits are often a little smaller than normal but knowledge of parthenocarpy has been used to produce seedless varieties of fruit such as navel oranges and satsumas. Parthenocarpy does not happen in all trees; for example, in oaks *Quercus* spp., if there is no fertilised seed, the acorn will not develop, as hormones produced by the fertilised seed are needed to stimulate the growth of the fruit. This makes good evolutionary sense as no energy is wasted growing a fruit that is of no use.

It should be borne in mind that, although all trees can produce fruit and seeds, they might not always do so. In some cases, this is a result of poor pollination. This may be because there is no other specimen of the same species close enough, or it needs to be a specific variety of that species. A good example of this is apple *cultivars* (a contraction of 'cultivated varieties'). Many of these human-produced varieties are self-incompatible: they cannot pollinate themselves and require a different variety nearby with which to cross-pollinate. For example, 'Granny Smith' produces most fruit when pollinated by 'Golden Delicious' but generally most cultivars will pollinate each other, providing their flowering periods overlap (Ramírez and Davenport 2013). Some trees may not even flower at all, or they may flower but not produce seeds and fruits. This is common in trees that are introduced into a different climate, typically those that are grown in cooler conditions than where they grow naturally. For example, the Japanese pagoda tree *Styphnolobium japonicum* (formally known as *Sophora japonica*) is native to China, despite the name, but is widely grown in Europe. In London it will flower regularly but in northern Britain, particularly in a cool summer and when shaded, it may never flower. Similarly, the silk tree *Albizia julibrissin*, from southwest Asia, will produce a riot of colour in a hot summer in southern Britain but is much less likely to flower in the cooler summers of northern Britain. In a similar way, small-leaved lime *Tilia cordata* now rarely sets seed in the UK, although it did 5000 years ago when the climate was significantly warmer (Pigott and Huntley 1981).

Cost of Reproduction

A growing tree has limited resources available to it, whether these are nutrients, such as nitrogen extracted from the soil, or carbohydrates produced by the plant through photosynthesis. Since they are limited, their use must be carefully budgeted. The most limiting of these budgets is usually the carbohydrates, and so the *carbon budget* is crucial. Carbon is 'fixed' by the *source* organs in the form of carbohydrates produced by photosynthesis in the leaves and young branches (see Chapter 3). The carbon is then transported to, and used in, *sink* organs, such as the roots and reproductive organs. The controls regulating which sinks get what (the *partitioning of resources*) are still poorly understood (Génard *et al.* 2008) but depend upon the number and position of the sinks, the strength of the sink (how much 'pull' it has), how much carbon might have been stored as well as the rate at which the carbon can be transported.

What we do know is that putting carbon into flowers and fruits comes at a cost to the tree because good seed years will generally slow the growth of the tree; this can be seen as a reduction in height, but more usually it shows itself in a reduction in the width of annual rings. For example, in beech *Fagus sylvatica*, ring width can be 20% smaller in years of heaviest seed production (Hacket-Pain *et al.* 2015) and in Scots pine *Pinus sylvestris*, an average load of cones reduces the amount of wood grown by an estimated 10–15% (Dick *et al.* 1990). Part of this reduction in growth is because the fruits, cones and seeds are acting as a powerful sink, dragging carbon away from other sinks. Moreover, in years of abundant flowers and fruits, the total area of leaves on a tree can be lower because of new branches being shorter, more buds dying (Ishihara and Kikuzawa 2009) and the leaves themselves being smaller (Innes 1992): the amount of carbon produced by the sources therefore is also reduced. There is also a physical issue in that the flowers and fruits replace leaves, further reducing the size of the source. As an example, male cones in lodgepole pine *Pinus contorta* reduce the number of needles on branches by 27–50%; female cones, hanging on branch ends, replace few needles (Dick *et al.* 1990). Flowers and fruits are not just greedy sinks, but often help towards their own costs. The green flower-bud scales (sepals) of apple *Malus* flowers can produce up to 15–33% of their own carbohydrate running costs. Similarly, green parts of fruits can photosynthesise, producing a percentage of their own costs, ranging from 2.3% in the green acorns of bur oak *Quercus macrocarpa* to 65% in the green samara wings of Norway maple *Acer platanoides* (Bazzaz *et al.* 1979). Leaves next to fruits may also be more productive, increasing their photosynthetic rate by up to 100%, seemingly because the strong sink close by creates a steep carbohydrate gradient, triggering extra production by these leaves. Finally, in some trees at least, part of the cost of a heavy seed crop is funded by carbohydrates stored up from the previous year(s) of excess. This *carbon reserve* can be particularly important early in the growing season, when lots of sinks around the tree (flower growth but also wood, shoot and leaf production) would otherwise stretch the incoming carbohydrate too thinly (Campioli *et al.* 2011). Thus, in beech *Fagus sylvatica*, around 10–20% of the wood in an annual growth ring is funded using stored carbohydrates (Skomarkova *et al.* 2006).

It is worth pointing out that although the above concentrates on carbon, flowering can also alter the distribution of nutrients. For example, Alla *et al.* (2012) point out that in several Mediterranean oak *Quercus* species, the amount of nitrogen is lower in the leaves on seed-bearing shoots because the fruits are a more powerful sink and outcompete the rest of the branch.

Figure 5.4 A branch of horse chestnut *Aesculus hippocastanum*. The terminal growing points die once they have flowered and fruited, so next year's growth will be from buds behind (a) leaving forks (b) where the flowers and fruits have previously been. This gives the branches of older trees a distinctive forked appearance. *Source:* Thomas (2014). Reproduced with permission of Cambridge University Press.

As reproduction is expensive, most trees delay the onset of flower production until maturity; in fact, the change from the juvenile to the mature growth phase is usually defined by the build-up of flowering. Just when that stage is reached partly depends on the type of tree. Pioneer trees, such as species of *Betula, Populus, Salix* and some *Pinus* species that rely on invading new areas for survival, produce seeds within the first decade or so. Others that grow in deeper shade will invest more in growing taller to establish themselves in the canopy before flowering, and may only produce flowers and seeds after 50–60 years.

A way of reducing the waiting time before flower production is through the use of cuttings (either rooted or used in grafting) which often retain a 'memory' of the older age of the plant they are taken from. Cuttings, thus, tend to grow more slowly, produce fewer branches, and flower and set seed at a much younger age. This can be quite useful for getting a good floral display soon after planting, and certainly flowering cherries will produce their first flowers just one or two years after grafting.

A final comment on the cost of reproduction is to do with the loss of growing points. A bud that produces only flowers will die when the flowering and fruiting is finished, and cannot be re-used in future years for vegetative growth. This is particularly important in species of *Acer, Aesculus, Cornus* and *Magnolia* because the fruits are produced at the end of the branch and so a new vegetative bud cannot be formed there (Figure 5.4). The following year, the shoot can only carry on growing from lateral buds, which changes the shape of the branch and can have a large impact on the overall shape of the tree, once it is old enough to flower.

Numbers Involved

It is usually the case that not every flower will result in a fruit. A common reason for this is that not all flowers are pollinated. Pollination rate will vary depending upon the weather, availability of pollinators (or how windy it is for wind-pollinated flowers) and how many trees of the same species are nearby to provide pollen. For example, pollination in white spruce *Picea glauca*, a common North American forestry tree, in areas with <10 trees was 38% lower than in areas with >100 trees (Gärtner *et al.* 2011). Overall, pollination rates may be as high as 75% of flowers being pollinated in *Pinus* spp., to less than 1 flower in 1000 in the tropical kapok tree *Ceiba pentandra*. However, this is not the complete answer to why few fruits are produced, because in many cases it is not unpollinated flowers that are dropped but rather young fruits that are aborted. This may come down to a lack of resources: there is not enough energy to keep them all. In these cases, rather than all half starve, the smallest, or the ones at the far end of the cluster, or the ones with seeds being attacked by insects, are actively shed in the same way that leaves are shed in autumn (see Chapter 3). A good example is the 'June drop' of apples, where excess small apples are deliberately shed by the tree.

Many trees, such as species of *Betula*, *Salix* and *Ulmus*, produce a more or less constant number of seeds and fruits each year, varying a little depending upon how good the growing season has been. Others produce a large number of seeds and fruits one year, followed by one or more years of very few. This *masting* behaviour is seen in many conifers and those angiosperms with large seeds, such as beech *Fagus sylvatica* and pedunculate oak *Quercus robur*. These two trees produce high seeding mast years every 3–4 and 2–3 years, respectively. The reason behind masting may be that a very large number of seeds requires a considerable amount of energy that must be stored up over several years. This still raises the question: why not produce fewer seeds more constantly every year? The underlying reason appears to be *predator satiation*. Large seeds, such as acorns, are widely sought after by herbivorous animals, so most seeds would be eaten if they produced a similar number each year. By producing a huge number occasionally, there will be more acorns than the squirrels, birds and other seed eaters could possibly consume so some should survive to germinate – the predators of the seeds are soon satiated and cannot eat everything. Moreover, in the years between, when very few acorns are produced, the seed predator populations are likely to dwindle from lack of food, so when the next mast year comes along they are even less likely to be able to eat all the seeds.

Tree seeds are generally quite large compared to non-woody plants. The smallest seeds, belonging to orchids, are like grains of dust weighing less than 2 µg each. Wind-blown seeds are bigger but still light enough to spread easily into newly disturbed areas (e.g. aspen *Populus tremula*, 125 µg). Trees that normally grow in woodlands, by contrast, tend to have heavier seeds, ranging from 0.07 g in European ash *Fraxinus excelsior* to 3.5 g in pedunculate oak *Quercus robur*. So why should tree seeds be so large? As with many aspects of trees design, seed size is a compromise between producing as many seeds as possible to increase the chance of a seed finding a suitable place to germinate and grow, and giving each seed the best start by supplying it with food and nutrients. Pioneer trees that invade open areas tend to have, not surprisingly, many small seeds to increase their chance of finding a suitable place to germinate and grow. Trees that grow in the more closed conditions of woodlands are at the other end of the spectrum, producing larger seeds to help each seedling at the start of its life. The large food supply in the seed allows the young root to grow rapidly downwards, through the upper layers of litter that can dry very quickly and easily lead to a seedling dehydrating if it tarries. The food supply also allows the shoot to grow as tall as possible before it needs to support itself by its own photosynthesis. This allows the seedling to get above much of the woodland floor vegetation that would otherwise compete with it for light.

Seed size tends to be fairly constant within a species and the quantity of seeds tends to vary much more than the size, which is not surprising if size is finely tuned to success in a particular environment. However, under very extreme conditions, seed size will vary. For example, in almonds *Prunus dulcis* growing in drought conditions in California, 50% of the flowers set seeds when hand-pollinated but they produced small seeds (0.8 g); whereas, in flowers where pollination was prevented by enclosing the trees in cages to stop insects getting at them, only 5% of the flowers produced seeds but they were double the weight at 1.6 g (Klein *et al.* 2015). It is as if the tree has an allocation of carbohydrate for each seed and if there are fewer seeds, each gets more of the food.

Flowering and Fruiting in Urban Landscapes

A huge amount has recently been written on the large number of benefits of urban trees, sometimes couched in terms of *ecosystem services*: those things an ecosystem provides that we find useful (Roy *et al.* 2012; Derkzen *et al.* 2015; Duinker *et al.* 2015; Janhäll 2015; Mullaney *et al.* 2015; Norton *et al.* 2015; Pitman *et al.* 2015; Wolf and Robbins 2015). Obviously, there are also ecosystem *disservices*, things we find harmful or less useful (von Döhrena and Haase 2015; Lyytimäki 2017).

In terms of the theme of this section, many trees are planted in urban areas for their attractive flowers. Flowering cherries (*Prunus* varieties) are an obvious example (Figure 5.5). As well as their aesthetic value, the flowers can also be of direct use to wildlife, particularly insects. Protecting biodiversity is of inherent value in itself, but the encouragement of pollinating insects also has human benefits in helping the production of urban crops in window-boxes, gardens and allotments. Pollination is a very important ecosystem service that we take for granted. As an example of its importance, more than $346 million was spent in the USA in 2012 on pollination services – bringing hives of bees to crops to ensure adequate pollination (Bond *et al.* 2014). It is also worth bearing in mind that trees may flower earlier in cities and towns because of the urban heat island effect – being warmer than surrounding rural areas. In South Korea, the flowering of trees and shrubs, such as *Forsythia koreana*, can be up to 9 days earlier for every 1 °C increase in temperature over rural areas (Jeong *et al.* 2011). This may not appear to be much, but in reality it is likely to have a significant mental boost for commuters seeing the first signs of spring on their commute to work. It is not just warmth that brings spring forward. For example, the flowering of common lime *Tilia × europaea* in Florence, Italy, was found to be 1.4 days earlier in spring for each 10% increase in impervious surfaces over the roots, because of a mix of water stress and increased root temperature (Massetti *et al.* 2015).

Fruits also have a similar positive benefit for wildlife, so much so that Belaire *et al.* (2014) found that in Chicago, trees with fruits or berries had a significant positive effect on native bird richness where bird feeders did not: planting trees that fruit is far

Figure 5.5 Flowering cherries can have an attractive impact on a landscape. In this case *Prunus* 'Kanzan' largely hides from view a rather untidy street market at Hirosaki Park, Aomori, Japan.

more beneficial than feeding birds in gardens. Fruit is also of direct benefit to humans providing that it is grown in a manner that encourages its use (this is discussed later in Tree Crops).

There are, of course, negative effects of flowering trees. There can be complaints over the mess created by many petals being shed (e.g. with cherries), blocking drains and being trampled into shops and buildings. Plus, pollen can trigger allergies or asthma, and the volatile organic compounds released by flowers can exacerbate pollution: for example, the pea-like flowers of yellow honey locust *Gleditsia triacanthos* 'Sunburst' have been found to release almost 4.5 times the quantity of monoterpenes as when just in leaf (Baghi *et al.* 2012). Monoterpenes are good for human health in small doses but contribute to the creation of pollutants including ozone. Some flowers can also be remarkably pungent and unpleasant, such as the male flowers of sweet chestnut *Castanea sativa*.

Fruits can similarly be a nuisance. Examples that commonly cause problems are the light fluffy seeds of poplars *Populus* and willows *Salix* spp. that seem to get everywhere including into eyes; the rancid butter smell of maidenhair tree *Ginkgo biloba* fruits; and the physical stumbling hazard of large woody fruits such as acorns and chestnuts. There are many incidents where people have felled magnificent horse chestnut *Aesculus hippocastanum* specimens because of the trouble caused by children with sticks trying to collect the conkers (largely a thing of the past now?). There have also been cases of fruit trees such as pears being felled in school grounds because of potential litigation issues if falling fruit should hit a child or parent. In some regions, slightly more extreme problems include bears coming into urban areas to forage for fruit (Lewis *et al.* 2015)!

Figure 5.6 Seedlings of European ash *Fraxinus excelsior* and suckers of wych elm *Ulmus glabra* growing profusely on top of an embankment near the village of Keele, UK. The local council has to repeatedly cut down the seedlings each summer to avoid obscuring sight lines for drivers – an expensive process.

A further problem of fruiting trees in urban areas is the knotty problem of seedlings. These are free trees but unless they are moved when very young they can be difficult to transplant without causing damage to the seedling and too much disturbance to their seedbed. In Europe and other temperate areas, some maples *Acer* spp., tree of heaven *Ailanthus altissima*, ash *Fraxinus excelsior* and similar species can cause problems by being prolific seed producers, resulting in saplings growing in places that are hard to access, with resultant problems when they become too large. Even when they are easy to access, they still take time and money to remove (Figure 5.6).

Flower and seed production can be problematic for the tree itself. Flowers can be a significant 'sink' of nutrients and carbohydrate. This can put the tree under stress, or add further stress to an already physiologically stretched tree that may be suffering as a result of challenging urban conditions. This can be exacerbated because a tree under stress will often flower and fruit heavily, as if it is making a final attempt to reproduce before it dies. This of course can 'tip' the tree over the edge, hastening its death.

Tree Crops

Most of the fruits and nuts we eat come from woody plants and have been cultivated for millennia. Traditionally, this has been in orchards in rural areas, but there is increasing demand for fruit production in urban areas, classed as *urban food forestry*, using fruit trees in designated urban orchards or in parks and community areas. In Burlington, Vermont (a city of some 42 000 people), it has been calculated that planting apples

(a)  (b)

Figure 5.7 Systems for training branches of fruit trees: (a) fan; and (b) espalier. These have the advantage of keeping the trees small, making maximum use of space (by keeping them tight to a wall or frame) and encouraging maximum fruit production which is easy to reach. Photos from Croxteth Hall, Liverpool.

Malus domestica on publicly accessible open space would result in enough fruit to meet the daily recommended minimum intake of fruit for the entire city's population (Clark and Nicholas 2013). They also identified 70 species of fruit trees that could be grown, with 30 being 'highly suitable'. There can be problems of pollutants contaminating fruit but most studies have found this to be below recommended levels of consumption. For example, heavy metals (such as cadmium and lead) have been found on fruit grown in Berlin, Germany, but levels were lower than in much commercially sold fruit (von Hoffen and Säumel 2014). Moreover, growing trees away from traffic in parks and orchards greatly reduces the problem, and the fruit is likely to be perfectly safe to eat.

Trees grown for fruit are often intensively pruned to encourage side shoots on which fruit is grown, and to ensure that the tree remains low enough for fruit to be easily picked. This is taken to extremes in trees trained as espaliers, where branches are trained to grow horizontally against a wall or on a trellis or other frame, or fans and cordons where the main trunk is trained at an angle (Figure 5.7): branches near the horizontal produce more fruit that those nearer the vertical.

Orchards can also be grown for the seed rather than the fruit. Seed orchards are commonly used for producing the seeds of forestry species in sufficient quantity to make their use viable.

Vegetative Reproduction

Most angiosperms and a few conifers can grow new shoots (*basal sprouts*; see Chapter 2) from the base of the trunk, either just below or just above the soil level. Young trees are usually better at this (when less than 15 cm diameter) and most trees will gradually lose the ability until, by the time they are more than 30 cm diameter, very few species can still produce prolific basal sprouts. Most species will sprout only when the top of the

(a) (b)

Figure 5.8 (a) Hazel *Corylus avellana* growing out of a garden hedge. (b) The base of the tree has a number of stems of different ages even though the crown of the tree is undamaged and healthy. This is a common occurrence in hazel and is referred to as 'self-coppicing'.

tree is damaged or otherwise compromised, but a few trees will produce basal sprouts on even healthy trees. Thus, hazel *Corylus avellana* tends to be shrubby with many shoots from below ground, all of different ages (Figure 5.8); this has been referred to as self-coppicing. Others, such as common lime *Tilia × europaea*, are readily identified by the great skirt of basal sprouts that surrounds a healthy specimen. In many ways, these shoots are supplementing the existing canopy or, if the tree is felled or snapped, just replacing the former canopy, so it may not be quite fair to accept this as reproduction because new trees are not produced. However, sprouting does produce a different *form* of tree – replacing the normal single trunk with multiple stems that may persist through to adulthood. Resprouting after the loss of the main trunk also allows the tree to persist by regrowing. In some areas where seed production is rare, resprouting may allow an individual to hang on long enough to reproduce by seed (de Lucena *et al.* 2015). A word of caution, however: if cutting a tree back to the ground to encourage new shoots, the cut should be as close to the ground as possible. This encourages the new shoots to come from as low as possible, making them more firmly attached to the stump, less susceptible to any heart rot in the original stem and more likely to develop a new, adventitious root system of their own (del Tredici 2001).

The most basic way for naturally producing new trees without seeds is for branches of trees that are abscised or broken to embed themselves in suitable patch of ground, allowing them to root and produce whole new trees. In species renowned for this, particularly willow *Salix* and poplar *Populus* spp., branches that are abscised retain live buds, and little of the nitrogen and phosphorous is resorbed out of the branch back into the tree. Dewitt and Reid (1992) give balsam poplar *Populus balsamifera* in Alberta, Canada, as an example: branches with a mean age of 6.7 years and 20.4 cm long were abscised in October before leaf fall and, because they held live buds, were capable of lodging in mud along rivers and producing new trees. *Populus* and *Salix* are dioecious (separate male and female trees – see above), so vegetative reproduction goes a long way to explaining why the trees along a particular river tend to be either predominantly male or female.

New trees can also arise from sprouts or *suckers* from roots (see Chapter 2). These can arise from stored *additional* buds that persist on roots for decades or from newly grown *reparative* buds. Some trees (e.g. sweetgum *Liquidambar styraciflua*) use just additional buds, others (e.g. American beech *Fagus grandifolia*) use just reparative buds while many (e.g. aspen *Populus tremuloides*) can use either (del Tredici 2001). Whichever type of bud is used, the new shoots are initially dependent upon the existing root system but will produce their own adventitious roots and so be capable of becoming a separate individual with, of course, the same genetic make-up as the parent. In this way, similar-looking *clones* build up in the same way as from apomictic seeds. Some trees growing in high light are likely to sucker throughout their life, even when healthy – Caucasian wingnut *Pterocarya fraxinifolia* and the invasive tree of heaven *Ailanthus altissima* are classic cases. However, most trees will normally only produce suckers if the top of the tree is damaged or diseased, or there is a dramatic change in the tree's environment, weakening the apical dominance that controls these root buds. Suckers can appear from anywhere along the main woody roots. As an example, an investigation of dozens of 'shoots' that had appeared in a prize lawn over a period of a few weeks revealed they were suckers of false acacia *Robinia pseudoacacia*, which could only have come from a tree that had been felled some 20 m away from the furthest sprout. The fact that this tree was on the other side of the house caused a few worries: no sprouts had previously been seen. Notably though, anything that stops roots spreading, such as streams, ditches or even woodland paths, will also prevent suckering beyond those boundaries (Jarni *et al.* 2015).

New trees can also start from above ground where low branches bend down to press into the ground. This is particularly common in conifers, especially on boggy soil where the underside of the branch is overgrown by moss, or where heavy snowfall bends and holds down branches for extended periods, as happens with hiba cedar *Thujopsis dolabrata* in Japan (Hitsuma *et al.* 2015). It can also happen in angiosperms that retain low heavy branches, such as beech *Fagus sylvatica* and sweet chestnut *Castanea sativa* (Figure 5.9). On the underside, they produce adventitious roots and either the branch will bend upwards to form a new canopy, or a side shoot on the original branch will do so. The best evidence for this having happened is that the branch is fatter on the side away from the original trunk (Figure 5.10). As with suckers, the new tree is initially connected to the parent root system, and may stay so, but equally the junction between the two root systems can break down and the new stem is then truly an independent tree. A similar thing can happen if a tree falls but maintains some root connection: the branches pushed hard against the ground can root, producing a thicket of new but connected trees (Figure 5.10).

Figure 5.9 A mature sweet chestnut *Castanea sativa* growing in a lawn with a lower branch (on the right) that has touched the ground and produced adventitious roots. The branch end has now become a tree in its own right. Both may be connected by a communal root system, but there is potential for the two trees to completely separate and become independent trees.

(a)   (b)

Figure 5.10 (a) An American lime *Tilia americana* that has fallen but maintained a connection with living roots. Side shoots that had buds in high light have taken over as the new leaders, and branches that were pushed against or into the ground have now rooted and are growing up as new trees. In this way, a tangled clone of new trees, all genetically identical with the parent, is growing up. (b) It is clear that the branch starting on the left of the picture has rooted, because the branch itself is much fatter near to the first new stem, and much fatter between the two new stems.

Growing Trees

Seeds and Their Origins

Seeds can either be bought from a supplier or collected directly from trees. If collected, it is important that any fleshy fruit is removed before planting because this often slows or prevents germination (Gosling 2007 gives good practical advice). An advantage of collecting seed is that you know what the adult tree looks like which may give an indication how the new trees will appear later in life (though see the start of this chapter on genetic variation in seeds). When planting native species, there is the question of provenance (the geographical origin of the seed). Some planting schemes will demand trees from local or at least national provenance, so it is as well to check provenance when buying seeds: seeds of many tree species native to the British Isles are often collected in mainland Europe and certainly in the past have been sold as 'native'. The question of provenance also applies to seed directly collected, as the original origin of the mother tree is usually unknown and most European countries have imported native trees for centuries.

Provenance is important, because it is widely recognised by national planting standards (see Hubert and Cottrell 2007, readily available online) that local wildlife is often better adapted at living with local trees. For example, hawthorn *Crataegus monogyna* imported into Britain can flower up to 5 weeks earlier than native provenances, and insects and birds that produce offspring synchronised to hawthorn flowering may be threatened by the earlier flowering (Jones *et al.* 2001). Moreover, the more tightly packed branches and dense thorns of the local provenances provide better protection for nesting birds. It is sometimes stated that we should also plant local provenance trees because they are best adapted to their environment, honed by millennia of evolution. As the climate is constantly changing, it is no more likely that seeds from an oak that established in the Little Ice Age of 1300–1850 are better adapted to its current climate than those from Hungary. In fact, it can be argued that using trees from other areas will help increase the genetic diversity of the local population, and because the rate at which populations can evolve (and therefore survive in a changing climate) is proportional to its genetic variability (Mather 1973), introducing new genes by out-of-provenance planting may be very beneficial. As Sackville Hamilton (2001) has said: 'Limited introduction of seed from other sources can produce new [genetic combinations] to be "tested" by natural selection.' This is fine in woodland settings where the strongest, most fit trees will survive, but less good in formal plantings where dead trees are a problem. There is therefore an argument for choosing tree provenances most suited to the climate and conditions of the planting area, even if they are not local. Indeed, given the longevity of trees, the source of seed we are now planting will become increasingly important because of climate change. Future predictions for the latter part of this century suggest that some parts of Europe will see marked changes in temperature and precipitation (Figure 5.11), so we will have to make difficult decisions about where we get seed from if we are looking for trees to survive through the next few centuries. These decisions are often based on matching climatic regions and choosing trees from the provenance that is most compatible. For example, if temperatures in central Scotland are due to rise by 3–3.5 °C by 2100 (Figure 5.11), we should be planting using seed from northern France, which currently has that climate. In this way, when temperatures rise

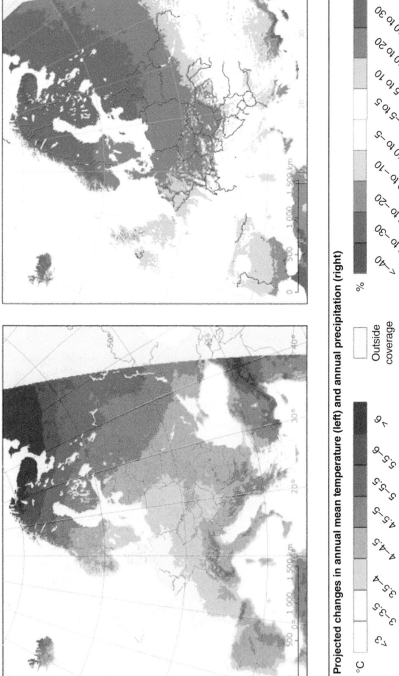

Projected changes in annual mean temperature (left) and annual precipitation (right)

°C

< 3 | 3–3.5 | 3.5–4 | 4–4.5 | 4.5–5 | 5–5.5 | 5.5–6 | > 6

Outside coverage

%

< −40 | −40 to −30 | −30 to −20 | −20 to −10 | −10 to −5 | −5 to 5 | 5 to 10 | 10 to 20 | 10 to 30 | > 30

Figure 5.11 Climate change may have a profound impact on parts of Europe over the next century. Here, projected changes are for 2071–2100, compared with 1971–2000, based on the average of a multimodel ensemble. All changes marked with a colour (i.e. not white) are statistically significant. Individual models from the EURO-CORDEX ensemble or high-resolution models for smaller regions may show different results. *Source:* Courtesy of European Environment Agency.

in Scotland, the trees being grown will be already adapted to the warmer conditions. This will undoubtedly present some challenging ecological problems, and there is no generic right answer (Vander Mijnsbruggea *et al.* 2010). It will certainly be important to maintain genetic variation and promote natural regeneration to help wildlife, adopt a portfolio of provenances alongside the current population, and plant trees from more southerly provenances (called *assisted migration*), in order to help cope with future climate change (Hubert and Cottrell 2007).

While dealing with provenance, we should also mention the common practice of collecting seeds locally and shipping them to other countries to grow into seedlings, before importing them back again. We can thus plant 'local' provenance trees, but only at the huge risk of importing pests and diseases. It is important that robust biosecurity is put in place to minimise this threat as a result of global trade. As an example, ash dieback caused by the fungus *Hymenoscyphus fraxineus* (originally described as *Chalara fraxinea*, the asexual stage) was most likely imported into the British Isles on seedlings brought in by nurseries from mainland Europe, although some spores probably also arrived by wind. The disease currently covers some 2 million km^2 from Scandinavia down to France and Italy, and is killing up to 90% of ash trees within infected areas (Thomas 2016).

Storing Seeds

Seeds can usually be dried before storage to increase their shelf life. Desiccation-tolerant seeds (described as *orthodox*) are easily stored, and will maintain their viability for up to 50 years when dried to 5% or even 2% moisture content and stored at 5 °C (Kozlowski and Pallardy 2002). Seeds of most temperate trees are orthodox (Table 5.1), including angiosperms, such as *Alnus*, *Betula*, *Fraxinus*, *Platanus* and *Prunus*, and gymnosperms, such as *Larix*, *Pinus*, *Pseudotsuga* and *Tsuga* (Bonner 1990).

However, seeds of some species are desiccation-sensitive and short-lived (termed *recalcitrant*), and rapidly lose viability if dried or stored for more than a few weeks or months: they are best planted as soon as they are collected. This includes species of *Acer, Aesculus, Castanea, Corylus, Fagus, Juglans* and *Quercus*, but also around 63% of tropical trees as the constant climate allows them to germinate as soon as they hit the ground (Khurana and Singh 2001). In reality, seeds are rarely entirely orthodox or recalcitrant and some species can be intermediate or *suborthodox*, such as *Carya, Corylus, Fagus* and *Juglans*, because they can be carefully dried and stored just for a short time (Bonner 1990).

Seed Dormancy

Seeds are very simple things. Inside the protective seed coat is an *embryo* – the young plant with the beginning of a shoot and root – and a food supply. The food may be stored around the embryo in a tissue called *endosperm*, but in most trees it is rapidly absorbed into the *cotyledons* (seed leaves) of the embryo.

Seeds (particularly orthodox seed) may be *dormant*; that is, they will not germinate until they have experienced the right pre-germination conditions. Many temperate and northern trees show a low degree of dormancy (sometimes called *shallow dormancy*; Table 5.1) and will germinate as soon as planted, but will often show faster, more complete and more

Table 5.1 Species that have different ease of storage (orthodox, intermediate and recalcitrant) and different degrees of dormancy. All 'hard' seeds (see text) are orthodox.

Dormancy type	Ease of storage		
	Orthodox	Intermediate	Recalcitrant
Hard	Broom *Cytisus scoparius*		
	Laburnum *Laburnum anagyroides*		
	False acacia *Robinia pseudoacacia*		
	Gorse *Ulex* spp.		
Shallow	Most conifers	Many firs *Abies* spp.	Horse chestnut *Aesculus hippocastanum*
	Alder *Alnus* spp.	Incense cedar *Calocedrus decurrens*	
	Birch *Betula* spp.	Cedar *Cedrus* spp.	Monkey puzzle *Aurucaria araucana*
		Lawson's cypress *Chamaecyparis lawsoniana*	Sweet chestnut *Castanea sativa*
		Western red cedar *Thuja plicata*	Poplar *Populus* spp.
			Willow *Salix* spp.
Deep	Most broadleaved trees	Norway maple *Acer platanoides*	Sycamore *Acer pseudoplatanus*
	Juniper *Juniperus communis*	Beech *Fagus sylvatica*	
	Macedonian pine *Pinus peuce*	Bay *Laurus nobilis*	
	Yew *Taxus baccata*		

Source: Gosling (2007). Contains public sector information licensed under the Open Government Licence v3.0.

uniform germination if the dormancy is removed. If dormancy is strong (or *deep*), it can almost prevent germination without suitable treatment. For example, removing the dormancy of Douglas-fir *Pseudotsuga menziesii* seeds increased germination at 20 °C from 5–7%, in untreated seeds, to 73% when treated (Taylor *et al.* 1993).

As shown in Figure 5.12, dormancy is usually broken by some combination of temperature and moisture. In some cases, the seeds need to experience dry warmth (20–35 °C), referred to as after-ripening, or need moist warmth (warm stratification). In temperate areas, dormancy is usually broken by *cold stratification*: a period of cool, moist conditions (normally between 1 and 15 °C) for weeks or a number of months. This ensures that seeds germinate in the spring rather than in a warm period in late autumn. Cold stratification needs are met either by planting seeds in the autumn, and so exposing them to winter conditions, or are met artificially by mixing the seeds with damp sand or organic matter and storing them in a sealed plastic bag in the fridge. Before stratification is carried out, however, it is as well to check what sort of stratification a species needs: if the wrong conditions are given, dormancy can be deepened (inducing *secondary dormancy*), making it harder to break in the future. Fortunately, stratification requirements for most species are well known and can be found in books, such as Dirr and Heuser (2006), Dirr (2009) or Kock *et al.* (2008), as well as via the Internet. If in doubt, it is best to sow some seeds to see if they are dormant before considering stratification.

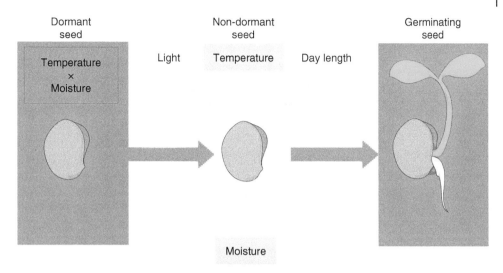

Figure 5.12 The main factors regulating the breaking of dormancy and the germination of seeds. Seeds can be non-dormant and ready to germinate straight away, or they may need a suitable combination of temperature and moisture to break the dormancy. For germination, temperature and moisture are also key factors, although the temperature requirements may be quite different from that needed to break dormancy. *Source:* Adapted from Walck *et al.* (2011). Reproduced with permission of John Wiley and Sons.

A special case of dormancy is seen in the seeds of the pea family (Fabaceae), such as false acacia *Robinia pseudoacacia*, Kentucky coffee tree *Gymnocladus dioica*, and species of *Cercis*, *Gleditsia* and *Laburnum*, which have *hard* seeds (Table 5.1). These have no physiological dormancy, but the hard seed coat prevents water entering the seed and thus stops germination. Stratification eventually weakens the hard coat, allowing germination but it can take a long time. In nature, the seed coat is usually cracked by abrasion, heat or by passing through the gut of an animal. Germination can be artificially sped up by mimicking these things in scarification by the cutting, sand-papering or otherwise physically breaking the seed coat, or dissolving it using acid.

Finally, some seeds are dormant because the embryo is too immature to germinate when the seed falls. European ash *Fraxinus excelsior* and holly *Ilex aquifolium* seeds are in this category. When they fall in the autumn, a small number (<5%) will germinate the following spring but the rest will wait until the next spring, 18 months after they have fallen from the tree. In cases such as these, a long period of stratification is the only answer.

Germination

As with dormancy, the main factors controlling germination are temperature and mois-ture. All tree species have an optimum germination temperature which will depend on the climate of their native origin; for most, the acceptable temperature range is wide and only becomes important if it is exceptionally cold or hot (again, an Internet search will help with these limits for any particular species), although temperature will likely become more crucial with climate change (Dürr *et al.* 2015). Similarly, moisture is also important: there needs to be enough for the seed to absorb (*imbibe*) to hydrate the contents and burst the seed coat, and to keep the emerging root and shoot alive. Above this threshold, the exact degree of water availability is not critical, and is much more likely to affect speed and uniformity of germination rather than total numbers. In a few species, light is also needed for germination. Day length is important to a small number of species: for example, the small-seeded downy birch *Betula pubescens* was long ago known to germinate better in 20-hour (light) days than in 4-hour (light) days (Black and Wareing 1955), which, in a natural environment, helps it to germinate at the right time of the year. For most seeds that need light, the actual amount is not crucial. Small seeds are more likely to need some exposure to some light to trigger germination (Milberg *et al.* 2000) because they have to be at or close to the surface of the soil to germinate, otherwise the young seedlings may not have enough energy reserves to get to the surface. Without light they will not germinate (referred to as *photodormancy*) or, more usually, germination is slow and erratic. To show how little light is needed, Black and Wareing (1955) found that the germination of downy birch was greatly improved by just 2 minutes' exposure to light after the seeds had absorbed water. The light level needed is also usually much lower than is needed for subsequent growth (downy birch needed just 3% of full sunlight), so even dim light in a greenhouse is sufficient. The importance of this in a nursery is that seeds of light-demanding species should always be sown on the surface of the soil. Most larger tree seeds do not require light for germination, however, and so can be safely buried in soil where they have better contact with the soil moisture.

Aside from the basic factors of being warm and moist, germination is fairly self-contained with very few other needs, and so it is possible to germinate most seeds on absorbent paper in a dish. The seed contains sufficient nutrients for germination; it is the subsequent growth and survival that is far more demanding of substrate, including its structure and fertility. However, it is emerging that the seeds from trees grown in experimentally high carbon dioxide levels show on average 55% increase in germination over those grown at current levels, presumed to be because the seeds have grown better (Marty and BassiriRad 2014).

Seedlings

When a seed germinates, the first thing to emerge is the young root (*radicle*) which will grow downwards, attracted by gravity, to ensure a supply of water. How the young shoot emerges depends upon the type of germination. In many small-seeded species (including those in *Pinus, Cedrus, Eucalyptus, Magnolia, Acer, Fagus, Fraxinus*), germination is *epigeal* where the cotyledons are pulled out of the seedcase and raised above ground, as can be seen in Figure 5.13(a). As the shoot emerges, the stem below the cotyledons (the *hypocotyl*) elongates and is curved over below the cotyledons to produce an *apical hook*, in effect dragging the cotyledons out by the base to give them protection against the soil. Once the cotyledons are above ground, they open up, turn green and begin photosynthesis. The young shoot grows above the cotyledons and will produce the first true leaves which will take over from the cotyledons normally after 1–3 months, at which point the cotyledons wither and fall off. The cotyledons are often oval or strap-like in shape, and the typical leaf shape only appears with the first true leaves.

(a)   (b)

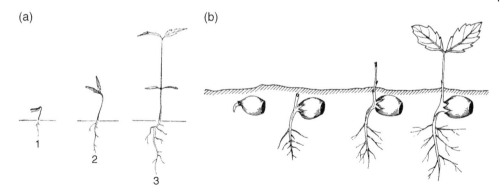

Figure 5.13 Germination can be either (a) epigeal or (b) hypogeal. In epigeal germination the cotyledons are raised above the soil surface and, when they emerge from the seedcase will turn green and photosynthesise. In hypogeal germination, the cotyledons remain inside the seed case at or below the soil surface. *Source:* Kozlowski and Pallardy (1997). Reproduced with permission of Elsevier.

In larger seeds (species of *Aesculus, Quercus, Juglans, Castanea* and many *Prunus*), germination is *hypogeal* and the cotyledons stay buried underground inside the seedcase. In this case, the shoot above the cotyledons (the *epicotyl*) forms the hook and emerges, and photosynthesis can only begin when the first true leaves are produced (Figure 5.13b). Epigeal germination would appear to be better because the seedling gets a head start by the cotyledons being photosynthetic. In hypogeal germination, this advantage is lost but there is an advantage to keeping the large nutritious cotyledons out of harm's way inside the seedcase, often underground, where they are less likely to be eaten.

The number of cotyledons a seedling has depends upon the type of plant. In angiosperms, the two big groups, the monocotyledons (palms, etc.) and dicotyledons, have either one (mono) or two (di) cotyledons by definition. Gymnosperms usually have more: in pines, typically around eight, but they can have more than 20.

As the roots develop, some species (such as in *Carya, Juglans, Pinus* and *Quercus*) produce a taproot growing down like a carrot (Figure 5.14). Others (such as *Acer, Fraxinus, Picea* and *Salix*) do not and have a fibrous root system. The taproot can eventually penetrate 1–2 m below the soil surface (or as far as the oxygen in the soil will allow), and is used primarily as a food store and a way of getting some fine roots deep into the soil very quickly, helping young trees with water uptake in dry periods. How root systems subsequently develop is covered in Chapter 4.

The weight of the roots tends to stay roughly equivalent to the weight of the young branches and leaves (without the trunk), so the root to shoot ratio stays fairly constant and close to 1. However, this varies between species: a classic study by Bazzaz and Miao (1993), looking at New England trees, found that red maple *Acer rubrum* had a root to shoot ratio of up to 1.7 (more weight of roots than shoots) while American ash *Fraxinus americana* was as low as 0.7 (less weight of roots than shoots). The ratio within a species also varies with nutrient level – more nutrients lead to a smaller proportion of roots (Ågren and Franklin 2003).

(a)

(b)

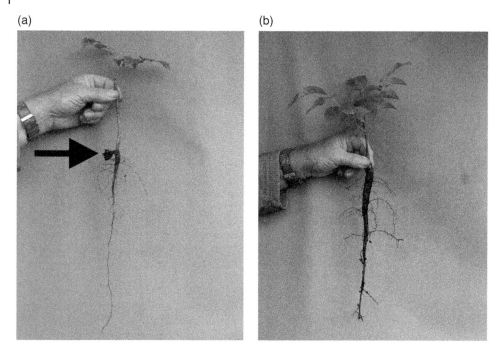

Figure 5.14 Common walnut *Juglans regia*, like many tree species, produces a taproot after germination. In both cases, the seedling is being held at the soil level. (a) Showing the root development of a 1-year-old seedling, the arrow indicates the remnants of the seed (walnut) which was buried by squirrels. Notice that in this case, the epicotyl has had to grow vertically about 10 cm before the true leaves could emerge above the soil. (b) Showing a 2-year-old walnut seedling with a slightly more developed taproot.

Tree Establishment – From Production to the Landscape

Mankind has been involved with the transplantation of live trees for thousands of years. Egyptian reliefs at Deir el-Bahri describe an eighteenth dynasty (c. 1550–1292 BC) expedition organised by Queen Hatshepsut to transplant incense trees (mainly *Boswellia* and *Commiphora* spp.) from a region in the horn of Africa to Egypt (Dixon 1969). It was hoped that by establishing these trees in Egypt, a more secure (and cheaper) supply of incense would be available for their burial rituals. There is also evidence that several thousands years ago the apple was being moved from its wild origins in the mountains of Kazakhstan westwards along the Silk Road and into Europe. No doubt other tree crops were also making their way along the major trade routes of the world, although it is often unclear if the trade was in seed or live plants.

The movement of trees for amenity purposes proliferated in the eighteenth, nineteenth and twentieth centuries as plant collectors sought to enrich the range of plant material available for parks and gardens (Ignatieva 2012). Hundreds of species and cultivars are now grown by nurseries and shared across regions with similar climates. For example, in temperate Europe, many species available from nurseries are native to temperate regions of North America, South America, Asia or Australasia. If used appropriately, this internationalisation of plant material can lead to diverse landscapes, delivering considerable amenity value and a wide range of ecosystem services. As an example, Mitchell (1974) suggested that in addition to the 35 or so trees native to the British Isles (the number depending upon what you classify as a tree or a shrub), an exploration of parks and gardens will readily reveal another 700 more species and cultivars. If botanical gardens and arboreta are included, the number of tree species grown in the British Isles is over 1700.

Despite our long history of cultivating and transplanting trees, transplant failure rates in urban environments are frequently 30–70% during the first few growing seasons (Gilbertson and Bradshaw 1985; Britt and Johnston 2008, Roman *et al.* 2014). Such high death rates of young trees must be considered unacceptable and underscores the importance of ensuring that best practices of establishing trees are followed. Comprehensive guidance on planting trees can be found in Watson and Himelick (2013) but some key considerations are discussed below.

Momentum of Tree Establishment

Tree establishment in any habitat or landscape is greatly aided by what might be called *developmental momentum*. If a young tree has a good quality root system with access to sufficient resources, then leaf expansion and shoot growth is relatively easily achieved. Providing the root system can maintain the supply of water and nutrient resources, the tree's crown will rapidly develop. This further increases the whole-tree carbon gain and helps to secure future development. However, loss of roots during transplanting, poor soil at the new site and/or failure of roots to couple (link) with the soil at the new site seriously reduce developmental momentum. If the root system is unable to supply water effectively, leaves fail to expand and carbon gain will be diminished as a consequence of a reduced leaf area. If the tree's need for water (transpirational demand) exceeds the supply of water from the roots, water deficits build up and there may be a loss of hydraulic conductivity (see Chapter 2), leading to a downward spiral of water shortage and reduced carbon fixation in the crown: developmental momentum is lost and tree growth is seriously supressed. Some of the effects of water deficits can be minimised by frequent irrigation but, if this is not positively managed, tree decline is inevitable. In very young trees, the difference in root system quality can make a huge difference in the tree's ability to build its crown, as shown by the young Père David's maple *Acer davidii* saplings in Figure 5.15. Aftercare immediately after planting trees into the landscape can be equally important to secure well-established trees. This is elegantly demonstrated by the contrast in growth between two sets of pedunculate oak *Quercus robur* in Figure 5.16. Those trees that were well-watered during establishment maintained developmental momentum, well after any supplementary irrigation had ceased. Once a good quality root system is established, the tree's long-term capacity for growth is much more secure. Where the root system fails to establish effectively immediately after transplanting, slow growth and tree decline is often apparent for years or even decades afterwards. This condition is widely termed *transplant stress* or *transplant shock*.

Good developmental momentum is dependent on four elements: *tree species selection, tree quality, rooting environment* and *arboricultural practices*. Failure to consider any one of these factors will reduce the performance of the tree within the landscape and

Figure 5.15 Two Père David's maple *Acer davidii* saplings both grown from seed and transplanted from 0.5-L pots into 3-L Air Pots. The sapling on the left started the growth season as the much shorter tree (approximately half the size shown in the photo) but had a well-developed root system at time of transplanting. The sapling on the right started as the taller sapling of the two but had a very poor root system at transplanting. Three months into the growth season, the sapling with the well-developed root system (left) has been able to fully expand its leaves and grow about 30 cm in height. The leaves are large and well-formed, ensuring effective photosynthesis that will foster further tree growth: it has developmental momentum. Conversely, the sapling with the poor root system (right) has failed to fully expand its leaves, leading to a severely reduce leaf area and negligible shoot elongation. Both saplings have been watered regularly and have access to the same volume of soil. This serves to illustrate the impact that a poor quality root system can have on subsequent tree development.

may lead to early tree mortality (Hirons and Percival 2012). These factors should be under the control of whoever is planning and carrying out tree planting. It should be pointed out, of course, that subsequent care of the trees is also very important. Evidence suggests that social factors, such as ongoing stewardship and controlling vandalism, help determine long-term survival (Ko *et al.* 2015). However, here we focus on the aspects of tree establishment that are important from a biological perspective.

Tree Species Selection

The constraints of the planting site are the most important factors to consider when choosing which species to plant (Figure 5.17): how the tree's future space requirement is likely to interact with other infrastructure (above and below ground); soil conditions; likely water and nutrient availability; as well as practical considerations, such as plant

(a)   (b)

Figure 5.16 The momentum of tree establishment can last for decades. Here, a line of penduculate oaks *Quercus robur* were planted in 1991 along a roadside next to the Swedish University of Agricultural Sciences (SLU), Alnarp. Half of the trees came under the jurisdiction of the university and half came under the jurisdiction of the local authority. (a) The trees on university land were well-watered during the establishment phase and are making a significant contribution to the landscape. (b) The trees on the local authority land were neglected during establishment and, although still surviving, are much smaller, making only a minor contribution to the landscape. They are very unlikely ever to achieve the stature of the trees that were cared for in their early years. *Source:* Courtesy of Henrik Sjöman.

availability and budget. It is also increasingly important to consider biological constraints, such as the vulnerability of trees to emerging pests and diseases, as well as the potential of a species to become invasive.

While it might be tempting to select a tree based on its abundant flowering or striking autumn colour, in many ways these more aesthetic attributes must be secondary to selecting a tree that can withstand the physical conditions presented at the planting site: the *ecophysiological factors* (Figure 5.17). Of primary importance in many temperate regions is the tree's cold-hardiness, but other characteristics, such as tolerance to water deficits (drought), can be just as important, particularly where soil volumes are constrained. Species must be selected that are appropriate to the climate of the planting site. This need not mean that the tree species is native to the planting area, but its natural range should have a comparable climate. It is also rewarding to try to match not just the climate, but also the tree's natural habitat. For example, paved urban environments are somewhat similar to warm mountain slopes with their shallow soils, exposed position and low humidity. Trees found naturally in these conditions often make excellent

Total Species Pool

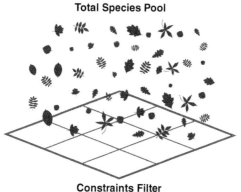

Constraints Filter
(Site, Biological, Practical)

Figure 5.17 The process of filtering out unsuitable tree species for urban environments using three main filters (see text for details). These help to refine a potential list of trees that are suitable for any particular planting site.

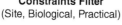

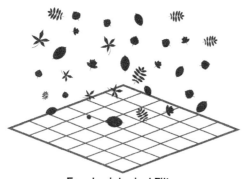

Ecophysiological Filter
(Local microclimate, Phenology, Tolerance)

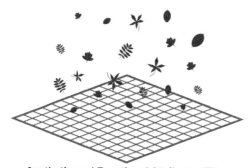

Aesthetic and Functional Attributes Filter
(Ornamental traits, Form, Mature size, Shelter, Shade)

Appropriate Species Pool

candidates for paved sites (Sjöman 2012; Sjöman *et al.* 2012). Looking at specific plant traits can also be very useful when identifying a species' potential for challenging urban sites. For example, leaf turgor loss (see Chapter 6) is useful to help identify species that are tolerant of dry urban sites (Sjöman *et al.* 2015). Tolerance of waterlogging is similarly useful for sites that are either prone to flooding or are designed to periodically flood as part of a sustainable urban drainage scheme. Whether a tree is a pioneer species adapted to open conditions or more adapted to humid, shady conditions will determine whether it would do well in an open street environment or in a large, dense planting scheme: forcing shade-tolerant trees into paved street environments will rarely prove successful.

Once these site constraints and ecophysiological factors have been used to filter out unsuitable species, aesthetic and 'functional attributes', such as shape, size, colour and flowering, can be used to hone the final pool of suitable trees (Figure 5.17). The key message is to ensure the tree will survive, before you decide what you want it to look like (see also Expert Box 5.1).

Expert Box 5.1 Selection of Trees for Urban Environments – Learning From Nature
Henrik Sjöman

Trees constitute an important element in the cityscape. The natural, graceful shapes of trees provide an architectural transition between human size and the scale of buildings and streets; over the ages, urban tree plantings have been regarded as a mirror of the prosperity and achievements of society. While beautification has traditionally been the main argument for planting trees in towns and cities, recent decades have seen numerous reports of a number of other beneficial effects provided by trees for the quality of life in urban areas. Trees help reduce the urban heat island intensity (King and Davis 2007), and thus decrease the need for energy for cooling buildings (Akbari *et al.* 2001). Urban trees are capable of reducing storm water run-off, and thereby reduce flooding and subsequent damage to property and infrastructure (McPherson *et al.* 1997). They act as noise filters and purify the air through capturing particulate matter, carbon dioxide, ozone and other air pollutants originating from traffic and industrial activities (Nowak *et al.* 2006). Urban trees also have an important role in recreation for the urban population, because they are an important element of green spaces in residential and commercial areas (Tyrväinen *et al.* 2007). However, these aesthetic, social and microclimatic ameliorations are only possible if the urban tree stock remains alive. Therefore, tree selection is one of the most important issues in creating sustainable urban forests and, thus, resilient urban environments. Today, we face two challenges in the selection of urban trees:

1) the development of reliable guidance in choosing the right tree for the right site and function, and;
2) using a broader catalogue of tree species.

Lack of Diversity of Urban Trees

From comprehensive tree inventories around the world, it is clear that a few tree species and genera dominate in many cities in the northern hemisphere. For example, in New York State, USA, Norway maple *Acer platanoides* represents over 20% of all street trees; in Helsinki, Finland, common lime *Tilia × europaea* makes up 44% of all street trees; and in Beijing, China, the pagoda tree *Styphnolobium japonicum* comprises 25% of all urban trees (Yang *et al.* 2005; Sjöman *et al.* 2010; Cowett and Bassuk 2014). Such poor diversity poses a major risk if a dominant species develops a severe disease or pest episode. Indeed, such events have already occurred: Dutch elm disease, caused by *Ophiostoma novo-ulmi*, had serious consequences for many cities in Europe and North America with large populations of elm *Ulmus* spp. Today, the legacy of Dutch elm disease is still apparent, with some urban areas lacking large, mature trees because removed elms were not replaced. This loss of a large proportion of the tree population is a major setback in view of the many ecosystem services associated with old and large trees – and it will take a long time before new trees can replace the former elm trees, even if suitable species are used as replacements. By far the best way to protect against the threats facing urban tree populations is to use as wide a diversity of species and genera as possible. It is recommended that this high diversity be evenly distributed throughout the entire city. Today, it is mainly city parks, arboreta or botanical gardens that contain a high and sustainable diversity of species, genera and families.

While recommendations on minimum levels of species diversity are important for increasing diversity, an essential consideration may be overlooked in the rush to diversify. Simply ordering new tree species and genotypes that are untested for the region is not the correct course, because the adaptability and longevity of species in stressful urban habitats must weigh heavily in the selection. Poor or incorrect choices may reduce the lifespan of trees and ultimately increase costs, as failed or failing trees must be removed or replaced. Therefore, knowledge of a species' capacity to grow in different habitats and climates remains vital.

Guidance in Selecting the Right Tree for the Right Site

The intention when choosing tree species is usually that they will stand on the site for many years, perhaps even in excess of 100 years, so it is critical to choose the best possible plant material for the specific site and climate. Based on the knowledge that many ecosystem services increase with tree size (e.g. Gomez-Munoz *et al.* 2010), choosing tree species that are capable of reaching maturity on the site must be a priority. When reviewing the literature for guidance in selecting trees for urban environments, it is clear that much of the information is based on the authors' own experiences and reflections rather than solid scientific experiments. This leads to rather varying descriptions where one author may state that the species is very tolerant for dry environments, while another author rates the same species as moisture demanding and sensitive to drought (Sjöman *et al.* 2015). This confusion is much more common for rare or non-traditional species where the experience of the tree outside of exclusive tree collections is scarce. However, knowledge of how these rare species grow in their natural habitats can give valuable guidance on how they will tolerate different urban habitats and climates, as well as the traits they will develop in order to compete successfully for resources.

Ecological Matching

The grandfather of modern arboriculture, Alex Shigo (1991), on the use and maintenance of city trees, said that: 'we must understand the tree as it grows in its natural site first. To try to treat a city tree without understanding the tree as it grows in its natural site is like drawing a data curve with only a y-axis and no base line!'

Choosing plants according to *fitness for site* reduces the need for intensive site management (Dunnet 2004). Some species have evolved an extensive plasticity and tolerance to a range of environmental conditions, while others have specialised in certain habitat types. Plants from habitats that share similar environmental constraints tend to

Figure EB5.1 When analysing a tree's capacity to grow in urban environments, its natural background or ecological heritage can give valuable guidance. Trees growing on steep, south-facing mountain slopes with shallow, rocky soil layers have developed traits that makes them tolerant of these types of conditions. The growing environment is similar to paved urban sites, so these species/genotypes are a better choice than those originating from moist river valleys, which have more in common with park environments. *Source:* Courtesy of Henrik Sjöman

share common traits or characteristics, which can be fully exploited when using the species in urban environments. For instance, steep, south-facing mountain slopes with shallow, rocky soil layers or warm and dry steppe environments represent distinct habitat types that have shaped the evolution of plants in the same way (Sjöman *et al.* 2010, 2012a). Consequently, in the search for species and genotypes tolerant of dry urban sites, information on the ecological strategy and performance of a species in different habitats can give valuable guidance (Figure EB5.1).

Failing to reflect on the species' (and genotype's) ecological heritage, its natural habitats and the traits that help the tree grow successfully in these habitats increases the likelihood of poor vitality in the landscape.

Succession – Guidance From Nature

The concept of succession can be described in simplified terms as the change in species distribution in one place over time (Picket *et al.* 2013). Today, this concept is seldom included in the selection process for urban trees, but the successional status of the species can be of critical importance for ensuring tree establishment and early development – the phase that often determines the long-term growth and survival of the tree. An urban square or courtyard built on concrete foundations has few similarities with a mature forest environment. Therefore, late-successional forest species find it much more difficult to establish in a warm square or enclosed courtyard, where the conditions resemble a much earlier stage of succession. Similarly, planting pioneer species in narrow, heavily shaded urban canyons will expose the trees to light levels that favour more shade-tolerant, late-successional species. Attempting to identify the phase of succession of an intended planting site can help in the search for plant material that possesses naturally developed strategies to deal with these conditions. It can also help in anticipating the initial maintenance measures needed.

Furthermore, paved urban environments, such as urban plazas, are represented in nature by warm mountain slopes with limited soil volume at an early phase of succession. This means that species such as black pine *Pinus nigra*, sessile oak *Quercus petraea*, goldenrain tree *Koelreuteria paniculata*, mahaleb cherry *Prunus mahaleb*, Russian olive *Elaeagnus angustifolia* and manna ash *Fraxinus ornus* are suitable trees, as they occur naturally in similar conditions and have developed strategies for coping with these conditions. However, if the planting site is 'improved' by ameliorating the rooting environment with structural soil, the site can be comparable to a natural scree slope with rooting conditions that provide good access to oxygen and a relatively large soil volume to hold water and nutrient resources. A number of species naturally grow in this type of environment and display very good long-term development. If the urban planting site is fully exposed to the sun, pioneer species are most suitable because they can cope with the open, exposed site with high evapotranspiration. Examples of such species are Italian alder *Alnus cordata*, Hungarian oak *Quercus frainetto*, Swedish whitebeam *Sorbus intermedia*, Sargent's cherry *Prunus sargentii*, zelkova *Zelkova serrata*, field maple *Acer campestre*, Japanese tree lilac *Syringa reticulata*, ginkgo *Ginkgo biloba* and European hackberry *Celtis australis*. Where the urban square planting site is shaded by neighbouring buildings for part of the day, late-successional species that occur naturally on scree slopes may be more suitable. These include hop hornbeam *Ostrya* spp., hornbeam *Carpinus* spp., elm *Ulmus* spp. and silver lime *Tilia tomentosa*, which can cope with the soil conditions and lower light quality, as the shady conditions also create a cooler, more humid site with lower evapotranspiration. Even on rich parkland sites, it is important to consider succession. On open sites where it may be desirable to create a windbreak, pioneer species from cool, rich forests should be selected, as they possess developmental strategies that facilitates rapid establishment. Examples of such species are silver maple *Acer saccharinum*, poplar *Populus* spp., many willow *Salix* spp., silver birch *Betula pendula*, alder *Alnus* spp. and Russian olive *Elaeagnus angustifolia*. Where established trees already exist on parkland, their mature crowns modify microenvironmental conditions, influencing the light and humidity levels on the site. These planting locations represent a later phase in forest succession and favour species such as western hemlock *Tsuga heterophylla*, fir *Abies* spp., sycamore *Acer pseudoplatanus*, small-leaved lime *Tilia cordata*, beech *Fagus* spp., yew *Taxus* spp. and western red cedar *Thuja plicata*.

Succession is a natural process that creates conditions that plants must adapt to. Therefore, if this ecological process is not considered when planting, it is likely that the species will be forced into a situation that they are not prepared for. Often, this results in poorly performing trees with a limited capacity to deliver important ecosystem services.

Future Vista

All planting design must, to some extent, be a compromise between what is desirable (artistic or creative vision) and what is possible (scientific reality) (Dunnet 2004). Even if it is possible to push the boundaries through technology, this often occurs at a considerable environmental cost that is not sustainable (e.g. continuous management costs). The great advantage for an ecologically informed basis for plantings is that it has the potential to achieve full creative vision with relatively little site modification. To succeed with long-term sustainable tree plantations in urban environments, a greater understanding of the trees' biological and ecological capacity to perform well in urban habitats is necessary, an understanding that can be gained by evaluating how they grow in their natural environments.

Knowledge of trees' natural habitats is of crucial importance when advocating less traditional tree species for planting, as it gives valuable guidance when long-term experience of the species in urban environments is scarce.

Tree Quality

Success in tree planting is strongly associated with high quality nursery stock. Tree handling during transport, as well as on the planting site, will also affect final plant quality. Robust biosecurity procedures during production and handling are becoming increasingly important in ensuring that planting nursery stock does not introduce new pests and diseases into our landscapes.

Tree quality can be very variable between tree nurseries so those buying trees need to evaluate nurseries, if necessary avoiding those that fail to deliver consistently high quality stock. Robust specifications should be made to provide clear, unequivocal guidance to nurseries and planting contractors to ensure high quality stock is procured and that its quality is maintained all the way to the planting site. Table 5.2 gives guidance to be used as the basis of robust planting specifications.

A number of nursery production practices can influence the establishment of trees. Perhaps of greatest significance is the preparation of the root system. Tree roots will naturally seek to explore new volumes of soil to exploit resource-rich patches and secure stability. Typically, this means that the root system rapidly becomes too extensive to be easily transplanted intact. For this reason, field-grown trees must either be regularly transplanted within the nursery, or root pruned so that a compact root system is formed.

Nursery Production

Root loss during transplantation from the nursery to the landscape is a major driver of early tree mortality. Therefore, practices and methods that maximise the fine root area taken with the tree at time of transplanting are essential for producing high quality amenity trees.

As seedlings grow in the seedbed, they will eventually need transplanting, either to their final home or into a new bed or container. In forestry, most seedlings are planted out as bare-root stock because it is easier and cheaper to transport and handle the large numbers involved. Even on a smaller scale, bare-root stock is cheaper, but comes with the disadvantage that the planting season is much shorter because it must be performed during the dormant season. Seedlings can be grown as $1+0$ or $2+0$ transplants (spending 1 or 2 years in the seedling bed before being planted out) but many are grown as $1+1$ or $2+1$ seedlings, being transplanted from the seedbed to another bed

Table 5.2 Key elements and criteria that may be used to generate robust planting specifications for work contracts and method statements. Adapted from Johnston and Hirons (2014).

Specification elements	Specification criteria
Tree characteristics before planting	• Specimen true to species/variety type • Graft compatibility (if appropriate) • Healthy with good vitality • Free from pests, disease or abiotic stress • Free from injury • Self-supporting with good stem taper • Stem–branch transition height (e.g. 180 cm clear stem) • Sound branch attachment and crown structure • Good pruning wound occlusion • Crown symmetry • High root-ball occupancy • Diversity in rooting direction, including to all quarters of the root-zone • Good root division • Extensive fibrous root system with root-ball diameter: trunk diameter (at 15 cm above soil surface) ratio of ~10:1 • Free from root defects (e.g. girdling roots)
Planting pit and root-zone	• Planting pit 2–3 times the diameter of the root-ball • Imported soil is of defined standard (e.g. BS 3882) • Low soil bulk density (e.g. ~1.2 g cm^{-3}) maintained in planting pit and root-zone • Potential rooting (soil) volume adequate for mature tree of species planted
Tree handling	• Transport should only be conducted in closed-canopy vehicles to prevent plant tissue desiccation • Stems should be protected from abrasion • Root-balls must not be allowed to dry out
Planting practice	• Hessian, wire baskets and other containers removed from root-ball and correctly disposed of • Tree planted at stem–root transition • An area 2–3 times the diameter of the root-ball should be cultivated around the base of the tree to a soil bulk density <1.2 g cm^{-3} • Tree upright and supported (where necessary) using above- or below-ground techniques; wooden posts or rails should never be in direct contact with the stem
Formative pruning	• Damaged branches removed using natural target pruning methodology (see Chapter 3) • Rubbing and crossing branches removed • Subordination of competing stems
Tree aftercare	• Mulch depth of 5–10 cm and to defined width. Stem to remain exposed and not buried by mulch • Mulch replenishment schedule defined • Irrigation schedule based on local soil variables (preferably soil matric potential) • Tree support to be placed as low as possible and not higher than one-third of the total height of the tree • Tree protection specified to meet the potential threats on the planting site • Tree protection and support to have defined timescale for evaluation and/or removal

Source: Dixon and Aldous (2014). Reproduced with permission of Springer.

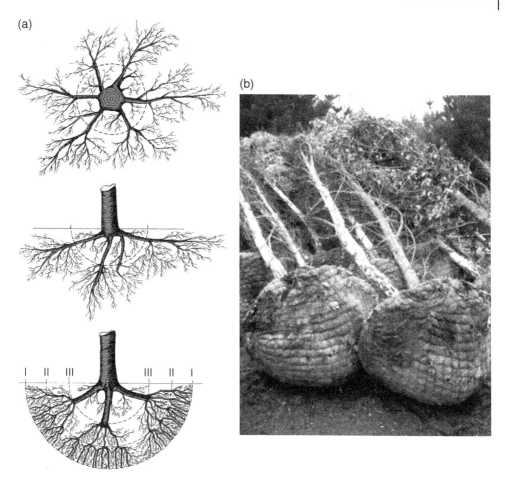

Figure 5.18 Root pruning or regular transplantation in the nursery is vital if a compact root system is to be created. Without root pruning, the majority of the root system can be left in the field at time of purchase (top and middle left [a]). Regular root pruning can create a compact root system (bottom left [a]) that can then be wrapped with hessian and a wire cage – rootballed (UK) balled and burlapped (USA) – prior to transportation to the planting site (right [b]). *Source:* (a) Courtesy of Keith Sacre.

for an extra year. The older seedlings (2 years or older) are often *undercut* (or horizontally root pruned) before transplanting. This involves cutting off any roots, including the taproot, at 15–20 cm below the surface. This often results in increased lateral root growth and the development of a more compact, fibrous root system that is better able to cope with transplanting. However, if trees are grown for extended periods in a nursery bed, the roots will naturally extend outwards to explore new volumes of soil. Typically, this means that the root system rapidly becomes too extensive to be easily transplanted intact. For this reason, *field grown* trees must either be regularly transplanted within the nursery or be root pruned so that a compact root system is formed (Figure 5.18). Despite such procedures, bare-rooted trees can lose the majority of their fine roots during transplanting (Watson and Himelick 1982), so planting them early in the dormant season is important to give the tree time to grow new fine roots before there is a high water demand from the leafy crown.

For ease of handling, and to give a much longer planting season, trees are often either container-grown (grown in a container) or containerised (transplanted into a container for the last year or so before planting out). Container-grown trees can be planted out at almost any time, as long as sufficient aftercare is provided, but they can be affected by several disadvantages. If trees are grown for too long in containers they become *root-bound*. In such cases, it is also common for roots reaching the edge of the container to bend sideways and start circling the pot. If these are not removed before planting out, these *circling* or *girdling roots* can cause the growing tree significant problems by restricting trunk growth, and by the roots failing to grow outwards to develop a healthy root system (Figure 5.19). Root pruning practices can help reduce the impact of root defects (Gilman 2012) but it is much better to buy trees with high quality roots systems in the first place. There are various ways of reducing circling roots: historically, a layer of copper sulfate was applied inside the pot to kill roots as they approached the sides, but a much more environmentally sustainable method is to use suitable pot designs. For example, an Air-Pot® has open holes around the sides and bottom, allowing the outer-most roots to dehydrate and 'air-prune' (Figure 5.20). These pots significantly reduce root defects and improve the quality of the root system (Amoroso *et al.* 2010; Gilman *et al.* 2010; Mariotti *et al.* 2015). Typically, air-pruned root systems are more fibrous and have few, if any, girdling roots (Figure 5.21), providing they are not left in the pot for excessive periods of time.

However, other approaches such as the use of white fabric containers (e.g. Barcham Light Pots™), which allow the transmission of diffuse light through, have also been shown to reduce root girdling (Grimshaw and Bayton 2010). Paper-based propagation cells (e.g. Jiffy® plugs or Ellepot) are also designed to facilitate air-pruning in seedlings and provide well-developed young root systems for growing on in the field or larger containers (Figure 5.22). Regardless of the container type used to grow trees, all trees grown in containers should be assigned a 'shelf-life', as even specialised containers will not postpone the development of root defects indefinitely. Root system defects will persist after planting so tree specifications requiring the absence of root defects provide useful safeguards against trees with poor quality root systems.

Another potential problem of container-grown or containerised trees is that when the tree is planted out, it has many roots concentrated in a small volume and so dehydration and root death is more likely unless the tree is regularly watered. However, the shape and size of the container can make a difference to the subsequent growth and survival of the tree. A study of container-grown stone pine *Pinus pinea* showed that the height and stem diameter of the seedlings was greater as container volume increased from 123 to 400 cm^3. Interestingly for a taproot species, it was diameter of the container that made most difference to seedling growth rather than depth. The ideal was a pot whose width was four times the depth (Dominguez-Lerena *et al.* 2006). However, trees from arid areas, such as cork oak *Quercus suber*, perform better if grown in 30-cm deep pots, rather than 18-cm deep pots. The longer taproot allows them to reach deeper soil water more quickly when planted out, aiding their survival (Chirino *et al.* 2008). Commercially, of course, there will always be a compromise between producing healthy trees and the cost of larger pots with more soil and higher transportation costs.

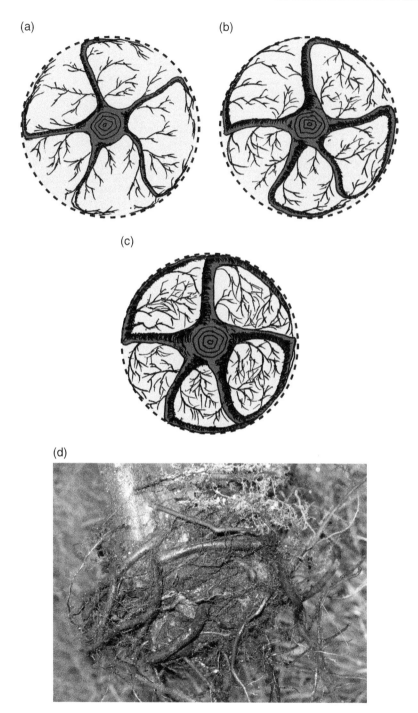

Figure 5.19 (a–c) Once the roots hit the inside wall of a container, their roots get deflected and begin circling. (d) Serious root defects can occur as a consequence. These will compromise the future stability of the tree and impede root development. *Source:* Courtesy of Keith Sacre.

(a)

(b)

Figure 5.20 Air-Pots®. Trees grown in these do not suffer from girdling roots that grow around the inside of a normal pot. (a) Manchurian alder *Alnus hirsuta* growing in an Air-Pot®; (b) a close-up of a lateral root that will be 'air-pruned'.

(a)

(b)

Figure 5.21 (a) Containerised production of dawn redwood *Metasequoia glyptostroboides* using an Air-Pot® (the black pipes are irrigation lines). (b) These create a fibrous root system free from significant defects. Photographed at Stairway Trees, Stair, UK.

A number of other nursery practices influence the quality of the trees they produce. High-density planting in the nursery can reduce stem taper, and hence the ability of the tree to be self-supporting (see Chapter 2). Shading can also reduce crown development and reduce the ability of the tree to photosynthesise, thus reducing transplant success (Sellmer and Kuhns 2007). Pruning can improve or reduce the quality of the tree, so it is important to have a clear vision of what the tree crown should look like prior to

(a)

(b)

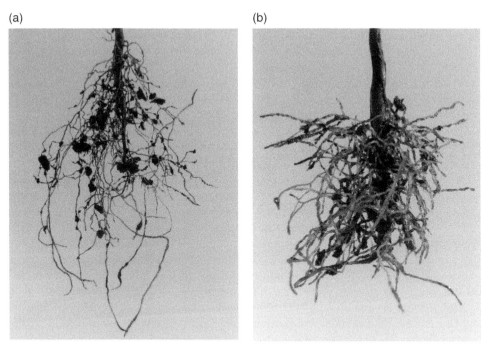

Figure 5.22 Young root systems of the Japanese snowbell tree *Styrax japonicus* grown in (a) an Air-Pot® propagation tray and (b) a Jiffy® paper propagation plug. In both images, the root systems have been washed to reveal a roots system free from defects that are ideal for growing on in larger containers or a nursery bed.

purchasing. Crown symmetry and a well-structured framework of branches with sound branch unions are important for future crown development.

Tree handling between the nursery and planting site can also have a significant effect on the subsequent quality of the tree. It is quite possible for an excellent tree to be severely damaged en-route to the planting site. Exposed roots, buds or leaves will rapidly desiccate in the back of an open vehicle, so a closed-canopy vehicle should always be used to transport trees. Other problems can arise from stem abrasion or the roots being allowed to dry out. Therefore, when writing a specification for a planting scheme, it is imperative to consider both the quality of nursery stock and tree handling procedures.

Rooting Environment

Soils in urban landscapes are generally thought of as highly disturbed, variable and of low fertility (Craul 1999). However, Pouyat *et al.* (2010) looked at entire urban landscapes and found that soils that are largely undisturbed or of high fertility can also be found in urban areas. For example, many parks and gardens provide excellent soil conditions for trees. Such diversity in soil quality means that those involved with tree establishment should have an understanding of how soil conditions can influence tree development.

The extent of soil compaction has particular significance for root development and, as discussed in Chapter 4, can readily limit tree vigour or even survival unless measures are taken to reduce it. Moreover, the soil volume needed for trees is very important (see Chapter 4). Despite the uncertainty surrounding absolute soil requirements, it is clear that soil volumes frequently found in urban environments are inadequate. A general principle of any tree planting scheme should therefore be to maximise the uncompacted soil volume available to roots.

On planting sites that are subjected to large-scale earth works, any topsoil that is excavated should ideally be kept aside and used in the final landscaping (Urban 2008). However, in some circumstances, it might be necessary to import soil on to a planting site. In such cases, it is useful to provide a soil specification that includes guidance on the physical, chemical and organic matter components of the soil, as well as acceptable contamination levels so that some quality control can take place as the rooting environment is constructed. Ensuring the soil is biologically healthy would also be of great value but, given the highly dynamic nature of biological interactions in soil, this is very hard to quantify. In practice, adding a good layer of mulch after planting will go a long way in aiding the development of good soil health. Further information on the impact of soil conditions on plant growth can be found in Gregory and Nortcliff (2013).

The benefits of mulching include reducing fluctuations of soil temperature and soil moisture; weed suppression; enriching the soil with nutrients; preventing soil erosion from heavy rains, regulation of pH and the cation exchange capacity in favour of the tree (see Chapter 4); suppressing plant diseases; and encouraging soil biological activity. Mulch achieves this, in part, because of its influence on the rhizosphere which contains a complex array of organisms vital for soil health (Buée *et al.* 2009). Mulch can also reduce the impact of soil contaminants such as those from pesticides and heavy metals. Unsurprisingly, the combined effect of all these factors is an increase in plant performance (Figure 5.23). In addition, mulch can help prevent mechanical damage to the tree trunk and act as a buffer in preventing de-icing salts from percolating into the rooting zone (Chalker-Scott 2007). Landscape mulches can be inorganic (crushed stone, crushed brick, gravel, polyethylene films) or organic (shredded leaves, wood bark, pine straw, recycled pallets or a combination). While inorganic mulches have some value in reducing weed competition, organic mulches offer wider benefits and are preferred for young trees where practicable. There is some evidence that mulches made just of chipped wood (xylem) will reduce the amount of nitrogen available to young trees: wood is low in nitrogen, so the microorganisms breaking it down will scavenge nitrogen from the soil for their own use and leave less for the trees to take up. This effect can be reduced by adding some bark and leaves to the mulch to raise the nitrogen level (Jackson *et al.* 2009).

Mulch is very often made from a mix of different species; however, there is increasing evidence that 'pure' mulches made from one species may offer an advantage over more general 'mixed' mulches. Percival *et al.* (2009) found that growth and yield of field-grown fruit trees, including the apple *Malus* 'Gala' and pear *Pyrus communis*, could be significantly increased by applying pure mulches. Growth and survival rates following containerisation of a transplant-sensitive species (beech *Fagus sylvatica*) were also significantly increased using appropriate pure mulches. Interestingly, the pure mulch did not have to be made of the same species to which it was applied. Mulches of hawthorn *Crataegus monogyna* and wild cherry *Prunus avium* were consistently better than those

Figure 5.23 Likely impacts of supplying good quality mulch to a tree root-zone. Triangles indicate whether the general trend of each factor increases or decreases with the addition of organic mulch. Dashed margins suggest that the relationships are not necessarily linear but represent a 'direction of travel'. *Source:* Hirons and Percival (2012). Contains public sector information licensed under the Open Government Licence v3.0, http://www.nationalarchives.gov.uk/doc/open-government-licence/version/3/and

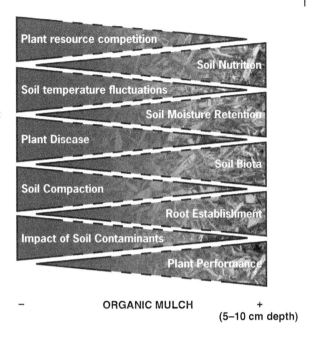

of silver birch *Betula pendula*, holm oak *Quercus ilex* and pedunculate oak *Q. robur*, which in turn were better than beech *Fagus sylvatica* (Percival *et al.* 2009). Although the reasons why some pure mulches are better than others are not yet fully understood, it may well relate to the sugar content of the mulch, allelochemicals (toxic compounds used in plant defence) produced by different mulch species, changes in soil organisms or, more likely, a combination of all three. Certainly, further investigation is warranted as the positive effect of mulches may be enhanced by only minor adjustments to management practices.

Mulch should be applied from the drip line to the trunk, 5–10 cm in depth. If this is not practical, mulch should be in a circle at least 0.3 m radius for small trees, 1 m for medium trees and 3 m for large trees. Mulch should not be placed against the trunk, as it will retain moisture that may result in disease around the base of the tree. It is the roots that require mulching, not the trunk.

Arboricultural Practices

Often, after an appropriate tree species has been selected for a planting site, high quality trees have been selected from the nursery and an excellent root environment has been prepared, a planted tree will still die as a result of poor arboricultural practices. Project management and specifications that provide precise expectations for all arboricultural procedures should enforce good practice. Practitioners and contractors should be accountable to these specifications, and audits should be used to monitor work standards.

To ensure that roots can readily develop into the new host soil, an area two to three times the diameter of the root-ball should be cultivated to a low soil bulk density (around $1.2\,\mathrm{g\,cm^{-3}}$) and the planting pit should be no deeper than the existing root-ball. High quality trees should have been ordered and correctly delivered as discussed earlier. Before putting the tree into the planting pit, it should be assessed against the criteria of Table 5.2. The tree crown and stems should be free from defects, such as graft incompatibility, and major damage, such as regions of missing bark or broken branches. The upper roots must not be more than a few centimetres below the root-ball's soil surface and a stem flare must be clearly visible. Once any root-ball packaging has been removed, any roots that circle over one-quarter of the root-ball should be removed, along with any other major root defects. Intervention at planting, if necessary, is preferable to leaving the tree with root defects that may compromise the long-term viability of the tree.

Good planting practice for amenity trees requires a circular or square hole to be cultivated 2–3 times the diameter of the root-ball, so that the tree's root system is unconstrained in its development into the new host soil (Figure 5.24). This process of root–soil coupling is vital if the tree is to avoid serious water and nutrient limitations shortly after transplanting. Planting the tree at the correct depth is also vital. Put simply, stems are designed to be above ground; roots are designed to be below ground. If

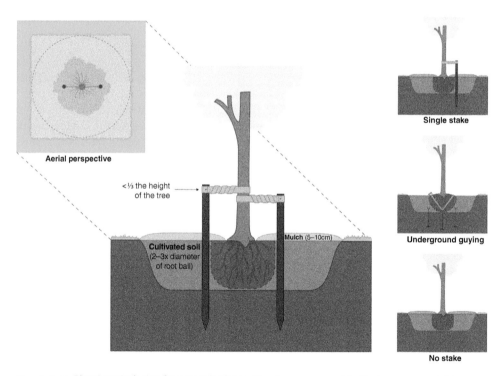

Figure 5.24 Planting pit design for a standard tree. Circular or square-sided holes are acceptable; cultivated soil should be in proportion to the size of the root-ball and should allow the tree to sit in the hole at the correct depth. If support is necessary, a number of different staking formats may be used for tree stability, including staking less than one-third the height of the tree) and underground guying. Mulch should be applied over the planting pit to a depth of 5–10 cm. *Source:* Hirons and Percival (2012). Contains public sector information licensed under the Open Government Licence v3.0, http://www.nationalarchives.gov.uk/doc/open-government-licence/version/3/.

the stem is buried, moisture levels around the stem often make it vulnerable to decay; if the roots are exposed, they rapidly dry out and die. To avoid either of these outcomes it is important to identify the base of the root–stem transition zone (often referred to as the stem or trunk flare) and make sure that this is level with the final soil grade. Consequently, the upper roots should be no more than a few centimetres below the soil. As the root-ball soil sometimes becomes raised around the base of the stem during nursery operations, it is important not to treat the root-ball soil level as a guide for planting depth. Always identify where the root–stem transition zone and upper roots are before finalising planting depth.

There are a number of different ways to physically support trees during establishment (Figure 5.24). Regardless of approach, the support system should allow stem and crown movement so that stem and root growth are stimulated. To aid this, support of stems should be as low as possible and certainly less than one-third the height of the tree (Appleton *et al.* 2008). Tree ties should seek to spread the load on the stem using a wide band (usually hessian or rubber) to reduce abrasion, and it must not restrict expansion of the stem. Below-ground root anchor systems allow full above-ground movement and help give the visual impression of an established tree, and potential trip hazards are avoided. All forms of support should be removed after new root growth adequately stabilises the tree. This can vary by tree species, size and soil type but, as a general guide, it should be possible to remove all support within 2 years of planting. It is essential to make sure that tree stakes and ties are not left on the tree indefinitely, as these frequently cause damage by restricting growth in tree girth or rubbing against the trunk.

The loss of roots at transplanting means that management of tree water deficits is often required (Pallardy 2008). In established trees, water deficit is often associated with periodic drought. However, newly planted trees in urban environments may need to cope with small soil volumes and relatively impermeable surfaces, as well as root loss. This reduces the amount of water that can be taken up by the tree and is an important reason why water deficits are frequently cited as a major cause of death of young landscape trees. In areas where newly planted trees are not irrigated, establishment has to rely on rain reaching the root zone. In drier climates and in situations where infiltration is disrupted by sealed surfaces or soil compaction, the moisture content of the root zone can be effectively decoupled from local rainfall. This means that young trees may suffer from water deficit even in areas with relatively high rainfall. In such cases, irrigation after planting can be critical to successful tree establishment. This can be facilitated by including an irrigation hose in the planting hole or using a tree irrigation bag (more detail on these and other ways of managing soil water availability can be found in Chapter 6).

On many planting sites, young trees require some form of protection to limit mechanical injury to the trunk caused by animals or machinery. In many urban areas where grass is being managed alongside trees, inept handling of strimmers and mowers frequently leads to damage to the base of a tree. This type of damage can be reduced by using strimmer guards and stakes. However, contractual clauses that safeguard against damage to newly planted trees by negligence of contractors should be included wherever possible. Targeted tree protection measures can then be specified in contracts and method statements. This should also be true of tree support systems: it is vital to include conditions of replacement and removal within any planting specification so that tree guards left in place do not cause damage to the tree as it grows.

Figure 5.25 Tree shelters, known as Tuley tubes in the UK, protecting young oak trees from deer (hence their height). As well as keeping deer and smaller browsing animals away they also act as greenhouses, keeping the saplings warmer and helping them grow faster.

Tree shelters or tree guards (also known as Tuley tubes in Britain; Figure 5.25) benefit the newly planted tree by keeping browsing animals away, providing a more favourable microclimate (particularly higher temperature), speeding growth and increasing survival. The original work assessing their effectiveness was carried out on sessile oak *Quercus petraea* transplants by their inventor, Graham Tuley (1985), who showed that after 3 years, trees in the shelters were 142 cm high compared to 45 cm using a mesh guard (which just kept browsers away) and 27 cm for those transplants with no protection. The average stem volumes were 118, 37 and 19 cm^3, respectively. More recently, a study on seeds of Chinese chestnut *Castanea mollissima*, American chestnut *C. dentata* and hybrids between the two planted on mining spoil in Kentucky, showed that germination was much higher (77–82%) when tree shelters were used compared to 1–12% with no shelters (Barton *et al.* 2015); without shelters the seeds were eaten by rodents. Survival of seedlings was also an order of magnitude higher when shelters were used. Shelters of various diameters can accommodate trees and shrubs with different spreads, and height can be specified depending upon whether rabbits (0.6 m) or deer (1.8 m) are the main problem (Trout and Brunt 2014). Shelters also have the advantage that they protect trees from herbicide used to create a weed-free zone over the roots. However, there are disadvantages in using shelters. Care should be taken that competing vegetation, such as grass, does not also grow better inside the tube and choke the young tree. In some climates, it can become too hot inside the tube, causing excessive moisture loss and death or stunting of the tree. In this case, white-coloured shelters help by reflecting sunlight (Oliet and Jacobs 2007). Shelters are also expensive (in large numbers), require maintenance and must be removed before they restrict tree growth or become an eyesore. The correct shelter also needs to be used. It is no good trying to stop deer eating the tree with a rabbit guard; however, a deer guard will stop rabbits. Spiral rabbit guards can be used where the aim is to stop rabbits and small mammals from eating the bark. Depending upon the site and the browsing animals, it may be cheaper to fence the planting area (Trout and Brunt 2014).

Formative pruning of newly planted trees can help achieve good branch structure right from the start and reduce future problems. It may be necessary to remove broken branches (from handling procedures) and, occasionally, branches that show serious conflict with others. However, it is essential that as much of the crown remain intact as possible, as a reduction in leaf area reduces carbon gain and therefore the energy resources available for root development. Further formative pruning can always take place, if necessary, once the tree is established. General guidance on pruning can be found in Chapter 3 and more comprehensive guidance can be found in standard arboricultural texts such as Gilman (2012) and Brown and Kirkham (2017).

References

Ågren, G.I. and Franklin, O. (2003) Root: shoot ratios, optimization and nitrogen productivity. *Annals of Botany*, 92: 795–800.

Akbari, H., Pomerantz, M. and Taha, H. (2001) Cool surfaces and shade trees to reduce energy use and improve air quality in urban areas. *Solar Energy*, 70: 295–310.

Alla, A.Q., Camarero, J.J., Maestro-Martínez, M. and Montserrat-Martí, G. (2012) Acorn production is linked to secondary growth but not to declining carbohydrate concentrations in current-year shoots of two oak species. *Trees*, 26: 841–850.

Amoroso, G., Frangim P., Piatti, R., Ferrini, F., Fini, A. and Faoro, M. (2010) Effect of container design on plant growth and root deformation of littleleaf linden and field elm. *HortScience*, 45: 1824–1829.

Appleton, B.L., Cannella, C.M., Wiseman, P.E. and Alvey, A.A. (2008) Tree stabilization: Current products and practices. *Arboriculture and Urban Forestry*, 34: 54–58.

Baghi, R. Helmig, D., Guenther, A., Duhl, T. and Daly, R. (2012) Contribution of flowering trees to urban atmospheric biogenic volatile organic compound emissions. *Biogeosciences*, 9: 3777–3785.

Barrett, S.C.H. and Hough, J. (2013) Sexual dimorphism in flowering plants. *Journal of Experimental Botany*, 63: 695–697.

Barton, C., Miller, J., Sena, K., Angel, P. and French, M. (2015) Evaluating the use of tree shelters for direct seeding of *Castanea* on a surface mine in Appalachia. *Forests*, 6: 3514–1527.

Bazzaz, F.A. and Miao, S.L. (1993) Successional status, seed size, and responses of tree seedlings to CO_2, light, and nutrients. *Ecology*, 74: 104–112.

Bazzaz, F.A., Carlson, R.W. and Harper, J.L. (1979) Contribution to reproductive effort by photosynthesis of flowers and fruits. *Nature*, 279: 554–555.

Belaire, J.A., Whelan, C.J. and Minor, E.S. (2014) Having our yards and sharing them too: the collective effects of yards on native bird species in an urban landscape. *Ecological Applications*, 24: 2132–2143.

Black, M. and Wareing, P.F. (1955) Growth studies in woody species VII. Photoperiodic control of germination in *Betula pubescens* Ehrh. *Physiologia Plantarum*, 8: 300–316.

Bond, J., Plattner, K. and Hunt, K. (2014) *Fruit and Tree Nuts Outlook: Economic Insight.* Situation and Outlook FTS-357SA, United States Department of Agriculture, Economic Research Service, Washington, DC, USA.

Bonner, F.T. (1990) Storage of seeds: Potential and limitations for germplasm conservation. *Forest Ecology and Management*, 35: 35–43.

Britt, C. and Johnston, M. (2008) *Trees in Towns II. A New Survey of Urban Trees in England and their Condition and Management.*, Department for Communities and Local Government, Wetherby, UK.

Brown, G.E. and Kirkham, T. (2017) *Essential Pruning Techniques: Trees, Shrubs and Conifers.* Timber Press, Portland, OR, USA.

Buée, M., Boer, W.D. and Martin, F. (2009) The rhizosphere zoo: An overview of plant-associated communities of microorganisms, including phages, bacteria, archaea, and fungi, and of some of their structuring factors. *Plant and Soil*, 321: 189–212.

Campioli, M., Gielen, B., Göckede, M., Papale, D., Bouriaud, O. and Granier, A. (2011) Temporal variability of the NPP-GPP ratio at seasonal and interannual time scales in a temperate beech forest. *Biogeosciences*, 8: 2481–2492.

Chalker-Scott, L. (2007) Viewpoint impact of mulches on landscape plants and the environment: A review. *Journal of Environmental Horticulture*, 25: 239–249.

Chirino, E., Vilagrosa, A., Hernández, E.I., Matos, A. and Vallejo, V.R. (2008) Effects of a deep container on morpho-functional characteristics and root colonization in *Quercus suber* L. seedlings for reforestation in Mediterranean climate. *Forest Ecology and Management*, 256: 779–785.

Clark, K.H. and Nicholas, K.A. (2013) Introducing urban food forestry: A multifunctional approach to increase food security and provide ecosystem services. *Landscape Ecology*, 28: 1649–1669.

Cowett, F.D. and Bassuk, N.L. (2014) Statewide assessment of street trees in New York State, USA. *Urban Forestry Urban Greening*, 13: 213–220.

Craul, P.J. (1999) *Urban Soils: Applications and Practices.* Wiley, New York, USA.

de Lucena, I.C., Amorim, R.S.S., de Lobo, F.A., Silva, R.B., Silva, T.A.S. and Matos, D.M.S. (2015) The effects of resource availability on sprouting: a key trait influencing the population dynamics of a tree species. *Trees*, 29: 1301–1312.

del Tredici, P. (2001) Sprouting in temperate trees: A morphological and ecological review. *Botanical Review*, 67: 121–140.

Derkzen, M.L., van Teeffelen, A.J.A. and Verburg, P.H. (2015) Quantifying urban ecosystem services based on high resolution data of urban green space: An assessment for Rotterdam, the Netherlands. *Journal of Applied Ecology*, 52: 1020–1032.

Dewit, L. and Reid, D.M. (1992) Branch abscission in balsam poplar (*Populus balsamifera*): characterization of the phenomenon and the influence of wind. *International Journal of Plant Science*, 153: 556–564.

Dick, J.McP., Leakey, R.R.B. and Jarvis, P.G. (1990) Influence of female cones on the vegetative growth of *Pinus contorta* trees. *Tree Physiology*, 6: 151–163.

Dirr, M.A and Heuser, C.W. Jr (2006) *The Reference Manual of Woody Plant Propagation: From Seed to Tissue Culture*, 2nd edition. Timber Press, Portland, OR, USA.

Dirr, M.A (2009) *Manual of Woody Landscape Plants: Their identification, Ornamental Characteristics, Culture, Propagation and Uses*, 6th edition. Stipes Publishing L.L.C., Champaign, IL, USA.

Dixon, D.M. (1969) The transplantation of Punt incense trees in Egypt. *Journal of Egyptian Archaeology*, 55: 55–65.

Dixon, G. and Aldous, D. (eds) (2014) *Horticulture: Plants for People and Places*, Volume 2. Springer, Dordrecht, The Netherlands, pp. 693–711.

Dominguez-Lerena, S., Herrero Sierra, N., Carrasco Manzano, I., Ocaña Bueno, L., Peñuelas Rubira, J.L. and Mexal, J.G. (2006) Container characteristics influence *Pinus pinea* seedling development in the nursery and field. *Forest Ecology and Management*, 221: 63–71.

Duinker, P.N., Ordóñez, C., Steenberg, J.W.N., Miller, K.H., Toni, S.A. and Nitoslawski, S.A. (2015) Trees in Canadian cities: Indispensable life form for urban sustainability. *Sustainability*, 7: 7379–7396.

Dunnet, N. (2004) The dynamic of plant communities. In: Dunnet, N. and Hitchmough, J. (eds) *The Dynamic Landscape*. Spon Press, London, UK, pp. 97–114.

Dürr, C., Dickie, J.B., Yang, X.-Y. and Pritchard, H.W. (2015) Ranges of critical temperature and water potential values for the germination of species worldwide: Contribution to a seed trait database. *Agricultural and Forest Meteorology*, 200: 222–232.

Génard, M., Dauzat, J., Franck, N., Lescourret, F., Moitrier, N., Vaast, P., *et al.* (2008) Carbon allocation in fruit trees: from theory to modelling. *Trees*, 22: 269–282.

Gilbertson, P. and Bradshaw, A.D. (1985) Tree survival in cities: The extent and nature of the problem. *Arboricultural Journal*, 9: 131–142.

Gilman, E.F. (2012) *An Illustrated Guide to Pruning*, 3rd edition. Delmar, Cengage Learning, NY, USA.

Gilman, E.F., Harchick, C. and Paz, M. (2010) Effect of container type on root form and growth of red maple. *Journal of Environmental Horticulture*, 28: 1–7.

Gómez-Muñoz, V.M., Porta-Gándara, M.A. and Fernández, J.L. (2010) Effect of treeshades in urban planning in hot-arid climatic regions. *Landscape and Urban Planning*, 94: 149–157.

Gosling, P. (2007) *Raising Trees and Shrubs from Seed: Practical Guide*. Forestry Commission, Edinburgh, UK.

Gregory, P.J. and Nortcliff, S. (2013) *Soil Conditions and Plant Growth*. Wiley-Blackwell, Oxford, UK.

Grimshaw, J. and Bayton, R. (2010) *New Trees: Recent Introductions to Cultivation*. Royal Botanic Gardens, Kew, Richmond, UK.

Gärtner, S.M., Lieffers, V.J. and Macdonald, S.E. (2011) Ecology and management of natural regeneration of white spruce in the boreal forest. *Environmental Reviews*, 19: 461–478.

Hacket-Pain, A.J., Friend, A.D., Lageard, J.G.A. and Thomas, P.A. (2015) The influence of masting phenomenon on growth–climate relationships in trees: explaining the influence of previous summers' climate on ring width. *Tree Physiology*, 35: 319–330.

Hirons, A.D. and Percival, G.C. (2012) Fundamentals of tree establishment. In: Johnston, M. and Percival, G.C. (eds) *Trees People and the Built Environment: Proceedings of the Urban Trees Research Conference 13–14 April 2011*. Forestry Commission, Edinburgh, UK, pp. 51–62.

Hitsuma, G., Morisawa, T. and Yagihashi, T. (2015) Orthotropic lateral branches contribute to shade tolerance and survival of *Thujopsis dolabrata* var. *hondai* saplings by altering crown architecture and promoting layering. *Botany*, 93: 353–360.

Hubert, J. and Cottrell, J. (2007). *The Role of Forest Genetic Resources in Helping British Forests Respond to Climate Change*. Information Note 86, Forestry Commission, Edinburgh, UK.

Ignatieva, M. (2012) Plant material for urban landscapes in the era of globalization: Roots, challenges and innovative solutions. In: Richter, M. and Weiland, U. (eds) *Applied Urban Ecology: A Global Framework*. Wiley-Blackwell, Oxford, UK, pp. 139–151.

Innes, J.L. (1992) Observations on the condition of beech (*Fagus sylvatica* L.) in Britain in 1990. *Forestry*, 65: 35–60.

Ishihara, M.I. and Kikuzawa, K. (2009) Annual and spatial variation in shoot demography associated with masting in *Betula grossa*: comparison between mature trees and saplings. *Annals of Botany*, 104: 1195–1205.

Jackson, B.E., Wright, R.D. and Alley, M.M. (2009). Comparison of fertilizer nitrogen availability, nitrogen immobilization, substrate carbon dioxide efflux, and nutrient leaching in peat-lite, pine bark, and pine tree substrates. *HortScience*, 44: 781–790.

Janhäll, S. (2015) Review on urban vegetation and particle air pollution: Deposition and dispersion. *Atmospheric Environment*, 105: 130–137.

Jarni, K., Jakše, J. and Brus, R. (2015) Vegetative propagation: Linear barriers and somatic mutation affect the genetic structure of a *Prunus avium* L. stand. *Forestry*, 88: 612–621.

Jeong, J.-H., Ho, C.-H., Linderholm, H.W., Jeong, S.-J., Chen, D. and Choi, Y.-S. (2011) Impact of urban warming on earlier spring flowering in Korea. *International Journal of Climatology*, 31: 1488–1497.

Johnston, M. and Hirons, A. (2014) Urban trees. In: Dixon, G. and Aldous, D. (eds) *Horticulture: Plants for People and Places*, Volume 2. Springer, Dordrecht, The Netherlands, pp. 693–711.

Jones, A.T., Hayes, M.J. and Sackville Hamilton, N.R. (2001) The effect of provenance on the performance of *Crataegus monogyna* in hedges. *Journal of Applied Ecology*, 38: 952–962.

Khurana, E. and Singh, J.S. (2001) Ecology of tree seed and seedlings: Implications for tropical forest conservation and restoration. *Current Science*, 80: 748–757.

King, V.J. and Davis, C. (2007) A case study of urban heat island in the Carolinas. *Environmental Hazards*, 7: 353–359.

Klein, A.-M., Hendrix, S.D., Clough, Y., Scofield, A. and Kremen, C. (2015) Interacting effects of pollination, water and nutrients on fruit tree performance. *Plant Biology*, 17: 201–208.

Ko, Y., Lee, J.H., McPherson, E.G. and Roman, L.A. (2015) Factors affecting long-term mortality of residential shade trees: Evidence from Sacramento, California. *Urban Forestry and Urban Greening*, 14: 500–507.

Kock, H., Aird, P., Ambrose, J. and Waldon, G. (2008) *Growing Trees from Seed: A Practical Guide to Growing Native Trees, Vines and Shrubs*. Firefly Books, Buffalo, NY, USA.

Kozlowski, T.T. and Pallardy, S.G. (1997) *Growth Control in Woody Plants*. Elsevier, Amsterdam, Netherlands.

Kozlowski, T.T. and Pallardy, S.G. (2002) Acclimation and adaptive responses of woody plants to environmental stresses. *Botanical Review*, 68: 270–334.

Kreyling, J., Buhk, C., Backhaus, S., Hallinger, M., Huber, G., Huber, L., *et al.* (2014) Local adaptations to frost in marginal and central populations of the dominant forest tree *Fagus sylvatica* L. as affected by temperature and extreme drought in common garden experiments. *Ecology and Evolution*, 4: 594–605.

Lewis, D.L., Baruch-Mordo, S., Wilson, K.R., Breck, S.W., Mao, J.S. and Broderick, J. (2015) Foraging ecology of black bears in urban environments: guidance for human-bear conflict mitigation. *Ecosphere*, 6: article141.

Lyytimäki, J. (2017) Disservices of urban trees. In: Ferrini, F., Van Den Bosch, C.C.K. and Fini, A. (eds.) *Routledge Handbook of Urban Forestry*. Routledge. Abingdon, UK, pp.164–176.

Mariotti, B., Maltoni, A., Jacobs, D.F. and Tani, A. (2015) Container effects on growth and biomass allocation in *Quercus robur* and *Juglans regia* seedlings. *Scandinavian Journal of Forest Research*, 30: 401–415.

Marty, C. and BassiriRad, H. (2014) Seed germination and rising atmospheric CO_2 concentration: a meta-analysis of parental and direct effects. *New Phytologist*, 202: 401–414.

Massetti, L., Petralli, M. and Orlandini, S. (2015) The effect of urban morphology on *Tilia × europaea* flowering. *Urban Forestry and Urban Greening*, 14: 187–193.

Mather, K. (1973) *Genetical Structure of Populations*. Chapman and Hall, London, UK.

McPherson, E.G., Nowak, D., Heisler, G., Grimmond, S., Souch, C., Grant, R., *et al.* (1997) Quantifying urban forest structure, function and value: The Chicago Urban Forest Climate Project. *Urban Ecosystem*, 1: 49–61.

Milberg, P., Andersson, L. and Thompson, K. (2000) Large-seeded species are less dependent on light for germination than small-seeded ones. *Seed Science Research*, 10: 99–104.

Mitchell, A. (1974) *Trees of Britain and Northern Europe*. HarperCollins, London, UK.

Mullaney, J., Lucke, T. and Trueman, S.J. (2015) A review of benefits and challenges in growing street trees in paved urban environments. *Landscape and Urban Planning*, 134: 157–166.

Norton, B.A. Coutts, A.M., Livesley, S.J., Harris, R.J., Hunter, A.M. and Williams, N.S.G. (2015) Planning for cooler cities: A framework to prioritise green infrastructure to mitigate high temperatures in urban landscapes. *Landscape and Urban Planning*, 134: 127–138.

Nowak, D., Crane, D. and Stevens, J. (2006) Air pollution removal by urban trees and shrubs in the United States. *Urban Forestry and Urban Greening*, 4: 115–123.

Oliet, J.A. and Jacobs, D.F. (2007) Microclimatic conditions and plant morpho-physiological development within a tree shelter environment during establishment of *Quercus ilex* seedlings. *Agricultural and Forest Meteorology*, 144: 58–72.

Pallardy, S.G. (2008) *Physiology of Woody Plants*, 3rd edition. Academic Press, San Diego, CA, USA.

Percival, G.C., Gklavakis, E. and Noviss, K. (2009) The influence of pure mulches on survival, growth and vitality of containerised and field planted trees. *Journal of Environmental Horticulture*, 27: 200–206.

Picket, S.T.A., Cadenasso, M.L. and Meiners, S.J. (2013) Vegetation dynamics. In: van der Maarel, E. and Franklin, J. (eds) *Vegetation Ecology*. Wiley-Blackwell, Chichester, UK, pp. 107–140.

Pigott, C.D. and Huntley, J.P. (1981) Factors controlling the distribution of *Tilia cordata* at the northern limits of its geographical range. III nature and causes of seed sterility. *New Phytologist*, 87: 817–839.

Pitman, S.D., Daniels, C.B. and Ely, M.E. (2015) Green infrastructure as life support: Urban nature and climate change. *Transactions of the Royal Society of South Australia*, 139: 97–112.

Pouyat, R.V., Szlavecz, K., Yesilonis, I.D., Groffman, P.M. and Schwarz, K. (2010) Chemical, physical and biological characteristics of urban soils. In: Aitkenhead-Peterson, J. and

Volder, A. (eds) *Urban Ecosystem Ecology*. Agronomy Monograph American Society of Agronomy, Crop Science Society of America, Soil Science Society of America, Madison, WI, USA, pp. 119–152.

Ramírez, F. and Davenport, T.L. (2013) Apple pollination: A review. *Scientia Horticulturae*, 162: 188–203.

Roman, L.A., Battles, J.J. and McBride, J.R. (2014) Determinants of establishment survival for residential trees in Sacramento County, CA. *Landscape and Urban Planning*, 129: 22–31.

Roy, S., Byrne, J. and Pickering, C. (2012) A systematic quantitative review of urban tree benefits, costs, and assessment methods across cities in different climatic zones. *Urban Forestry and Urban Greening*, 11: 351–363.

Sackville Hamilton, N.R. (2001) Is local provenance important in habitat creation? A reply. *Journal of Applied Ecology*, 38: 1374–1376.

Sellmer, J.C. and Kuhns, L.J. (2007) Guide to selecting and specifying nursery stock. In: Kuser, J. (ed.) *Urban and Community Forestry in the Northeast*, 2nd edition. Springer, Dordrecht, The Netherlands, pp. 199–219

Shigo, A.L. (1991) *Modern Arboriculture*. Shigo and Trees Associates, Durham, NC, USA.

Sjöman, H. (2012) Trees for tough urban sites. Doctoral Thesis, Swedish University of Agricultural Sciences, Alnarp, Sweden.

Sjöman, H., Gunnarsson, A., Pauleit, S. and Bothmer, R. (2012) Selection approach of urban trees for inner city environments: learning from nature. *Arboriculture and Urban Forestry*, 38: 194–204.

Sjöman, H., Hirons, A.D. and Bassuk, N.L. (2015) Urban forest resilience through tree selection: Variation in drought tolerance in *Acer*. *Urban Forestry and Urban Greening*, 14: 858–865.

Sjöman, H., Nielsen, A.B. and Oprea, A. (2012a) Trees for urban environments in northern parts of Central Europe: A dendroecological study in north-east Romania and Republic of Moldavia. *Urban Ecosystem*, 15: 267–281.

Sjöman, H., Nielsen, A.B., Pauleit, S. and Olsson, M. (2010) Habitat studies identifying potential trees for urban paved environments: A case study from Qinling Mt., China. *Arboriculture and Urban Forestry*, 36: 261–271.

Sjöman, H., Östberg, J. and Bühler, O. (2012b) Diversity and distribution of the urban tree population in ten major Nordic cities. *Urban Forestry and Urban Greening*, 11: 31–39.

Skomarkova, M.V., Vaganov, E.A., Mund, M., Knohl, A., Linke, P., Boerner, A., *et al.* (2006) Inter-annual and seasonal variability of radial growth, wood density and carbon isotope ratios in tree rings of beech (*Fagus sylvatica*) growing in Germany and Italy. *Trees*, 20: 571–586.

Tal, O. (2011) Flowering phenological pattern in crowns of four temperate deciduous tree species and its reproductive implications. *Plant Biology*, 13 (Suppl. 1): 62–70.

Thomas, P.A. (2014) *Trees: Their Natural History*, 2nd edition. Cambridge University Press, Cambridge, UK.

Thomas, P.A. (2016) Biological Flora of the British Isles: *Fraxinus excelsior*. *Journal of Ecology*, 104: 1158–1209.

Trout, R. and Brunt, A. (2014) *Protection of Trees from Mammal Damage*. BPG Note 12, Best Practice Guidance for Land Regeneration, Forest Research, Farnham, UK.

Tuley, G. (1985) The growth of young oak trees in shelters. *Forestry*, 58: 181–194.

Tyrväinen, L., Makinen, L. and Schipperijn, J. (2007) Tools for mapping social values for urban woodlands and of other green spaces. *Landscape and Urban Planning*, 79: 5–19.

Urban, J. (2008) *Up By Roots: Healthy Soils and Trees in the Built Environment.* International Society of Arboriculture, Champaign, IL, USA.

Walck, J.L., Hidayati, S.N., Dixon, K.W., Thompson, K. and Poschlod, P. (2011) Climate change and plant regeneration from seed. *Global Change Biology*, 17: 2145–2161.

Vander Mijnsbrugge, K., Bischoff, A. and Smith, B. (2010) A question of origin: Where and how to collect seed for ecological restoration. *Basic and Applied Ecology*, 11: 300–311.

Watson, G.W. and Himelick, E.B. (1982) Root distribution of nursery trees and its relationship to transplanting success. *Journal of Arboriculture*, 8: 225–229.

Watson, G.W. and Himelick, E.B. (2013) *The Practical Science of Planting Trees.* International Society of Arboriculture, Champaign, France.

Wolf, K.L. and Robbins, A.S.T. (2015) Metro nature, environmental health, and economic value. *Environmental Health Perspectives*, 123: 390–398.

von Döhrena, P. and Haase, D. (2015) Ecosystem disservices research: A review of the state of the art with a focus on cities. *Ecological Indicators*, 52: 490–497.

von Hoffen, L.P. and Säumel, I. (2014) Orchards for edible cities: Cadmium and lead content in nuts, berries, pome and stone fruits harvested within the inner city neighbourhoods in Berlin, Germany. *Ecotoxicology and Environmental Safety*, 101: 233–239.

Yang, Y., McBride, J., Zhou, J. and Sun, Z. (2005) The urban forest in Beijing and its role in air pollution reduction. *Urban Forestry and Urban Greening*, 3: 65–78.

6

Tree Water Relations

Water is Fundamental to Tree Development

Providing temperatures are suitable for growth, water is the factor that most constrains the development and growth of all plants, including trees. Consequently, the availability of water is critically important in determining their relative success in different environments. This is not surprising when you consider that water is not only a major constituent of plants, but it is involved with almost every physiological process.

Non-woody plant parts are made up of 70–95% water; even wood (when fresh) is made up of about 50% water. Water held with the cells maintains the stiffness of the cell (cell turgor) and provides the substrate for biological activity, including key processes such as photosynthesis. It provides the solvent in which gases, minerals and other compounds can be transported from cell to cell or over longer distances between different parts of the tree. Indeed, growth can only occur if the positive turgor pressure (the internal pressure of cells) achieved in well-hydrated cells provides a driving force for cellular enlargement.

In addition to the large amount of water held in a tree, water is required in huge quantities just for the tree to function. Plants are able to incorporate the vast majority of absorbed minerals, such as nitrogen, phosphorous and potassium, into new tissues, but only a tiny fraction (1–5%) of water that enters the tree is retained in biomass. Most of the water taken in by the tree will be lost back to the atmosphere by transpiration (the evaporation of water from plant surfaces). This apparent profligacy in water use is an unavoidable consequence of photosynthesis. Stomata in the leaf must be open to enable access to carbon dioxide from the atmosphere, but in so doing they provide a gateway for water to be lost from the leaf. However, this water should not simply be seen as a waste because the evaporation of water provides the pulling force that draws water and minerals up the tree from the soil. Regardless of how you look at it, trees need large volumes of water (see Chapter 4). Consequently, understanding how trees maintain their water supply and respond to variable water availability is of central importance to all those managing trees or seeking to understand how climate and environment affect tree performance.

Applied Tree Biology, First Edition. Andrew D. Hirons and Peter A. Thomas.
© 2018 John Wiley & Sons Ltd. Published 2018 by John Wiley & Sons Ltd.

Importance of Water Potential

A hugely valuable, unifying concept used to describe the status of water in the soil, plant and atmosphere is that of *water potential*.

Box 6.1 gives the technical definition of water potential but, more simply, water potential can be thought of as a pressure difference, with water moving from a place with a higher pressure to a place where it is lower. In plants, such movement is often in response to a suction (referred to as tension), and because suction is below atmospheric pressure, it has a negative value. In this case water will move towards the place with the greatest suction (i.e. the most negative pressure).

In most circumstances, water will move down this *water potential gradient*. Therefore, by assessing the differences in water potential between different parts of the soil–plant–atmosphere continuum (SPAC), it is possible to predict the direction water will move in. For example, water will be released from the soil to the root if the water potential of the root (e.g. −0.1 MPa) is lower than the soil water potential (e.g. −0.01 MPa): the water moves towards the more negative pressure of −0.1 MPa. The water will move from the root (−0.1 MPa) to the shoot (e.g. −1.5 MPa) as long as the shoot water potential is lower (more negative) than the root water potential (see Ascent of Sap from Roots to Shoots).

Potential is written as the Greek letter psi Ψ with a subscript letter to indicate what sort of potential it is. Whilst water potential (Ψ_w) inside any part of a tree is often presented as a single value, it is actually made up of a series of other potentials: *osmotic potential* (Ψ_π), *pressure potential* (Ψ_p) and *gravitational potential* (Ψ_g).

The osmotic potential is always negative, as it is a measure of the amount of substances dissolved in the water (technically called *solutes*) held in the solution that act to suck water towards them; the higher the solute concentration, the lower (more negative) the osmotic potential. The pressure potential is derived from the positive pressure inside cells caused by water pressing the cell membrane against the internal cell walls (turgor pressure), or from the tension (negative pressure) caused by evaporation of water. The gravitational potential is important in tall trees but, as it only varies by 0.1 MPa per 10 m in height, it is often ignored in plants that are not very tall.

Soils also have their own water potentials. In soils, the *matric potential* (Ψ_m) describes how tightly the water is held by the soil particles, and is the most critical component of soil water potential. The matric potential always has a negative value, as the forces at

---

Box 6.1 Technical Definition of Water Potential

Water potential is derived from a calculation of the chemical potential of water in a particular part of the system. Whilst this is measured in joules per mole of water compared with pure free water at atmospheric pressure and a temperature of 298 Kelvin (~25 °C), this value is converted to pressure units, normally megapascals (MPa). The exact derivation of these units is quite complex; interested readers can consult Kramer and Boyer (1995) or Jones (2013) for a comprehensive explanation.

For ease of conversion, 0.1 MPa is equivalent to 1 bar, which is in turn roughly equivalent to 1 atmospheric pressure.

play tend to want to hold water on to the soil particles or colloids. The overall soil water potential (the matric potential plus the sum of the other potentials) may be slightly negative or even positive.

In saline soils, the osmotic potential of the soil solution acts to further reduce the soil water potential (it becomes more negative), making it harder for the roots to access water at a given soil water content. This is exacerbated as the soil dries and the salts become more concentrated (the osmotic potential becomes even more negative). Once the soil water potential becomes lower than the root water potential, then conditions exist whereby water may be drawn out of the root and into the soil (reverse osmosis). Although this is not a major problem in moist, humid environments that experience plenty of rainfall, spray (or run-off) from salt-treated roads can result in saline soils along roadsides. If these salts are not adequately flushed through the soil profile, the osmotic effect can seriously inhibit water uptake during the growth season, even in apparently well-watered conditions. Incorrect use of fertilisers can also create soils with low osmotic potentials that can limit water uptake.

Trees Experience Soil Water Potential, Not Soil Water Content

It is easy to think that the soil water availability simply depends on the quantity of water in the soil. Of course, this has to be partly true: moist soils provide easier access to water than dry soils. However, soils are very variable in texture, pore size, organic content and compaction (see Chapter 4) and, as a consequence, the total volume of water retained by the soil and the way in which water is released from the soil is very different from soil to soil.

The best way to visualise how soil will release water as it dries is to plot soil water potential (MPa) against the soil water content (typically presented as a volume, m^3 water per m^3 soil, or simply as a percentage of soil volume) in a *soil water release curve* (Figure 6.1a). In a drying soil, water will be available to the plant until some minimum soil water potential threshold is met, often referred to as the *permanent wilting point* (PWP). This relates to the water potential in the plant, where leaves reach their *turgor loss point* (Ψ_{P0}) (i.e. they irreversibly wilt) and are unable to recover. (Note that plants with wilted leaves may recover after watering because they have not reached the *permanent* wilting point.) In agricultural crops, this PWP is widely considered to be −1.5 MPa, but it can be much lower in temperate trees (−2.0 to less than −4.0 MPa) and even lower in some very drought-tolerant trees of the Mediterranean or other arid areas. Water will cease to become accessible by the plant when the soil water potential is lower than the turgor loss point of the species in question. Therefore, the quantity of water that is available to the plant corresponds to the water content between when the soil is full of water at field capacity,[1] and when the soil has little water and the plant reaches the turgor loss point.

---

1 Field capacity is the water content after the soil becomes saturated, minus the water drained away under the influence of gravity. Typically, the water potential of soils at field capacity is between −0.01 and 0.03 MPa.

When you compare two contrasting soil types, it is easy to see that measurement of the soil water content gives you rather limited information on the availability of soil water. For example, in Figure 6.1(a), the field capacity of the sand is at approximately 10% soil water content whilst in the loam, the extra silt, clay and organic matter increases

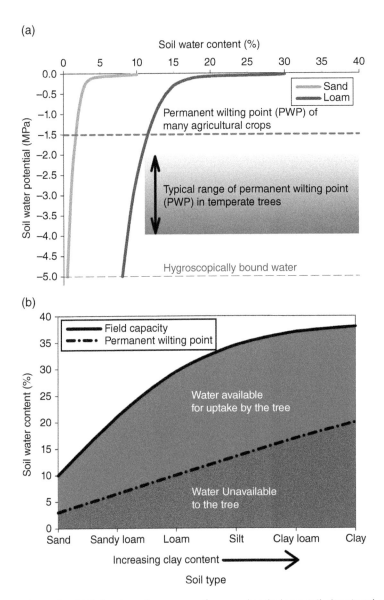

Figure 6.1 (a) Soil water release curves for a sand and a loam soil, showing the typical relationship between the soil water content and the soil water potential (the ease with which a plant can extract water from the soil). The turgor loss point (below which plants cannot grow) for many agricultural crops is taken as −1.5 MPa, and for many temperate trees it ranges between −2 and −4 MPa. At soil water potential below −5 MPa, water is hygroscopically bound to the soil so tightly that it is completely unavailable to plants. (b) A general relationship between soil water content and soil type.

the water retained in the soil to a little over 30%. The rate of decline in soil water potential also differs between the soils as they dry. At 10% soil water, all the water in the sandy soil is available to the plant, whilst the loam at 10% soil water has no water available to roots because its water potential is lower than the PWP: that is, the soil is holding the water too tightly for the plant to be able to remove any. Additionally, the sandy soil may only be able to hold 5% of its volume as available water, whilst the loam may be able to hold 15–20% of its volume as available water. These characteristics can make substantial differences to the volume of soil that trees require and the duration of time that trees can survive without rainfall or irrigation (see Chapter 4 for a discussion of this).

The way in which soil releases water differs widely with soil type, so the first challenge in managing soil water is to understand this relationship in the soil with which you are working. The construction of water release curves requires specialist equipment so it may be best to seek expert analysis from a professional laboratory. Once the soil water release curve is established, it is relatively easy to use a soil moisture probe to estimate the soil water content, and therefore predict the soil water potential. However, if a precise water release curve is unavailable, more general relationships between soil water content and soil type can help inform those trying to estimate the amount of water available to plants (Figure 6.1b).

Managing Soil Water Availability

Approaches to managing soil water availability will depend to a great extent on the context. For landscape trees, the first priority should be to ensure that opportunities are taken to minimise any major physical restrictions to root development. Large soil volumes are of very limited value if they are compacted and physically restrict root development (see Chapter 4). It may be that reducing soil compaction will substantially increase the rootable soil volume and therefore the availability of soil water. Equally, the prevention of soil compaction to protect rootable soil volumes can be just as important. Where it is possible to design rooting environments, soil volumes should be maximised within the constraints of other below-ground infrastructure (again, see Chapter 4).

In some situations, it may be possible to increase the water-holding capacity of the soil using amendments. For example, sandy soils are likely to benefit from the addition of organic matter and silt and clay particles. There is evidence that *biochar* from non-woody plants added to sandy soils does a similar job (Basso *et al.* 2012). Other factors important in determining how to manage soil water availability are the scale of the site, the number of trees, the potential value of the trees (or their crops) and, ultimately, the budget available for irrigation.

In recently planted landscape trees, before roots have had time to grow extensively, the small soil volume occupied by roots means that the available water is very rapidly depleted. Therefore, water deficits are a major challenge to tree establishment, even in humid areas with relatively high levels of rainfall. Impermeable surfaces will further compound the problems young landscape trees have in accessing water if rainfall does not adequately recharge the soil water. Supplementary irrigation will almost always be a good thing for young landscape trees, providing the soil does not become waterlogged.

(a)

(b)

Figure 6.2 (a) An irrigation tube being installed in a tree planting pit. (b) A watering bag placed around a recently planted tree in Copenhagen, Denmark. The bag has a porous base that slowly releases water to the root ball and surrounding soil over a number of hours. This helps reduce surface run-off and ensures deeper soil water recharge. Here, a Treegator® bag is being used, but a number of different brands and designs are available.

A number of mechanisms can be used to deliver water to the root-ball. In some tree pit designs, an irrigation pipe may be pre-installed into the planting pit (Figure 6.2a). In these cases it is simply a matter of connecting a water source to this pipe and delivering a prescribed volume of water to the root environment. In other cases, watering bags can be used (Figure 6.2b). These bags typically hold around 75 L which is slowly released over a number of hours to help ensure that the water sinks in, rather than running off across the surface. This will reduce the frequency that trees need to be irrigated. For high value trees, hydrogels (usually super-absorbent polyacrylate) can also be mixed into the soil backfilled around planted trees. There is some evidence that, at least in the short term, they can improve survival and growth (Orikiriza *et al.* 2013), but the extra expense must be factored in. In some more managed landscapes, sprinkler systems may be in place and, of course, a hose-pipe can be used to water a tree. However, it is also important to remember that saturated soils can be just as bad for the tree as dry soils. If managed incorrectly, there are risks associated with all of these methods:

- Irrigation pipes must be expertly installed to ensure that water is delivered to the actual root-ball and not just the surrounding soil.
- Watering bags must not be left around the stem for long periods of time as they cause high levels of moisture around the lower stem (extended use of watering bags may also discourage root development out into a wider soil volume).
- Manual watering can lead to excessive surface run-off, or superficial surface wetting.
- Irrigation systems using timers are not responsive to actual tree water demand.

Figure 6.3 A permeable geotextile barrier reduces soil evaporation and, importantly, weed competition in a field trial based at the Swedish University of Agriculture (SLU), Alnarp, Sweden.

To encourage root development beyond the root-ball, it is better to irrigate with larger volumes of water less often than with smaller volumes more often. Small, frequent irrigation often encourages roots to develop higher in the soil profile, making them more vulnerable to soil drying once irrigation is removed. Larger irrigation volumes recharge the soil water to a greater depth and so encourage root development in a larger volume of soil. In turn, this can slow the impact of tree water deficits during dry periods because the tree roots occupy a greater volume of soil, and deeper soil water is less prone to evaporation or uptake from shallow-rooted herbaceous competitors, such as grass. The use of mulch or a geotextile barrier to limit soil evaporation and competing herbaceous plants is always helpful when trying to establish young trees, particularly with regards to managing available soil water (Figure 6.3).

Commercial operations may need to irrigate trees to produce high-quality plants or profitable crop yields. With increasing pressures on regional water resources, water abstraction rights may only be granted if the grower can demonstrate sustainable water-management practices. Even then, in some dry regions, the water available for irrigation may not be adequate to provide irrigation throughout the year. Precise scheduling of irrigation and periods of *deficit irrigation* (delivering less water than the tree is losing by evapotranspiration[2]) may be necessary to preserve this precious resource.

---

2 Evapotranspiration (ET) is a measure of total plant water use, including *evaporation* from the soil surface over the roots and *transpiration* from within the leaves. Potential evapotranspiration (ET_p) is predicted from environmental variables, such as temperature and humidity, and what is known about the resistance to water loss offered by the plant.

Box 6.2 Types of Soil Moisture Sensors

The two main types of soil moisture sensor are those that measure soil water content by volume and those that measure the availability of soil water to the plant (soil water potential) (see Trees Experience Soil Water Potential, Not Soil Water Content for the distinction). Sensors based on time domain reflectometry (TDR), neutron-attenuation and measurements of soil conductance of a current (based on dielectric properties) will estimate soil volumetric content. If calibrated correctly, these work well over a wide range of soil moisture contents, are reliable and relatively maintenance free. Tensiometers directly measure soil water potential, but have the limitations of being quite labour intensive to maintain and they only work in a narrow range of soil moisture (0 to −0.1 MPa). Porous matrix sensors estimate soil water potential from dielectric properties and can operate across the plant available range, albeit with a small loss in accuracy compared to tensiometers. Regardless of the sensor used, irrigation scheduling decisions should always be made with reference to the soil water potential because this most closely represents the availability of water to the root (see main text).

Although many irrigation decisions still rely on the tree manager's intuition, there are now a number of ways to schedule irrigation more precisely to ensure sustainable practice. These are typically based on feedback from soil moisture sensors, the measurement of plant water status, or on an estimation of evapotranspiration (ET) or potential evapotranspiration (ET_p).

A range of soil moisture sensors are available commercially (Box 6.2). The major limitation of using these to help in irrigation decisions is that the volume of soil measured tends to be very small compared with the soil volume from which roots extract water. Differences in the soil texture, drainage properties and root absorption rate can lead to variable soil moisture over small distances, so sampling in just a few places may mean making an irrigation decision based on atypical soil conditions. More sensors can be used to overcome this, but this is invariably limited by time and cost. In tree nurseries growing many species, the selection of a suitable reference tree under which to monitor soil moisture is particularly challenging because the demands of one type of tree may not be the same for other species or sizes of tree. Nevertheless, when installed and calibrated correctly, soil moisture sensors assist in assessing when to irrigate and how much water to apply. Figure 6.4 shows the various components needed in an irrigation system that uses soil moisture sensors. This system is used for scheduling irrigation for containerised trees but there is no reason why a similar approach could not be used for high value landscape trees.

It is possible to judge the need for watering by looking carefully at a tree. This might be as simple as watching for wilting or could involve measuring water potential, stomatal conductance, sap flow or crown temperature. The main drawback is that not all species behave in the same way. Consequently, it is important that the response of a species to water shortage is known so that a suitable *bioindicator* can be used to help schedule irrigation events. A general disadvantage to using plant-based approaches is that they do not give any information on how much water needs to be added to the soil (Jones 2004).

In some scenarios, it may be possible to use other plants as *biological sensors* to help guide irrigation decisions. For example, the wilting point of a sunflower has been

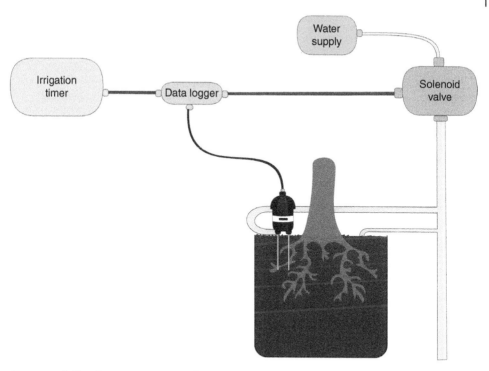

Figure 6.4 Soil moisture sensors pass information on the moisture status of the soil to a data-logger. When this is integrated with an irrigation timer and a solenoid valve, irrigation scheduling is very responsive to the demands of the tree. However, it only delivers feedback from one root system, so it may not fairly represent the needs of other plants on the same irrigation line.

reliably established to be −1.5 MPa. This is very likely to be higher (less negative) than the wilting point of tree species so the wilting of a sunflower planted close to a tree could be used as an early warning that irrigation for the tree may be necessary. This low-tech surrogate sensor for soil drying also has the advantages of being cheap and responsive to local climatic conditions. In gardens with a wide range of species, well-established plants that are known to be sensitive to drying soils can be used to inform irrigation decisions for the rest of the garden. In this way, paying close attention to the indicator species can be an efficient way of guiding decisions for a much larger group of plants. Clearly, some precision is lost using this technique, but no system is perfect and this type of approach is both cheap and useful.

Irrigation may also be scheduled by calculating ET using the standard Penman–Monteith equation (for details see Allen *et al*. 1999). Used appropriately, this approach can give very good information on how much irrigation is required to replace that lost via ET. However, it does rely somewhat on the uniformity of plant material, as large diversity in plant size and species can reduce accuracy. As a result, this technique is most useful for scheduling irrigation in orchards.

Variation across soils, species, tree size and climate make precise recommendations for the irrigation of trees difficult without knowledge of all these factors. However, Table 6.1 gives some important general principles for managing soil water availability.

Table 6.1 Important principles when managing soil water availability for trees.

Principle	Considerations
Maximise the volume of soil that roots have access to	<ul><li>Ensure that potential rooting volume is not compacted</li><li>Provide root paths through other below-ground infrastructure to 'breakout zones' or new volumes of soil</li><li>Prevent future soil compaction</li></ul>
Improve the soil water holding capacity where possible	<ul><li>In very sandy soils, consider adding silt, clay, organic matter or biochar to increase water retention</li><li>Ensure any ameliorants actually increase available soil water and do not lock up soil water</li></ul>
Understand the water release characteristics of the soil	<ul><li>Remember that it is soil water potential that the tree experiences, not soil water content</li><li>A small investment in a laboratory analysis to get a soil water release curve can substantively increase the confidence in irrigation scheduling decisions</li><li>Accuracy in the 0 to −5 MPa range is most important</li></ul>
Reduce soil evaporation and competition from other vegetation	<ul><li>Ensure that water applied to the tree is taken up by the tree, and not lost via evaporation. Therefore, consider the time of day water is applied: avoid irrigating in the middle of the day when evaporative demand is highest</li><li>Organic mulches have numerous benefits, including reducing soil evaporation</li><li>Geotextile membranes can be useful in reducing evaporation from the soil, but they offer no nutritional value and do not readily biodegrade</li><li>Keep a circle around the tree (ideally to the dripline) clear of competing vegetation</li></ul>
Ensure water that is being applied gets to the roots	<ul><li>Pre-installed irrigation infrastructure may hydrate soil beyond the absorbing roots, especially on recently planted trees</li><li>Minimise surface run-off by slowly wetting the soil, rather than delivering high volumes of water very rapidly</li></ul>
Decide on timing of irrigation based on the tree's requirement for water	<ul><li>If the timing of irrigation events is not underpinned by tree physiology, then inefficient water use should be expected</li><li>Irrigation should be responsive to the requirements of the tree, not an arbitrary maintenance schedule</li></ul>
Avoid over-application of water	<ul><li>Saturated soils are very low oxygen soils and can be very damaging to trees without specialist adaptations</li><li>Excessive irrigation can cause the leaching of nutrients and reduce the fertility of the soil</li><li>Water is a precious commodity; use it sustainably</li></ul>
Ensure that the osmotic potential of the soil solution does not hinder root absorption	<ul><li>Excessive salt or fertiliser use can make it much harder for roots to absorb water</li><li>Flushing (leaching) soils is the best way to reduce the affect of saline soils (assuming water is available to do this)</li></ul>

Fine Roots are Critical for Water Absorption

It is the intricately branched fine roots and their root hairs that are responsible for the vast majority of water absorption from the soil. Whilst the larger, woody roots are capable of taking up some water, the resistance to water absorption caused by lignin and suberin impregnated cells is marked. Indeed, part of the role of secondary growth and suberised cell walls is to prevent water being lost from the roots system to a dry soil. It stands to reason therefore that the structures within older roots that inhibit water being leaked back to the soil are incompatible with efficient water absorption. For this reason, the removal or loss of fine roots can have catastrophic consequences for the delivery of water to a transpiring crown. So the preservation, protection and promotion of fine roots should be at the core of any tree health-care programme. Even in soils with plenty of available water, trees must still have sufficient root surface area to absorb enough water to supply it to the crown.

Uptake of water from the soil to the roots occurs along gradients of decreasing water potential. During the day, evaporation from the leaves creates a tension (negative pressure or 'suction') in the column of water extending from the leaf to the root. In turn, this acts to reduce the water potential of the root and water is drawn into the root, providing the soil water potential remains higher (less negative) than that of the root. This mechanism for water uptake is most important in transpiring trees. However, at night and during other periods of very low transpiration, water can still be drawn into the root by osmosis. To maintain this form of uptake, roots produce various osmotically active substances to keep a water potential gradient between the root and the soil. This active process uses energy so is only efficient in well-aerated soils when suitable temperatures exist around the roots. Therefore, warm, well-drained soils provide better conditions for this active uptake of water than cold, waterlogged soils.

The active production of ions and compounds, such as sugars, by the xylem parenchyma also appears to be the source of root pressure that can push water several metres up the tree. For the tree, this positive root pressure, sometimes in combination with stem pressure, is likely to be very useful in making sure that the vessels and/or tracheids start the growth season full of sap and not gas bubbles. It may also be a way of speeding the supply of sugars from the roots to the growing points on the trunk. In some species, particularly within the genera *Acer, Betula, Juglans* and *Ostrya*, this positive pressure within the xylem causes stems to 'bleed' when cut. For this reason, it is best to avoid pruning these trees in spring.

To enter a root, water must move through the outer cortex of the root and through the endodermis (Figure 6.5), before reaching the xylem of the root (more detail on root structure can be found at the beginning of Chapter 4). There are three different pathways that water can move along: water may pass *between* the cells of the cortex (the *apoplast* pathway) before they reach the endodermis; make its way *through* the cells (the *symplast* pathway); or pass through the *transmembrane* pathway (all are explained further in Figure 6.5). Inevitably, the relative importance of these alternative pathways varies somewhat between species, the nature of the driving force for uptake, root maturity and the surrounding soil environment. However, it is clear that roots are able to exert a high degree of control through the active adjustment of cell osmotic potentials, and the use of specialised water channels known as aquaporins. Research is still unravelling the

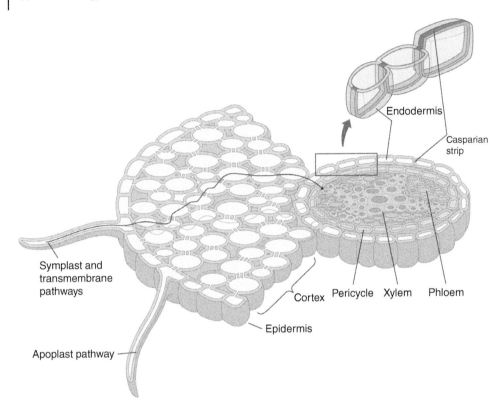

Figure 6.5 Alternative pathways for water and nutrient uptake by the root. In the *symplastic* pathway, water and nutrients cross the plasma membrane into a cell, and so move *through* the cells along the interconnected symplast (the inner surfaces of the cell or plasma membrane) via channels between cells known as plasmodesmata. In the *apoplastic* pathway, water and nutrients move *between* cells by following the gaps between cells and along the outside of cell walls until they reach the endodermis, at which point they must cross a plasma membrane into a cell before it can be taken into the xylem. A further route is known as the *transmembrane* or *transcellular* pathway, where water has to cross two cell membranes as well as the cell wall between two adjacent cells. Specialised water channels, known as *aquaporins*, mediate this transcellular pathway. Regardless of how the water crosses the root cortex, water must pass through the endodermis before entering the xylem. The Casparian strip in the endodermis is a corky, suberised layer that ensures that nothing enters the centre of the root without going through a cell. In this way the root has control over everything that enters. *Source:* Taiz and Zeiger (2010). Reproduced with permission of Oxford University Press.

relative importance of these alternative pathways for water uptake and the resistances that they confer to water movement through the plant. However, whilst these finer questions regarding root water uptake deserve scientific attention, for those trying to manage trees it is the preservation of fine roots and the provision of a high quality root environment that will make the greatest difference to the uptake of water.

For water to be efficiently absorbed into the root system, fine roots must be in contact with moist soil. Well-drained, uncompacted soils help ensure sufficient oxygen around the roots, and also tend to be warmer. Adding mulch can also help buffer temperature extremes, which is of particular importance in spring when night-time temperatures still regularly fall below 5 °C and limit root growth (see Chapter 4).

Hydraulic Redistribution

The tree's coarse roots are important, not just for linking the fine roots to the trunk, but also for the movement of water within the soil and even into the soil. Careful measurement of soil moisture and sap flow has yielded some fascinating insights into the way trees respond to variable water availability. At night, when transpirational demand is negligible and stomata close, the ascent of sap from the roots to the crown is halted. At this point, small remaining water potential gradients between different parts of the root system and between the crown and the roots can induce sap flow. This *hydraulic redistribution* (HR) (Burgess *et al.* 1998) can take place in a number of different forms (Figure 6.6), but it is a vital process for many trees. Importantly, HR not only moves water around the tree, it can also release water back into the soil so that other vegetation and soil organisms can benefit.

Where shallow soil layers are drier than deeper soil layers, water can flow from the deeper roots up to shallow roots via a process termed *hydraulic lift* (HL). *Lateral redistribution* (LR) can also occur where water is moved horizontally through roots found at the same depth but experiencing differing water potentials (water flowing from wetter to drier areas). This may occur naturally in trees at the edge of a group that have part of the root system sheltered under a canopy and other roots in an open environment or it may be brought about by localised irrigation. After rain (or irrigation), when deeper soil is drier than shallow soil layers, *downward hydraulic redistribution* (DHR) can occur to aid the water recharge of deeper soil compartments. In very humid conditions, such as those caused in fog or drizzle where the soil does not really experience rewetting, *foliar uptake* (FU), the absorption of water through the leaves, can move

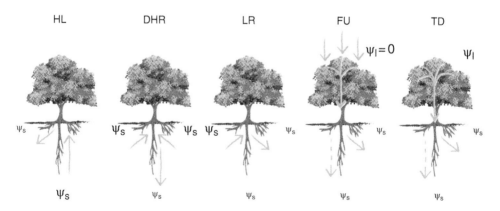

Figure 6.6 Alternative types of hydraulic redistribution in trees. Hydraulic lift (HL) brings water from deeper roots to the shallower roots. Downward hydraulic redistribution (DHR) moves water from shallower roots to deeper roots. Lateral redistribution (LR) moves water horizontally through roots of similar depth. Foliar uptake (FU) occurs when water moves from a very moist atmosphere through the crown and stems to the root system. In very dry circumstances, tissue dehydration (TD) can also occur. Soil and plant water potentials are shown by the symbol Ψ: different sizes of symbol indicate different sizes of water potential, with the bigger symbol representing the higher water potential (i.e. greater moisture). Arrows indicate the direction of water movement and dashed arrows indicate alternative pathways for water movement. *Source:* Adapted from Prieto *et al.* (2012). Reproduced with permission of John Wiley and Sons.

water from the crown, down the stem and into the roots. At the end of prolonged periods of water deficit, when all soil compartments are dry, the root system acts as a competing sink for the remaining water within the tree, and *tissue dehydration* (TD) of the crown can occur (Nadezhdina *et al.* 2010).

Intuitively, if a tree is rooted into deep soil compartments that remain moist through dry periods, the transpirational demand can simply be supplied from deeper roots. What then are the benefits of this redistribution of water around the root system? First, shallow lateral roots do have a number of advantages over deep roots. They offer lower resistance for water uptake into the crown so the leaves find it easier to draw water up from these shallow roots than they might from deeper roots. Secondly, shallow lateral roots are also much better placed to intercept rainfall than deep roots. Therefore, there is an advantage in keeping the lateral roots alive during dry periods, particularly keeping the fine roots healthy (Bauerle *et al.* 2008). If these shallow roots become dysfunctional through embolism, lose contact with the soil or die, they cannot take up rainfall. As seen in Figure 6.6, HL can also result in the release of water into the surrounding soil environment. This increase in soil water can improve nutrient availability either directly, by nutrients dissolving into the soil solution, or by increasing the activity of soil microorganisms, such as mycorrhizae. Needless to say, increases in water and nutrient availability can lead to a cascade of effects that can improve the performance of individual trees, as well as having larger-scale effects across whole ecological communities (Prieto *et al.* 2012).

During periods of plentiful rainfall, the preservation of lateral roots also means that the tree can recharge deep soil compartments through downward hydraulic redistribution. This effectively locks water away where it is less vulnerable to surface evaporation and competition from other vegetation. Then, during dry periods, the recharged deeper soil compartments release water back to the roots, so that the effects of water deficit are less pronounced and the growing season can be extended. Indeed, in a velvet mesquite *Prosopis velutina* savannah in Arizona, USA, this deep-water recharge during a wet period was able to provide 16–49% of the tree's water requirements throughout the dry season (Scott *et al.* 2008).

Lateral redistribution is unlikely to be able to extend the growing season in the same way that DHR can, but it can help areas of the crown survive in trees with a highly sectored vascular system (see Chapter 2). Here, LR will help maintain the water supply to a larger proportion of the crown when soil moisture availability would otherwise be low in some parts of the rooting zone.

Foliar uptake only occurs when the soil is dry and the atmosphere is saturated by fog or drizzle. The significance of fog is seen in coastal redwood *Sequoia sempervirens* in its natural Californian environment. During the frequent heavy fogs coming off the Pacific Ocean, foliar uptake (where sap flows in the direction of the roots) accounts for 5–7% of the water demanded by the crown. However, it is likely that some of the water taken up via the leaves is used to rehydrate plant tissues near the height limit for water transport up the xylem, thereby providing an important source of water to the crown but not a substantial source of water to the roots and soil (Burgess and Dawson 2004). Extra water can be acquired, however, without it entering the leaves. Intercepted fog will drip off the leaves and stem to reach the soil. In coastal redwood forests, this accounts for around one-third of the annual water input into the forest (Dawson 1998). In other coastal environments, such as the laurel forests of Tenerife in the Canary Islands, fog collection supplies up to 20 times that received from rainfall (Thomas 2014).

Whilst it is clear that HR is an important process, particularly in regions that experience extended periods of drought, it is difficult to manipulate through land management. Planting species known for HR certainly has long-term merit but, as with so many ecosystem services provided by trees, their greatest contribution is only seen when they reach maturity. Indeed, the fact that HR occurs most effectively in mature root systems underscores the importance of protecting mature trees in the landscape, especially in water limited environments where HR is likely to be an important component of the local hydrological cycle.

Understandably, the significance of HR to trees growing in urban environments has not been investigated to the same extent as to those growing in natural environments but, given that soil water is likely to be very variable under urban trees as hard landscapes and soil compaction alter infiltration and drainage, urban tree roots will certainly experience variation in soil moisture and water potential gradients across their root-zone. HR is likely to be a feature of urban trees and may well be critical to the survival of some individuals: it is easy to see how this strategy could be used to take advantage of a water leak from a damaged pipe (providing there is no waterlogging). Thus, it is preferable to design root paths (trenches) from areas with impermeable surfaces to areas of open ground (break-out zones), to provide future opportunities for HR. Roots growing in these break-out zones can then move water and nutrients to portions of the root system in less favourable conditions, and the whole tree is likely to fare better.

Ascent of Sap from Roots to Shoots

The ascent of sap within trees has intrigued scientists for centuries: just how do trees manage to move water up over 100 m in height? If plants had only managed to reach a few metres in height, then it might be possible to explain water movement as capillary rise in the very narrow xylem conduits, or the positive force caused by root pressure. However, these forces cannot explain water movement through trees that are tens of metres high.

The origins of what is now referred to as the *cohesion–tension* (CT) *theory* can be traced back hundreds of years to the insight provided by an English clergyman, Stephen Hales (1677–1761), who suggested in his book *Vegetable Staticks* (Hales 1727) that 'sap…is probably carried up to great heights in those vessels by the vigorous undulations of the sun's warmth'. Although our understanding is much more developed now, Hales was essentially right that evaporation of water from the leaves provides the driving force for the ascent of sap. It is, perhaps, too generous to give Hales the credit for what we now understand as the CT theory, this is usually reserved for Dixon and Joly (1895), but a number of scientists have been involved in its refinement since (see Brown 2013). Perhaps the most complete review can be found in Tyree and Zimmermann (2002).

In simple terms, water evaporates from inside the leaf (transpiration), creating a *tension* (negative pressure) in the mesophyll cells inside the leaf. Put another way, a water potential gradient is generated between the moist cells inside the leaf and the comparatively dry air outside, causing water loss. Strong *cohesive* forces hold the water molecules together, helped by adhesive forces between the water molecules and the cell walls, so the tension acts to pull water towards the drier cells from adjacent wetter cells. This tension is transmitted through the mesophyll, into the xylem, and all the way down

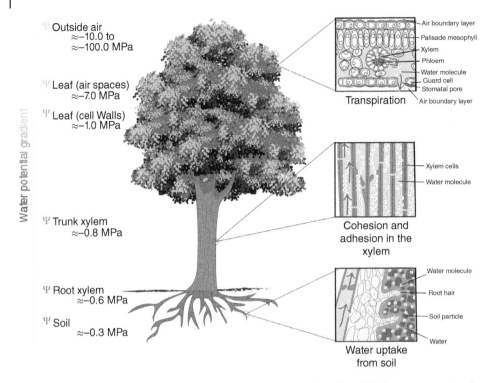

Figure 6.7 Sap moves up through a tree down a water potential gradient (Ψ): the more negative the value, the greater the suction or tension. Water is lost from the leaves to the relatively dry air which generates tension within a continuous column of sap that extends from the leaves to the roots, which then acts to pull water up through the xylem. The reduction in the amount of water in the roots (the reduced root water potential) passively draws water into the root and causes soil water to move towards the root down a water potential gradient. Water potential values shown on the left are indicative only.

a continuous column of sap held within the tracheids and/or vessels. Cohesive forces within the sap act to pull sap up from the roots to replace that lost from the leaves. In turn, this action reduces the amount of water inside the roots, reducing the water potential and thus drawing soil water into the root. Subsequently, water moves towards the root down a gradient of decreasing water potential within the soil (Figure 6.7). The continuous system of water from the evaporating surfaces in the leaves to the absorbing surfaces of the roots is known as the soil-plant-air-continuum (SPAC).

One of the most remarkable things about the ascent of sap is that the whole process simply relies on physics: it does not require any energy from the tree to lift sap from the deepest roots to its uppermost leaves; if it did, the energy demands of getting water to any height would have prevented trees from ever evolving. Sap is pulled up the tree by the evaporation of water from the leaf (causing tension), the incredible tensile strength of water (from cohesion) and the extraordinary ability of wood (xylem) to withstand these forces. However, this is not without its limits. Where the supply of water is not able to keep pace with transpiration demands, tension within the sap becomes ever greater. Eventually, under high tension, water columns will break and gases will be drawn into the tracheid or vessel via a pit in the cell wall; embolism will occur (see

Chapter 2). Embolism can disrupt the water supply to the leaves and so may lead to hydraulic failure, so numerous mechanisms and adaptations have evolved to help prevent this from occurring (again, see Chapter 2).

Transpiration

More than 95% of the water ascending the tree is lost via transpiration (McElrone *et al.* 2013). This is the process by which water evaporates from the plant and moves to the atmosphere. While this may seem very wasteful, the tree has little choice. In order to allow carbon dioxide and oxygen to diffuse in and out, the tree has to be 'leaky', a key side effect of which is loss of water. On the positive side, this transpiration stream helps cool heated leaves and is one of the main ways of delivering minerals that are dissolved in the water to the growing points of the tree. However, trees have ways of regulating this loss.

Although water can evaporate from any internal surfaces that come into contact with the air, and also the entire outer surface of the plant (hence, trees still lose some water when they have no leaves; look back to Figure 4.28), most of the water lost from the plant is lost via the leaves. Water evaporates from the moist internal surfaces of the leaf, and so the air spaces in the leaf mesophyll contain a higher concentration of water vapour relative to the dry air surrounding the plant. Thus, a *vapour concentration gradient* between the interior of the leaf and the outside air causes water vapour to move, via diffusion, from the inside of the leaf to a *boundary layer* of unstirred air surrounding the leaf, and then into the atmosphere. The air inside the leaf is typically saturated with water vapour, and the air outside the leaf contains less vapour, so the magnitude of this gradient in vapour concentration is described by its *vapour pressure deficit* (VPD).

However, it is more practical to think of the difference in vapour concentration as a difference in water potential. This can then be readily compared with other measures of water potential in the tree (e.g. shoot or leaf water potential). The conversion is quite complex (see Nobel 2009) as it depends upon relative humidity and temperature, but a range of air water potentials are given in Table 6.2. What should be clear is that in all but the most humid atmospheres (i.e. close to 100% relative humidity), the water potential of the air is extremely low and drives water loss from the leaves.

Resistance to Water Loss

Transpiration is essentially a process of evaporation, but there is substantial resistance to evaporation provided by the leaves. Leaf cuticles with their embedded waxes provide the greatest resistance to water loss. This resistance is not apparently related to the thickness of the cuticle, as commonly assumed (Kerstiens 1996): thicker cuticles do not lead to reduced water loss. It is therefore likely that the chemical composition of the waxes have a dominant role in regulating cuticular water loss. The boundary layer of very still air surrounding the leaf also provides some resistance to water loss because the water vapour has to diffuse further to escape, but this varies quite substantially with leaf size and wind speed. In fact, unless the leaf has dense leaf hairs (or similar) that increase

Table 6.2 The water potential of the air (Ψ_{air}) at different levels of relative humidity (%) and temperature (°C). Relative humidity describes the degree of saturation in the air as a percentage of the maximum possible saturation at a given temperature. The more negative the water potential, the greater the forces leading to evaporation of water from inside the leaves.

Relative humidity (%)	Ψ_{air} (MPa) at different temperatures (°C)				
	10	15	20	25	30
100	0	0	0	0	0
99.5	−0.65	−0.67	−0.68	−0.69	−0.70
99	−1.31	−1.33	−1.36	−1.38	−1.40
98	−2.64	−2.68	−2.73	−2.77	−2.81
95	−6.69	−6.81	−6.92	−7.04	−7.14
90	−13.75	−13.99	−14.22	−14.45	−14.66
80	−29.13	−29.63	−30.11	−30.61	−31.06
70	−46.56	−47.36	−48.14	−48.94	−49.65
50	−90.50	−92.04	−93.55	−95.11	−96.50
30	−157.2	−159.9	−162.5	−165.2	−167.6
10	−300.6	−305.8	−310.8	−316.0	−320.6

Source: Lambers et al. (2008). Reproduced with permission of Springer.

the boundary layer effect, as the wind speed rises above $2\,\text{m}\,\text{s}^{-1}$ (7.2 km per hour/4.5 mph) the boundary layer resistance becomes negligible. For this reason, open-grown trees, those growing in exposed locations and those growing in wind tunnels caused by urban canyons are likely to have lower resistance to water loss than those growing in more sheltered environments. Neither the cuticle nor the boundary layer resistances provide the tree with any short-term control over water loss: this role is left to the stomata.

Stomata, tiny pores in the leaf surface, provide the gateway between the internal leaf environment (including those moist evaporating surfaces) and the atmosphere. Stomata have two guard cells that act as doors controlling the size of the pore. When they are closed, stomata are quite resistant to water loss; when they are open they conduct water. This can be measured as *stomatal conductance*.[3] However, as noted above, the main role of stomata is to allow carbon dioxide into the leaf. Consequently, it is almost as if stomata have two masters: one telling them to make sure they do not lose too much water; the other telling them to make sure that they do not run short of that all important raw ingredient for photosynthesis. They have to walk the metaphorical tightrope, so it should come as no surprise that the regulation of their opening is complex.

When water is abundant, the advantage of maintaining a good supply of carbon dioxide for photosynthesis is greater than the disadvantage of losing water from the leaves. The cooling effect of transpiration can be vital in preventing leaf temperatures from reaching damaging levels. Therefore, during the day when light is available for

---

3 Stomatal conductance is a measure of water loss through the stomata, typically measured in millimoles of water lost per square metre of leaf area, per second ($\text{mmol}\,\text{m}^{-2}\,\text{s}^{-1}$).

Figure 6.8 Daily changes in transpiration with decreasing soil moisture (curves 1–5). The dotted line indicates potential evaporation, arrows indicate stomatal movements and the green area shows where transpiration is only through the cuticle. (1) Unrestricted transpiration; (2) limitation to transpiration in the middle of the day; (3) full closure of stomata at midday; (4) complete cessation of stomatal transpiration by persistent closure of stomata (only cuticular transpiration continues); (5) even further reduced cuticular transpiration as the result of membrane shrinkage. *Source:* Adapted from Stocker (1956). Reproduced with permission of Springer.

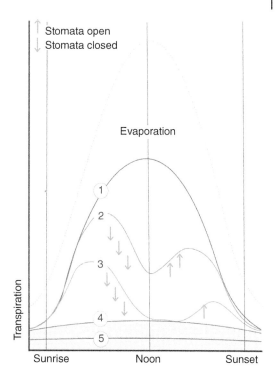

photosynthesis, the stomata tend to open. If minor water deficits develop they often partially close around midday to prevent excessive water loss, then open again to allow photosynthesis in the afternoon. At night, when there is no photosynthesis, the stomata tend to close to prevent unnecessary water loss (Figure 6.8). However, some night-time water loss may still occur, particularly on warm, dry nights. For example, night-time flow of water through the trunk of coastal redwood *Sequoia sempervirens* can be 10–40% that of the daytime flow (Dawson *et al.* 2007). This might actually be really useful to provide water and nutrients to parts of the tree that did not receive much sap flow during the day. If night-time transpiration does not occur, the water potential of the shoot (or leaf) pre-dawn (Ψ_{pd}) provides a good surrogate measurement for the soil water potential experienced by the tree as, in the absence of transpiration, these water potentials equilibrate. A decline in Ψ_{pd} is therefore good evidence of soil drying across a substantial portion of the root system.

The real challenge for stomata comes when these two 'masters' start competing in those situations where water supply cannot keep up with water demand. Here, prodigious water loss must be controlled if leaf dehydration and lasting damage to the tree is to be avoided.

To control against dehydration, stomata close in response to a wide range of variables so that they are able to provide a compromise between carbon gain (photosynthesis) and water loss. Roots provide hydraulic signals as their ability to supply water is diminished and the root water potential declines (Kramer and Boyer 1995). They can also provide chemical signals, such as the hormone abscisic acid (ABA), which is produced when the root experiences drying soil and transported via the sap to the leaves where it promotes stomatal closure (Davies and Zhang 1991). A range of variables, therefore, have the potential to impact stomatal aperture, but it seems that the hydraulic factors are usually dominant in forest trees (Augé *et al.* 2000).

Coordination between all the controlling variables helps to maintain the *water balance* of the tree. Only if the rates of water uptake, conduction and loss are adjusted to each other can a satisfactory water balance be maintained. Indeed, the difference between absorption and transpiration, measured over a given interval of time, gives a good idea of how well the water balance is maintained. The balance becomes negative as soon as the absorption of water is unable to meet the requirements of transpiration. If the stomata partially close and the rate of absorption remains unchanged, the balance can be restored.

During the day, the water balance almost always becomes negative as the supply of water struggles to match the demand from transpiration. The balance is restored in the evening or overnight, providing there is sufficient water in the soil. If soil water is not replenished by rainfall or irrigation, the water balance of the tree may not entirely recover overnight, so that the deficit accumulates from day to day. Inevitably, if the supply of water continues to fall behind the transpirational demand, then serious water deficits can develop. Ultimately, leaves may wilt, embolism may become widespread in the xylem and hydraulic failure can lead to the tree dying.

Species, tree size, rooting environment and climate can all have a profound effect on the volume of water a tree uses. Although this makes estimating the water use of trees complex, understanding the volume of water trees use can help answer important questions relating to forest hydrology, as well as the soil volumes required to support landscape trees. Chapter 4 discusses this in more detail.

References

Allen, R.G., Pereira, L.S., Raes, D. and Smith, M. (1999) *FAO Irrigation and Drainage Paper 56: Crop Evapotranspiration – Guidelines for computing crop water requirements.* Food and Agriculture Organisation, Rome, Italy.

Augé, R.M., Green, C.D., Stodola, A.J.W., Saxton, A.M., Olinick, J.B. and Evans, R.M. (2000) Correlations of stomatal conductance with hydraulic and chemical factors in several deciduous tree species in a natural habitat. *New Phytologist*, 145: 483–500.

Basso, A.S., Miguez, F.E., Laird, D.A., Horton, R. and Westgate, M. (2012) Assessing potential of biochar for increasing water-holding capacity of sandy soils. *Global Change Biology: Bioenergy*, 5: 132–143.

Bauerle, T.L., Richards, J.H., Smart, D.R. and Eissenstat, D.M. (2008) Importance of internal hydraulic redistribution for prolonging the lifespan of roots in dry soil. *Plant Cell and Environment*, 31: 177–186.

Brown, H. (2013) The theory of the rise of sap in trees: some historical and conceptual remarks. *Physics in Perspective*, 15: 320–358.

Burgess S.O., Adams M.A., Turner N.C. and Ong C.K. (1998) The redistribution of soil water by tree root systems. *Oecologia*, 115: 306–311.

Burgess, S.S.O. and Dawson, T.E. (2004) The contribution of fog to the water relations of *Sequoia sempervirens* (D. Don): foliar uptake and prevention of dehydration. *Plant, Cell and Environment*, 27: 1023–1034.

Davies, W.J. and Zhang, J.H. (1991) Root signals and the regulation of growth and development of plants in drying soil. *Annual Review of Plant Physiology and Plant Molecular Biology*, 42: 55–76.

Dawson, T.E. (1998) Fog in the California redwood forest: Ecosystem inputs and use by plants. *Oecologia*, 117: 476–485.

Dawson, T.E., Burgess, S.S.O., Tu, K.P., Oliveira, R.S., Santiago, L.S., Fisher, J.B., *et al.* (2007) Nighttime transpiration in woody plants from contrasting ecosystems. *Tree Physiology*, 27: 561–575.

Dixon, H.H. and Joly, J. (1895) On the ascent of sap. *Philosophical Transactions of the Royal Society of London*, 186: 563–576.

Hales, S. (1727) *Vegetable Staticks*. W. and J. Innys and T. Woodward. London, UK.

Jones, H.G. (2004) Irrigation scheduling: advantages and pitfalls of plant-based methods. *Journal of Experimental Botany*, 55: 2427–2436.

Jones, H.G. (2013) *Plants and Microclimate*, 3rd edition. Cambridge University Press, Cambridge, UK.

Kerstiens, G. (1996) Signalling across the divide: A wider perspective of cuticular structure-function relationships. *Trends in Plant Science*, 1: 125–129.

Kramer, P.J. and Boyer, J.S. (1995) *Water Relations of Plants and Soils*. Academic Press, New York, USA.

Lambers, H., Stuart Chaplin III, F. and Pons, T.L. (2008) *Plant Physiological Ecology*, 2nd edition. Springer, Berlin, Germany.

McElrone, A.J., Choat, B., Gambetta, G.A. and Brodersen, C.R. (2013) Water uptake and transport in vascular plants. *Nature Education Knowledge*, 4: article 6.

Nadezhdina, N., David, T.S., David, J.S., Ferreira, M.I., Dohnal, M., Tesař, M., *et al.* (2010) Trees never rest: the multiple facets of hydraulic redistribution. *Ecohydrology*, 3: 431–444.

Nobel, P.S. (2009) *Physiochemical and Environmental Plant Physiology*, 4th edition. Academic Press, San Diego, CA, USA.

Orikiriza, L.J.B., Agabam, H., Eilu, G., Kabasa, J.D., Worbs, M. and Hüttermann, A. (2013) Effects of hydrogels on tree seedling performance in temperate soils before and after water stress. *Journal of Environmental Protection*, 4: 713–721.

Prieto, I., Armas, C. and Pugnaire, F.I. (2012) Water release through plant roots: new insights into its consequences at the plant and ecosystem level. *New Phytologist*, 193: 830–841.

Scott, R.L., Cable, W.L. and Hultine, K.R. (2008) The ecohydrologic significance of hydraulic redistribution in a semiarid savanna. *Water Resources Research*, 44: article W02440.

Stocker, O. (1956) Die Abhängigkeit der transpiration von den Umweltfaktoren. In *Pflanze und Wasser/Water Relations of Plants*. Springer, Berlin, Germany, pp. 436–488.

Taiz, L. and Zeiger, E. (2010) *Plant Physiology*, 5th edition. Sinauer Associates, Sunderland, USA.

Thomas, P.A. (2014) *Trees: Their Natural History*, 2nd edition. Cambridge University Press, Cambridge, UK.

Tyree, M.T. and Zimmermann, M.H. (2002) *Xylem Structure and the Ascent of Sap*, 2nd edition. Springer, New York, USA.

7

Tree Carbon Relations

Through the process of *photosynthesis*, energy from sunlight (Figure 7.1) is used to combine carbon dioxide and water, to produce carbon compounds (carbohydrates) and release oxygen. Energy held in these carbohydrates can then be used for growth, metabolism (the running costs of the tree), reproduction and defence. This energy is released in a controlled manner through *respiration*. Respiration uses oxygen and carbohydrates while releasing carbon dioxide and water: the reverse of photosynthesis. Readers interested in comprehensive details of photosynthesis and respiration should consult standard plant biology texts, such as Taiz *et al.* (2014) and Mauseth (2016).

The tree produces carbohydrates but tree physiologists often talk about the amount of carbon *fixed* by photosynthesis. The amount of carbon fixed by a tree, minus its running costs (i.e. respiration), gives the net gain in carbon compounds. This *net carbon assimilation* through photosynthesis is essential if the tree is to grow. If carbon gain exceeds carbon losses to respiration, growth is possible. Conversely, if losses of carbohydrates via respiration exceed the supply of carbon from photosynthesis, growth must be funded using stored carbohydrates (if available). Once these are depleted, no growth can occur and any living tissues within the tree will ultimately die through carbon starvation. An increase in tree biomass is therefore a clear expression of a *positive carbon balance*. The size of a tree is also used to evaluate how much carbon has been sequestered or locked up from the atmosphere.

As well as fuelling all metabolic processes within the tree, carbon assimilated by photosynthesis provides the building blocks for key *structural carbohydrates*, such as lignin and cellulose. Although it varies considerably between species, about 40–60% of the dry mass of trees is made up of carbon (Thomas and Martin 2012). Some carbohydrates will also be given to other organisms such as mycorrhizae (see Chapter 9). Any carbon compounds left over from running and building the tree will be stored for later use. The proportion of *non-structural carbohydrates* (NSC)[1] in tree tissues represents carbon reserves or carbon storage that can be used for future growth (Chaplin *et al.* 1990). This potentially mobile pool of NSC is an indicator of a tree's 'fuelling' status as it reflects the balance between depletion and accumulation of carbon within the tree (Körner 2003).

———

1 Non-structural carbohydrates include low molecular weight sugars (e.g. glucose, fructose, sucrose); more complex sugars known as oligosaccharides (e.g. raffinose and stachyose); starch; sugar alcohols (e.g. cyclitols, hexitols and sorbitol); and lipids.

Applied Tree Biology, First Edition. Andrew D. Hirons and Peter A. Thomas.
© 2018 John Wiley & Sons Ltd. Published 2018 by John Wiley & Sons Ltd.

Figure 7.1 Sunlight is crucial to photosynthesis and, hence, tree health and survival. Trees have many ways of collecting as much light as they need; these are covered in this chapter. Here, the light is coming through a gap in the trees in Kakamega Forest Reserve, Kenya.

Of course, a positive carbon balance is the cumulative outcome of almost every aspect of tree morphology, anatomy, physiology and phenology. Roots must acquire resources for photosynthesis, and stems must provide safe passage for materials transported to and from the leaves which, in turn, must be held in a position that intercepts sufficient light. Leaves must balance the uptake of carbon dioxide with water loss. They must time their emergence to maximise the growing season without risking environmental damage, or be constructed to endure the annual fluctuations in temperature and water availability. Differences in morphology, anatomy, physiology and phenology therefore arise from alternative ways of acquiring the same resources to secure carbon profit. For this reason, some species can achieve a positive carbon gain on relatively impoverished sites but, when they are planted on more fertile sites, fail to compete effectively with vegetation adapted to resource-rich sites. Conversely, species used to growing on fertile sites will often perform poorly on more marginal sites because they cannot cope with the environmental constraints imposed on them.

A tree well-adapted to its environment will be able to gain sufficient carbon for the maintenance of its living cells, export to closely associated organisms and storage for future requirements.

Carbon Moves from Source to Sink via the Phloem

The movement of carbohydrates (carbon) occurs from *sources* (suppliers) to *sinks* (users) associated with growth, metabolism, export or storage. During active periods of photosynthesis, the most important source of carbon is from the leaves although, during other periods, carbohydrates stored within the roots or stems form significant sources.

Developing tissues as well as all other living cells across the entire plant body act as sinks because they need carbon to live and grow.

This creates a complex network where thousands of independent sources must supply the myriad of competing sinks within the tree. The exact nature and strength of these sinks will depend on the time of year and the relative demand of particular tissues (and cells) for carbon. For example, at the start of the growth season, roots are often the strongest sink of carbon as they start to grow; later, rapidly growing shoots and the cambia become strong sinks; during flowering, the reproductive structures are important sinks. Symbiotic organisms, such as bacteria or fungi on roots, are also often significant sinks. In addition, the need to allocate resources to defence can occur throughout the year (Kozlowski 1992).

Generally, the sources closest to the sink will supply that particular site. Therefore, in spring, the first leaves to emerge on a shoot will provide carbon for subsequent leaves; foliage on smaller branches will tend to contribute carbon to the larger branches they are attached to; and reproductive structures will be predominantly funded by the branches on which they grow. Nevertheless, long-distance transport of carbon is also possible, when and where needed.

For all but the shortest distances across cells and tissues, the movement or *translocation* of sugars occurs via the phloem (look back at Box 2.1). In both gymnosperms and angiosperms, this occurs through tubes called *sieve elements*. The active[2] loading and unloading of sugars from the sieve elements is controlled by associated *companion cells* (Beck 2010). In this way, the sieve element–companion cell complex collects sugars from the source (e.g. leaf mesophyll), transports it and unloads it at the target sink. Although the actual long distance transport is believed to be passive, this movement is driven by an osmotic gradient between the source and the sink created by the active loading and unloading of sugars from the phloem: the *pressure flow hypothesis* originally proposed by Ernst Munch in 1927 (Mauseth 2016).

Phloem (sometimes called the inner bark) is found just beneath the outer bark so injuries to the bark are very often associated with damage to the phloem. An intact vascular cambium may be able to repair the wound over time, providing it is not too extensive. However, large wounds on the surface of stems, especially in cases of ring-barking (the removal of bark all the way around the stem), will prevent sugars from being transported between sources and sinks on either side of the injury. For example, ring-barking the main stem will prevent sugars moving from the crown to the roots. This will lead to the death of the tree as the roots die from carbon starvation once their reserves are exhausted.

Light and Other Environmental Variables That Influence Photosynthesis

A number of factors can limit photosynthesis and carbon gain. Carbon dioxide in the atmosphere is rarely limiting so the factors most important for controlling photosynthesis are light, temperature, nutrients and water. Here, we only summarise the ways

---

2 'Active' in physiology means that the process requires energy to make it happen. A 'passive' process, by contrast, needs no energy input from the plant.

> ### Box 7.1 Light and Photosynthesis
>
> Although we commonly understand photosynthesis to be driven by sunlight, it is more precise to say that it is driven by solar radiation between 400 and 700 nm in wavelength: referred to as *photosynthetically active radiation* (PAR). In practice, the availability of light is measured by plant biologists as photosynthetic photon flux density (PPFD, μmol m^{-2} s^{-1}), because this is a good measure of the quantity of solar radiation relevant to photosynthesis.
>
> Strictly speaking, PPFD measures the number of photons or quanta arriving at the plant with the units of μmol of quanta per square metre per second (μmol *quanta* m^{-2} s^{-1}), which by convention is abbreviated to μmol m^{-2} s^{-1}. To add confusion, photosynthetic carbon assimilation is measured in μmol *carbon dioxide* per m^2 leaf area per second (μmol CO_2 m^{-2} s^{-1}), which, unfortunately, is also abbreviated to μmol m^{-2} s^{-1}. Care is therefore needed!

these factors affect photosynthesis but readers interested in detailed information should consult Flexas *et al.* (2012b). Of these factors, light is usually the most important (Box 7.1).

Numerous factors affect the proportion of light that reaches tree crowns. Latitude, climate (especially cloudiness), time of day and neighbouring vegetation (or objects) that cast shade on to leaves all affect the quantity or quality of light available for photosynthesis. In forest environments, the canopy modifies the quantity and quality of light for any plants growing in the understorey. Within a leafless temperate deciduous forest, 50–80% of full sunlight reaches the forest floor. When the canopy closes, this is often reduced to less than 10% and, in dense beech *Fagus* forests, can be as little as 0.2–0.4% (Barnes *et al.* 1998). This means that trees growing in the open or in a dominant position within the forest canopy may grow in 100% full sunlight (a PPFD greater than 2000 μmol m^{-2} s^{-1}), while some understorey species may have to cope with less than 0.25% full sunlight (around 5 μmol m^{-2} s^{-1}). Generalisations are difficult to make as light transmission through forests is highly complex (see Chapter 3), but the crown architecture of the dominant species will influence how much light reaches the understorey. On windy days, more light may get through because of the movement of the branches above; on cloudy days, the relative percentage of light reaching the ground is higher than on clear days because the cloudy sky radiates light from all directions. Light quality also changes in the understorey as leaves within the forest canopy absorb much of the red and blue light that is useful for photosynthesis (Valladares *et al.* 2012).

Tree species differ in their precise response to light but all leaves increase their photosynthetic rate with more light, up to a point (Figure 7.2). This *light response curve* provides useful information on the photosynthetic performance of leaves in different light conditions. Light response curves have three main parts: the initial slope of the curve where photosynthesis increases linearly with light; the curved part where photosynthesis starts to saturate with light and slow down; and a plateau region where photosynthesis is light saturated and does not increase any further, no matter how much light is given. In very low light levels, the net photosynthetic rates tend to be negative as leaf respiration uses carbon at a greater rate than it is being assimilated. As light levels begin to increase, net photosynthesis reaches zero when carbon assimilation matches the respiratory demand: this is known as the *light compensation point*.

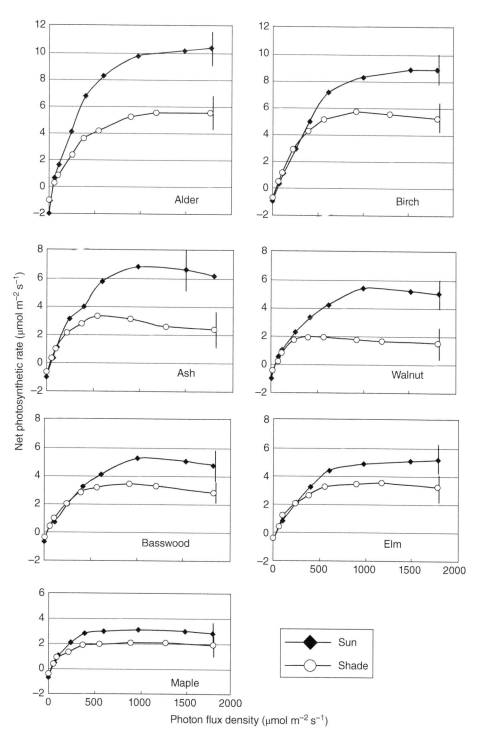

Figure 7.2 Light response curves for leaves from seven canopy tree species from a temperate forest in Japan. Early-successional or pioneer species (grey alder *Alnus hirsuta* also called *A. incana*, and white birch *Betula platyphylla* var. *japonica*) are light saturated at around 1000–1200 μmol m^{-2} s^{-1} for sun leaves and 800–900 μmol m^{-2} s^{-1} for shade leaves. Saturating light was lower in late-successional species (basswood *Tilia japonica* and mono or painted maple *Acer mono*) at around 400–500 μmol m^{-2} s^{-1}. Mid-successional species (Japanese elm *Ulmus davidiana* var. *japonica*) and gap-phase species (Japanese ash *Fraxinus mandshurica* var. *japonica* and Japanese walnut *Juglans ailanthifolia*) were intermediate. Full sunlight is around 2000 μmol m^{-2} s^{-1}. At high light intensities, some species exhibited signs of photoinhibition, as shown by a declining carbon gain under high light (see Coping With Too Much Light). *Source:* Koike *et al.* (2001). Reproduced with permission of Oxford University Press.

Importantly, the light response curve can differ quite substantially between species, particularly between early- and late-successional species, as shown for a range of temperate tree species from Japan in Figure 7.2. Sun leaves (see Chapter 3) have higher levels of photosynthesis than shade leaves given the same quantity of light, regardless of species, but these differences tend to be greater in early-successional species. Early-successional species have a higher light saturation point than late-successional species, with mid-successional and gap-phase species occupying an intermediate position (Koike *et al.* 2001).

Coping With Low Light

Trees have developed a range of morphological, physiological and biochemical responses to both low and high light environments. However, adaptations that increase the performance of a tree in low light are incompatible with those that increase its performance in high light. Consequently, tree species tend to specialise in either high- or low-light environments (Valladares and Niinemets 2008). In other words, shade-tolerant species tend not to be able to perform very well in high light and species adapted to high light environments will not be able to survive in environments where light is scarce. Inevitably, some species will also occupy intermediate positions and be able to cope with a range of light environments, but they are unlikely to thrive in deep shade or high light.

Leaves adapted to low light levels typically have thinner mesophyll (see Chapter 3) so have a lower leaf mass per unit of leaf area, resulting in cheaper leaf construction. The chlorophyll also adapts to absorb more efficiently the particular quality of light within the understorey and protective pigments, such as carotenoids are not produced in high quantities because the photoprotection they give is less valuable in shade. These adaptations, combined with other cellular level adjustments, result in leaves that are able to harvest light more efficiently than leaves acclimated to high light (Niinemets 2007). The advantage of this over the long term is clear: trees growing under a canopy of other trees can still use the small amount of light available to make incremental advances towards the the canopy so that, for example, shade-tolerant, late-successional species can eventually achieve dominance in the forest (Figure 7.3).

Within a tree crown and under the forest canopy, short bursts of light, known as sunflecks, can give saturating light levels for brief periods from fractions of a second (as the canopy moves in the wind) to many minutes (as gaps in the canopy allow light to enter) (Figure 7.4). On clear days, these sunflecks can account for 10–80% of the light available for photosynthesis so the efficiency of their utilisation can have an important effect on daily and seasonal carbon gain (Pearcy 2007). In very low light conditions, it takes some time for stomata to open and the photosynthetic processes to respond, which leads to an induction period of between 10 and 30 minutes, depending on the species, before photosynthesis is fully functional (Pearcy 1988). However, if a fully induced leaf is shaded for a few minutes and then exposed to a subsequent sunfleck, the leaf is already primed and it can regain maximal levels of photosynthesis in a few seconds. Some species can also be induced for longer than the period of illumination and can carry on fixing carbon (photosynthesising) for a brief period after the sunfleck has passed, which can greatly increase carbon gain if sunflecks are short-lived but frequent (Way and Pearcy 2012). This induction process can be extremely important for species that experience highly variable amounts of light under forest canopies. In general, shade-tolerant

Figure 7.3 Shade-tolerant western hemlock *Tsuga heterophylla* can be seen to make a steady incremental advance towards the canopy as they are quite capable of surviving in the low light conditions of the understorey for many decades.

Figure 7.4 The amount of light in the forest understorey can be very dynamic. Sunflecks, lasting from a few seconds to several minutes can make up a large portion of the light available for photosynthesis. The diverse light environment can be seen in this Kauri forest in New Zealand.

species show faster photosynthetic induction and are capable of remaining induced for longer, so are more efficient at utilising sunflecks in the understorey than are high-light (shade-intolerant) species (Valladares *et al.* 2012).

Coping With Too Much Light

Canopy trees and those growing in open environments experience light levels that are orders of magnitude greater than understorey species (e.g. more than $2000\,\mu\text{mol}\,\text{m}^{-2}\,\text{s}^{-1}$ and less than $20\,\mu\text{mol}\,\text{m}^{-2}\,\text{s}^{-1}$, respectively). Photosynthesis will increase with more light until a certain point is reached (the light saturation point), beyond which extra light leads to no further increase in photosynthesis. This extra light above and beyond what photosynthesis can use can be very problematic for leaves as they struggle to cope with all the energy it provides. This can actually lead to *photoinhibition*: the reduction in photosynthesis as a consequence of high light (Figure 7.2).

A solution is to increase the maximum level of photosynthesis to help dissipate the extra light energy (Niinemets 2007), but if other environmental limitations act to close the stomata (such as a lack of water) then access to carbon dioxide becomes limited and photosynthesis slows. In such cases, *photorespiration* can help deal with excess light energy. This light-dependent respiration uses up some of the excess light energy within the leaf but at a cost, because it also uses up some of the carbon gained through photosynthesis. Leaves acclimated to high-light environments also dissipate excess light energy as heat through evaporation of water and having increased levels of yellow–orange carotenoid pigments, especially those in the subgroup xanthophyll, such as violaxanthin, antheraxanthin and zeaxanthin (Pearcy 2007). These remain in the leaf through its life and lead to yellow leaves in autumn, as the carotenoids are revealed once the chlorophyll is broken down and reabsorbed.

In young leaves of many tree species, red pigments, predominantly anthocyanins, also seem to have a photoprotective role early in leaf development (Figure 7.5). They are thought to reduce the likelihood of photoinhibition while the photosynthetic machinery gets up and running (García-Plazaola and Flexas 2012); it may also help hide young

(a)   (b)

Figure 7.5 Red pigments, predominantly anthocyanins, seem to have a protective role against too much light in young leaves. Here young (a) European ash *Fraxinus excelsior* and (b) poplars *Populus* spp. grown in open, high light environments show red pigments in newly emerged leaves.

leaves from insects because they cannot see long-wave red light and so the leaves will appear dark or possibly dead (Dominy *et al.* 2002). As the leaves mature and chlorophyll is added, the anthocyanins are broken down and the redness tends to disappear rapidly. However, during leaf senescence, they often start to accumulate again to help create a more stable environment for the dismantling of photosynthetic apparatus and resorption of leaf nutrients (Field *et al.* 2001; Schaberg *et al.* 2008). Again, they also may help prevent insects (particularly aphids) laying their eggs on the branches, saving next year's leaves from being attacked (White 2009). On a more aesthetic level, these helpful red pigments also result in some spectacular autumn colours.

If these mechanisms are unable to dissipate the excess energy, it can lead to the production of reactive oxygen species: harmful chemically reactive compounds that contain oxygen. Photoprotective pigments such as carotenoids may be able to quench some of these compounds but any that are left will result in damage to the vital photosynthetic machinery. Repair of damage may be possible over time but obviously not without some cost and a temporary reduction in the efficiency of photosynthesis.

Of course, reducing the amount of light that is intercepted by the leaves can reduce these problems. Leaves can adjust their angle to hang down and so spill light (a well-known feature of some eucalyptus forests; see Figure 3.16). In individual crowns, this can also lead to more favourable light levels within the crown that can help reduce the impact of self-shading, as well as reduce potential damage in the upper leaves. Close observation of tree crowns in high light will often reveal leaves inclined at a steeper angle around the upper margins of the crown and more horizontally held leaves in lower positions. The amount of light captured inside the leaf can be reduced by changes in the pigment composition but also by the chloroplasts moving within the cell, resulting in them being stacked one above the other so that they shade each other (Valladares *et al.* 2012). Some species also possess features such as leaf hairs and epicuticular waxes that increase light reflectance and, by implication, reduce absorbance (García-Plazaola and Flexas 2012).

Practical Implications of the Light Environment and Shade Tolerance

The different light requirements of trees have two important practical implications for those managing trees. First, consideration of a tree's natural light environment is critical when considering planting locations. Although shade-tolerant trees certainly have a long-term advantage under low-light conditions, in full sun they can struggle to cope with excess light energy and suffer from photoinhibition. Therefore, planting very shade-tolerant species in the open can often be less than satisfactory as they simply do not thrive in such high light; many curators of collections will testify to the problems of bringing understorey trees out into the open. Equally, placing a pioneer species adapted to high-light conditions underneath an existing canopy or in the deep shade of a building can be similarly unrewarding.

Secondly, pruning that removes the outer portion of the crown (crown reductions), exposing shade-grown leaves to high light, will lead to photoinhibition in these leaves. Thus, as well as removing potentially productive leaves from the outer crown, it is likely that crown reductions reduce the productivity of the remaining leaves. Mature leaves do have a limited ability to adjust to altered light levels (by increasing protective pigments or adjusting the proportion of photosynthetic proteins) but most trees can only really adjust to new light levels by growing new leaves that are better adapted to these new light levels. For this reason, evergreen species that keep their leaves for a number of years find it particularly difficult to adjust to long-term changes in light conditions as it may be a number of years before the entire crown can be replaced. Consequently, when making recommendations for planting and pruning, consideration of the light environment is essential if tree health and performance is a priority.

Other Key Factors Influencing Photosynthesis – Temperature, Nutrition and Water

Temperature is important as it affects both photosynthesis and respiration. Its impact can be quite complex because temperature varies throughout the day (diurnally), between days and across the seasons. In forests as well as urban environments, there are gradients in temperature through the canopy, and these vary as a result of exposure and proximity to buildings or reflective surfaces. Even on the scale of a single leaf, temperature gradients can affect photosynthesis. Nevertheless, it is possible to identify some general patterns.

Whilst some of the key biochemical processes of photosynthesis tend to work best at around 35–40 °C (Dreyer *et al.* 2001), photosynthesis itself typically peaks at around 25 °C in temperate trees (Warren *et al.* 2012). The reason for this cooler optimum temperature is because the rate of respiration increases more rapidly than photosynthesis as temperature rises. This means that above around 25 °C, the gains made by increases in photosynthesis are eroded by the increase in cellular respiration.

Trees adapted to warmer or colder climates may, of course, have different optimum temperatures. Boreal trees have a lower optimal temperature for photosynthesis whilst Mediterranean and (sub)tropical trees typically have a slightly higher optimal temperature. Even within a single species, adaptation to temperature can be different. For example, red maple *Acer rubrum* from Florida was found to have higher growth and photosynthesis than a northern *A. rubrum* from Minnesota at moderate temperatures (33/25 °C day/night; Weston and Bauerle 2007).

Extreme temperatures can slow growth and be damaging. Above around 25 °C the net photosynthesis of most trees starts to diminish. At around 40 °C, temperate trees start to use more carbon in respiration than they gain in photosynthesis (and so make a net loss of carbon), as various key processes cease to work properly and damaging compounds, such as reactive oxygen species are produced in abundance (Warren *et al.* 2012).

Low temperatures do not affect light absorption but the rate of photosynthesis is reduced in the cold as the leaves are less able to dissipate excess light energy. This leads to an increase in photoinhibition on bright but cold days. Low temperatures can also reduce the export of sugars from the leaf, the build up of which can further slow the rate of photosynthesis. However, shorter and cooler nights can help the tree as they reduce the amount of night-time respiration where carbon loss greatly exceeds carbon gain. The corollary of this is that long warm nights increase night-time respiration, and so reduce the overall carbon gain that is possible. Indeed, in urban environments where night-time temperatures are raised by heat released from hard surfaces and buildings, night-time respiration may markedly increase compared with trees growing in adjacent rural areas.

As a general rule, in temperate trees net photosynthesis is likely to be minimal at less than 5°C, and typically is reduced to zero at 0°C. However, for cold-adapted boreal or high altitude trees, leaves can frequently reach 30% of their photosynthetic capacity at 0°C and 50–70% at 5°C (Körner 2012). Bear in mind that sunlight can warm leaves to 10–20°C above the ambient air temperature, so photosynthesis can happen even when the air temperature would suggest it is too cold (Vogel 2012). As a result, clear, still days during late autumn, winter and early spring can be critical for the carbon gain of broad-leaved evergreen species living in the understorey (Miyazawa and Kikuzawa 2005).

There is, perhaps, little that can be done to influence the temperature around the tree crown (without great expense) but it is, nonetheless, important to have an idea of the typical range of temperatures under which positive carbon gain is possible. When selecting trees, it can be very helpful to know the climatic origins of the particular cultivar as this can influence the photosynthetic performance of the tree if the new planting location is very different from its ecological origin. This is one of the many reasons why the nursery trade should keep accurate data on the precise origins of their plants. Unfortunately, this information has often been lost over time so it can be difficult to work out the climatic needs of one particular cultivar over another.

Adequate nutrition is also vital for photosynthesis although, inevitably, some nutrients have a more important effect on carbon gain than others (see Chapter 8). As a major constituent of chloroplasts and photosynthetic enzymes, nitrogen is certainly the most important nutrient for photosynthesis and a deficiency is often associated with reduced carbon gain. A number of other nutrients also exert an influence over photosynthesis (see Morales and Warren 2012), but many deficiencies are likely to affect growth as much by reducing shoot development and leaf area (Marschner 2012) as by affecting photosynthesis.

When water is in short supply, and the tree is under a mild water deficit, stomata close to restrict water loss from the leaves (see Chapter 6). This can limit photosynthesis because acquiring carbon dioxide becomes much harder. The significance of this largely depends on the duration of reduced water availability. Sustained periods of water deficit, particularly during periods of high temperature and high light, are likely to lead to increased photoinhibition, which further reduces net photosynthesis (Lambers *et al.* 2008). With repeated dry periods over a growth season, the rate of recovery after rainfall (or irrigation) will greatly affect carbon gain. Recovery usually depends on the duration of the water deficit and may take anything from a day to several weeks (Flexas *et al.* 2012a). Rapid recovery from water deficits may therefore exert an important influence on the performance of trees on sites that experience frequent drying and rewetting events. In managed landscapes, ensuring good soil water availability (see Chapter 6) is vital if prolonged periods of reduced photosynthesis are to be avoided.

Species Differ Widely in Their Leaf Photosynthetic Capacity

It would be easy to assume that species with the highest level of net leaf photosynthesis (total carbon uptake by photosynthesis minus respiratory losses) would be the most competitive in any given environment. However, this is not the case. The global mean for net photosynthesis under non-limiting conditions is around 9 µmol of carbon per

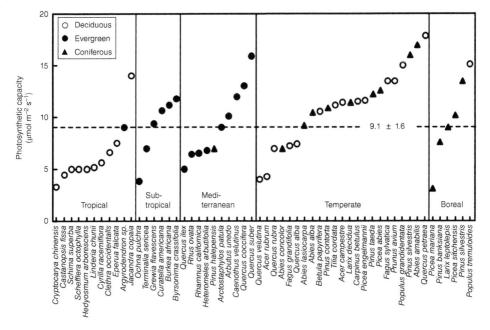

Figure 7.6 The variation in the amount of carbon fixed per square metre of leaf per second (photosynthetic capacity) in forest tree species across different climatic zones. *Source:* Körner (2005). Reproduced with permission of Springer.

square metre of leaf area per second ($\mu mol\,m^{-2}\,s^{-1}$) (Figure 7.6), but rates have been shown to vary across species by an order of magnitude between 3 and 30 $\mu mol\,m^{-2}\,s^{-1}$ (Ceulemans and Saugier 1991): even higher than shown in Figure 7.6.

Within temperate tree species, net photosynthetic rates are usually higher in deciduous broadleaved species (an average of 12 $\mu mol\,m^{-2}\,s^{-1}$), compared to evergreen broadleaved species (~11 $\mu mol\,m^{-2}\,s^{-1}$) and coniferous species (~8 $\mu mol\,m^{-2}\,s^{-1}$) (Warren *et al.* 2012). These differences seem not to be related to climate or leaf morphology but with how long the leaves live: the longer they live, the lower the net photosynthetic rate per second (Wright *et al.* 2004). Many environments favour leaf longevity, where the construction costs associated with leaves cannot readily be paid back within a single growth period (see Chapter 3). Typically, this results in more of the leaf nutrients being allocated to structural support and protective chemicals, to ensure the leaf is robust enough to cope with the year-round environmental variation and resist herbivory. As a result, a smaller proportion of the leaf nutrients can be used in photosynthesis. This leads to a reduced photosynthetic capacity per unit area of leaf, and is further compounded by reduced mesophyll conductance to carbon dioxide (caused by tightly packed cells with thick cell walls) that often goes with the ability to cope with tough environmental conditions (Niinemets and Sack 2006). Nevertheless, these slightly reduced rates of photosynthesis work well for evergreen broadleaves and conifers, as individual leaves live for many years and have a longer time to pay back the cost of their production.

Inevitably, the longer the leaf lifespan, the more leaf area is accumulated, so evergreen species tend to have a higher leaf area index (LAI: leaf area per unit of ground area).[3] This helps compensate for their lower leaf photosynthetic capacity. Species with more open crowns tend to have a higher leaf photosynthetic capacity. These nuances between leaf longevity, the photosynthetic capacity of leaves and LAI partly explain why stands of temperate forest species with varying LAI show similar annual rates of carbon gain (Warren *et al*. 2012).

Shade tolerance, at least in temperate deciduous trees, also affects photosynthesis. Early-successional, shade-intolerant trees generally have higher rates of photosynthesis and respiration than late-successional shade-tolerant species. In open, non-shaded environments, this allows early successional species to grow faster and consistently win the race for canopy dominance. However, in shady environments, shade-tolerant trees are able to achieve higher carbon gain at a lower light level because the leaves have a lower light compensation point; they also conserve carbon resources by having a relatively low respiration rate (Craine 2009).

Differences in photosynthetic capacity of drought-tolerant and drought-intolerant species are less easy to categorise. Some drought-tolerant species with robust leaves capable of surviving low internal water potential may, under normal conditions, be less able to take up carbon dioxide, and so show reduced photosynthetic rates (Niinemets and Sack 2006). Alternatively, there is evidence that some drought-tolerant species, such as the evergreen blue oak *Quercus douglasii*, native to south-west North America, can achieve very high leaf photosynthetic rates ($20–24\,\mu\mathrm{mol\,m^{-2}\,s^{-1}}$) during the early part of the growth season when soil moisture is not limiting. This compensates for low levels of photosynthesis in the latter part of the growth season when access to water is very limited and temperatures are high (Xu and Baldocchi 2003). Species are therefore likely to be very well adapted to their natural environments, including temporal variations in soil moisture and temperature. This underscores the relevance of understanding the natural environment a tree comes from, especially if exotic species are being used in amenity plantings.

The Big Picture – Carbon Gain Over the Years

Carbon gained by photosynthesis (carbohydrates) can vary within a tree's crown, as well as between different individual trees and between different species. Ultimately, the health of any tree is affected by how much carbon it can fix each year over its life.

Perhaps the most important factor in determining carbon gain each year is the time span over which high levels of photosynthesis are possible. This usually ties into the length of the growing season. In deciduous trees, the timing of leaf emergence and senescence (leaf phenology; see Chapter 3) is important because trees will lose out if they are slow to get going or too quick to lose their leaves. A few more days or

---

3 Leaf area index (LAI) is typically measured in $\mathrm{m^2}$ of leaf area per $\mathrm{m^2}$ of ground ($\mathrm{m^2\,m^{-2}}$). Temperate trees average around $6\,\mathrm{m^2\,m^{-2}}$ but some conifers can reach more than $10\,\mathrm{m^2\,m^{-2}}$.

weeks of positive carbon gain at either end of the growth season can make a substantial difference to long-term tree performance, but this must be balanced against the risk of frost damage to leaves. Understorey trees often have to deal with very short growing seasons: growth is most easily achieved before the canopy closes as only marginal gains can be made throughout the remainder of the growth season (see Figure 3.20).

A benefit of a long growing season is that in parts of the world where the growing season is long, as in warm temperate regions, even fairly modest daily carbon gains will build up over the season to a significant amount. Conversely, where favourable environmental conditions for photosynthesis are short-lived, as in cold temperate, boreal regions or high altitude habitats, even high daily rates of photosynthesis may yield only modest annual carbon gains (Larcher 2003).

The number of consecutive frost-free days provides a reasonable estimate of the growing season (Figure 7.7). Tropical regions have no frosts and potential for a year-long growing season. However, it is important to note that other factors, such as water availability, may shorten the growing season, particularly in sub-tropical climates. Warm temperate and Mediterranean regions typically have >250 consecutive days without frost, whilst in temperate environments it is around 200 days. Boreal and Arctic environments tend to have <150 days without frost, and higher latitudes may only have a few weeks where high levels of photosynthesis are possible. Trees close to the altitudinal tree line may have similarly short growing periods, as can be seen by the impact of mountain ranges on the number of consecutive frost-free days in Figure 7.7. Of course, within these generalisations the actual growing season may be shorter because of too much or too little water, too high or too low temperatures or other aspects of the environment that limit growth.

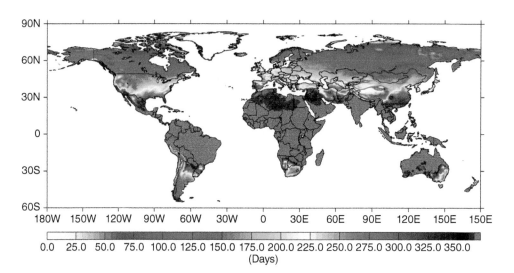

Figure 7.7 Average annual number of consecutive frost free (frost being <0°C) days using data from 1986–2015 from ERA-Interim Reanalysis (Dee *et al*. 2011). Plotted by Linda Hirons (National Centre for Atmospheric Science, University of Reading, UK).

Carbon Dynamics in Trees: Production, Use and Storage

One of the challenges faced by trees during their annual life cycle is that the carbon required by the tree is not always synchronised with its production by photosynthesis. Sometimes, growth and respiration occur at times when photosynthesis is not possible or when the supply for carbon is not able to meet demand. For example, a very dry period may effectively stop photosynthesis but the tree has to keep on respiring to stay alive. In the same way, a deciduous tree has to grow leaves in spring before photosynthesis can begin. In both cases, trees meet these demands by accumulating non-structural carbohydrates (NSCs), in the form of soluble sugars (mainly sucrose) and starch, which provide a carbon buffer during periods when supply does not quite meet demand (Chaplin *et al.* 1990). These carbon compounds are stored in the xylem parenchyma, often in the roots, before being reallocated via the phloem to carbon sinks or used to build the biomass of the tree (Placová and Jansen 2015). Put simply, trees have a carbon savings plan to insure them against periods when they need a little more carbon than can currently be supplied.

This accumulation of carbon is not, as was once assumed, simply a passive event whereby a little extra carbon is saved during periods of high production; it is, in reality, a highly regulated process. Storage can be prioritised (i.e. made an active sink for carbohydrates) to the extent that trees can reduce growth in the short term to fund growth and resilience to stress in the longer term (Sala *et al.* 2012).

It is difficult to accurately assess (and interpret) the NSC in trees because levels vary across the growing season(s) and between different storage compartments within the tree (i.e. course roots, stem, branches and leaves). In many temperate trees, the growth of new roots, wood and foliage in spring will deplete stores of sugar in the branches and trunk (Hoch *et al.* 2003; Richardson *et al.* 2013), as is shown in Figure 7.8. Interestingly, whilst sugar reserves are in decline, starch levels in most species increase at the same time (at least in stems). Production of starch (a comparatively immobile and inactive form of carbon) may seem counter-intuitive during a period of peak demand. However, the conversion of sugars to starch in stems will keep sugar concentrations low, maintain a strong sink strength and help draw sugars out of the leaves. This stops the sugars building up in the leaves, thus minimising the potential for down-regulation of photosynthesis (Richardson *et al.* 2013). Again, these findings suggest that the NSC pool is regulated to aid the long-term performance of the whole tree, and it should not be viewed as a passive pool of surplus carbon that simply accumulates and depletes at the whim of relative supply and demand.

It is clear from research using carbon isotopes (carbon atoms that have a slightly different mass and so can be tracked) that the carbon used for new growth relies on a mix of previously stored carbon and that which is recently produced (<1 year old). Indeed, some food reserves (NSCs) of red maple *Acer rubrum* and eastern hemlock *Tsuga canadensis* were found to be more than a decade old (Richardson *et al.* 2013). When the age profile of NSCs are closely examined, it appears that new growth is supported by a mix of some older and some recently assimilated carbon, but recently produced NSCs do not mix extensively with carbon reserves stored in the older growth rings. This suggests that during the course of normal tree growth, some older reserves are drawn upon to fund new growth, but recently produced NSCs do not get stored in the older growth rings as they become increasingly difficult to access as the stem ages.

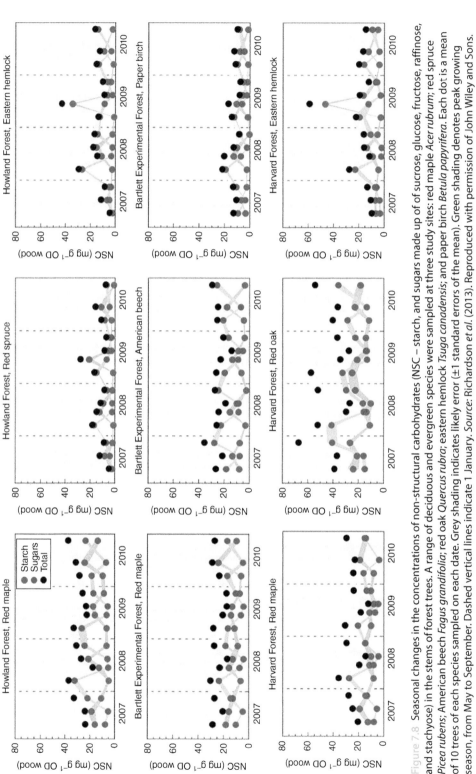

Figure 7.8 Seasonal changes in the concentrations of non-structural carbohydrates (NSC – starch, and sugars made up of of sucrose, glucose, fructose, raffinose, and stachyose) in the stems of forest trees. A range of deciduous and evergreen species were sampled at three study sites: red maple *Acer rubrum*; red spruce *Picea rubens*; American beech *Fagus grandifolia*; red oak *Quercus rubra*; eastern hemlock *Tsuga canadensis*; and paper birch *Betula papyrifera*. Each dot is a mean of 10 trees of each species sampled on each date. Grey shading indicates likely error (±1 standard errors of the mean). Green shading denotes peak growing season, from May to September. Dashed vertical lines indicate 1 January. *Source:* Richardson *et al.* (2013). Reproduced with permission of John Wiley and Sons.

The maintenance of these older carbon supplies seems to be more important for growth responses after major disturbances in roots (Vargas *et al.* 2009) and shoots (Carbone *et al.* 2013) than for current growth. Therefore, it seems that rather than a single pool of NSCs in trees, two pools exist: a *fast* and a *slow cycling pool* (Richardson *et al.* 2013, 2015). The fast pool responds to current demand for carbon whilst the slow pool can be drawn upon at times of more exceptional stress or disturbance. Thus, trees appear to have an instant savings account (fast pool) to service normal growth and maintenance across the growing season; they also have a long-term savings account (slow cycling pool) which can be mobilised during periods of prolonged stress or major disturbance.

A further detail emerging from the isotope studies is that trees operate a *last in, first out* policy with respect to carbon: the carbon most recently added to the NSC pool is the first to be used when demand requires it. Consequently, stressed trees will progressively use up older and older carbon reserves. Theoretically, this will causes the mean age of the carbon reserves to become older as the more recently assimilated carbon is utilised first. Therefore, in addition to the total proportion of NSC in tree tissues, the age distribution of these carbon pools can provide useful insights into the health of the tree by confirming if the slow cycling pool is being depleted (Richardson *et al.* 2013).

In healthy mature trees it is unlikely that carbon supply really limits tree development to any great extent (Fatichi *et al.* 2014; Körner 2015), even when stressed by, for example, drought; see Box 7.2. In both temperate (Hoch *et al.* 2003) and tropical (Würth *et al.* 2005) trees, high NSC pools have been found regardless of season, habitat or climate.

Box 7.2 Can Trees Starve to Death When Facing Drought?

McDowell *et al.* (2008) suggest that a tree's response to drought – closure of stomata – will lead to carbon starvation, particularly in *isohydric* species that readily close their stomata to maintain a high water content (*iso* = equal, *hydric* = water) rather than coping with internal water loss (*anisohydric* species). If the stomata remain closed, the tree's hydraulic conductivity is maintained, but eventually the tree will run out of carbon reserves –sugars and starch – and will literally starve to death. The problem comes in proving this. Seedlings and saplings are most likely to have limited reserves and so run out, but there is also some evidence of this happening in very old trees. However, many trees that die during drought still have large amounts of carbohydrates stored inside, so surely this disproves the argument that trees starve to death? Not necessarily. It may be that some of these reserves are unusable because, for example, the drought disrupts enzyme production or transport and the reserves become unattainable: although there are reserves left, these are unusable so the tree does indeed starve to death.

Others (such as Sala *et al.* 2011) argue that trees have mechanisms to hold themselves back from the brink of starvation. As photosynthesis declines because of drought, the tree will reduce growth and respiration to conserve carbohydrates. This is usually accompanied by a loss of energy-demanding tissues by, for example, shedding leaves or fine roots. Of course, this compounds the problem, reducing photosynthesis and water uptake, but will allow the tree to survive by ticking over until conditions improve.

It is obviously a complex set of problems to tease out and the debates continue. The bottom line is that drought and other physical problems, like high temperatures, do stress a tree, weakening it in ways that frequently accelerate tree mortality.

Under non-limiting conditions, trees can therefore be assumed to store more carbon than needed to meet the typical demands over time. Potentially, at least, these reserves could be used to sustain tree development in the absence of photosynthesis for quite a long time and are almost certainly capable of seeing the tree through several challenging growing seasons. Hoch *et al.* (2003) estimated that healthy temperate trees generally store enough carbon reserves to be able to replace the entire crown four times. Even masting years of beech *Fagus sylvatica* (when exceptionally large numbers of seeds are produced) did not substantially reduce NSC pools. Similarly, no carbohydrate depletion was observed in olive trees *Olea europaea* in years of high fruit production (Bustan *et al.* 2011). Thus, there is a growing body of evidence that suggests healthy trees, in natural ecosystems and in production systems, are not carbon limited in their development.

What does this say about old or large trees? Is the decline in growth in these trees a result of limited carbon availability, caused either by a progressive increase in the size of carbon sinks or a decrease in carbon sources? Do trees really grow so big and have to maintain so much living tissue that it becomes difficult to keep all the cells respiring? The best support for an increase in carbon demand resulting in a depleted carbon pool comes from a light-demanding sapling of eastern white pine *Pinus strobus* being grown in deep shade and not from a large tree (Machado and Reich 2006). However, some changes to allocation patterns have been observed between young and mature trees. Young sessile oak *Quercus petraea* showed preferential allocation of carbon to growth until the end of wood formation, whilst mature trees started to allocate carbon to storage shortly after budburst (Gilson *et al.* 2014). It would thus seem likely that even *old* trees have sufficient carbon reserves. However, old trees eventually begin to senesce, brought about by a whole range of internal and external factors, including rot (Brutovska *et al.* 2013; Salguero-Gomez *et al.* 2013; Thomas 2013). At this point, the problem is not initially one of too little stored carbon but that these reserves may become increasingly unavailable (as explained in Box 7.2), so the tree can have an adequately full larder but cannot access it. In this way, the carbon reserves effectively become disconnected from the carbon sinks so may not be able to be moved to the points of need around the tree. As the tree begins to die back, the crown becomes smaller, roots die and the amount of sound wood declines, so both the amount of carbon fixed and the amount stored will start to decline and NSC can begin to limit the defensive capabilities of the tree. In such old trees, the limited carbon storage *is* important. Clearly, understanding the precise mechanisms of tree size and age-related decline remains a live area of debate and research.

How Do Trees Die?

Despite there being many reasons for tree death, there are only really two principal mechanisms for tree mortality: carbon starvation and hydraulic failure (Figure 7.9). Carbon starvation occurs when the available carbon pool becomes depleted to the extent that it can no longer support basic metabolism (e.g. respiration). The size of this carbon pool is related to the development history of the tree: it will be greater for trees growing in excellent conditions than for those growing in impoverished conditions. Inevitably, over the life of a tree, the carbon pool will fluctuate on a daily, seasonal and

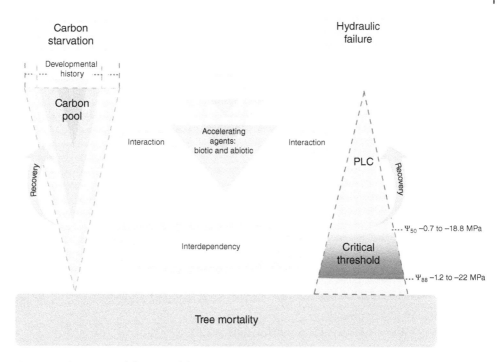

Figure 7.9 A conceptual diagram of the two dominant mechanisms of tree mortality: carbon starvation and hydraulic failure. PLC, percentage loss of conductivity; Ψ_{50}, water potential at 50% loss of conductivity; Ψ_{88}, water potential at 88% of conductivity. See text for a full explanation.

annual basis. A depleted pool can recover if conditions for carbon assimilation (photosynthesis) substantially improve, such as when a heavily shaded tree gains access to more light. However, there will a point where the carbon pool depletes below a certain threshold and the tree can no longer survive.

Hydraulic failure occurs when the percentage loss of conductivity (PLC) caused by embolism within the xylem (see Chapter 2) increases to such an extent that water can no longer be delivered to the leaves, photosynthesis is halted and the ability to move carbohydrates around the tree is also compromised. For this reason, there is a degree of interdependency between carbon starvation and hydraulic failure. The rate of the loss of conductivity is dependent on species, and highly variable. For example, the water potential inducing a 50% loss of hydraulic conductivity (Ψ_{50}) in trees varies between 0.7 and 18.8 MPa (again, see Chapter 2). At some point between Ψ_{50} and around 88% loss of conductivity (Ψ_{88}), trees will pass through a critical threshold, will be unable to recover, and mortality from hydraulic failure will be inevitable. However, providing the tree does not reach this critical threshold, it will be able to recover, for example by producing new functioning xylem (as happens seasonally in temperate ring porous trees) or by refilling embolised conduits.

Although tree mortality essentially follows one of these two principal mechanisms, in the real world there are a great many things that can influence their progression. These are termed *accelerating agents* and are either biotic or abiotic. For example, insect pests may accelerate the decline of the carbon pool as the tree makes a carbon-intensive defensive response; extreme temperatures and limited water availability will reduce photosynthesis and limit carbon gain; vascular wilt diseases may cause the tree to block its xylem and accelerate hydraulic failure; water deficits caused by climatic drought or loss of roots similarly accelerate hydraulic failure. So, while the stimulus for tree mortality can be highly variable, it is usually related to carbon starvation, hydraulic failure or the interdependency between the two. A notable exception to this is direct chemical phytotoxicity that can cause tree death by other means, but this is rare in natural environments.

Improving the Carbon Balance in Landscape Trees

Despite evidence that trees are not limited in their development by carbon in healthy mature forest trees, it is important not to be complacent about the carbon balance in amenity trees. Growing conditions in urban environments often differ greatly from the natural forest environment, so it is likely that the carbon pools do not accumulate to the same extent. Therefore, every attempt should be taken to enhance the carbon status of amenity trees, where possible.

A key management focus is to promote photosynthesis. Primarily, this involves planting in an excellent rooting environment that provides adequate oxygen, water and nutrition. Assuming there is no major constraint on light availability, this will mean that each leaf will be performing close to its photosynthetic capacity. A high-quality root environment will also prevent premature senescence of leaves, ensuring the longest growing season possible and maximising carbon gain. Many trees on challenging sites can be seen to shed their leaves earlier than their more fortunate brethren on moist, fertile sites.

It is also important to consider how crown management, particularly pruning, might influence the carbon gain of the whole tree, not just by reducing the photosynthetic leaf area but also by increasing photoinhibition in the remaining leaves. Clearly, there are many circumstances where the reasons for pruning a tree are more persuasive than the potential loss of productivity, but it is important that management decisions are well informed and take full consideration of the biological effects on the tree.

Management practices that help reduce carbon losses can also help. Reducing the night-time temperature around the tree to depress respiration rates is never going to be a viable option. However, it is worth considering that wounding, including that induced by pruning, will require extra carbon allocated to defence and will therefore reduce the carbon pool. Again, evaluating the requirement for pruning is relevant here.

Annual Carbon Dynamics of the Tree and the Timing of Arboricultural Work

Providing that the tree is healthy, for most species the timing of arboricultural operations should not be driven by the annual carbon dynamics of the tree (Hirons 2012). However, there is evidence that the seasonal depression in the carbon pool is greater in two groups of trees. Ring porous trees need to produce new xylem prior to leaf emergence (see Chapter 2), so they necessarily use stored carbohydrates to do this before photosynthesis starts in spring. Species that flower profusely prior to leaf emergence (e.g. many flowering cherries) similarly use stored carbohydrates. Needing to respond to a significant pruning event in early spring would likely disadvantage both these groups of trees because their carbon reserves will have been depleted as a result of other stresses. Pruning in early summer, prior to any substantial seasonal soil water deficits, would help to reduce the impact of pruning in these cases (Hirons 2012).

A tree's water status is much more significant than its carbon status when considering the timing of arboricultural operations. When a tree is pruned during a period of relatively low water potential (see Chapter 6), serious loss of hydraulic conductivity can occur as a result of embolism within the xylem. This reduces the tree's ability to supply water to the leaves for photosynthesis. Furthermore, if conductivity is not quickly restored, hydraulic connections within the tree may be lost, which in turn may isolate stored carbon pools from various carbon sinks around the tree. Hence, there is an important interaction between the loss of hydraulic function, caused by mechanical injury or water deficits, and the ability of the tree to supply carbon to parts of the tree that need it for growth and defence.

References

Barnes, B.B., Zak, D.R., Denton, S.R. and Spurr, S.H. (1998) *Forest Ecology*, 4th edition. Wiley, New York, USA.

Beck, C.B. (2010) *An Introduction to Plant Structure and Development: Plant Anatomy for the Twenty-first Century*, 2nd edition. Cambridge University Press, Cambridge, UK.

Brutovská, E., Sámelová, A., Dušička, J. and Micieta, K. (2013) Ageing of trees: application of general ageing theories. *Ageing Research Reviews*, 12: 855–866.

Bustan, A., Avni, A., Lavee S., Zipori, I., Yeselson, Y., Schaffer, A.A., *et al.* (2011) Role of carbohydrate reserves in yield production of intensively cultivated oil olive (*Olea europaea* L.) trees. *Tree Physiology*, 31: 519–530.

Carbone, M.S., Czimczik, C.I., Keenan, T.F., Murakami, P.F., Pederson, N., Schaberg, P.G., *et al.* (2013) Age, allocation and availability of nonstructural carbon in mature red maple trees. *New Phytologist*, 200: 1145–1155.

Ceulemans, R.J. and Saugier, B. (1991) Photosynthesis. In: Raghavendra, A.S. (ed.) *Physiology of Trees*. Wiley, New York, USA, pp. 21–50.

Chaplin, F.S. III, Schulze, E. and Mooney, H.A. (1990) The ecology and economics of storage in plants. *Annual Review of Ecology and Systematics*, 21: 423–447.

Craine, J.M. (2009) *Resource Strategies of Wild Plants*. Princeton University Press. Princeton, USA.

Dee, D.P., Uppala, S.M., Simmons, A.J., Berrisford, P., Poli, P., Kobayashi, S., *et al.* (2011) The ERA-Interim reanalysis: Configuration and performance of the data assimilation system. *Quarterly Journal of the Royal Meteorological Society*, 137: 553–597.

Dominy, N.J., Lucas, P.W., Ramsden, L.W., Rib-Hernandez, P., Stoner, K.E. and Turner, I.M. (2002) Why are young leaves red? *Oikos*, 98: 163–176.

Dreyer, E., Le Roux, X., Montpied, P. Daudet, F.A. and Masson, F. (2001) Temperature response of leaf photosynthetic capacity in seedlings from seven temperate tree species. *Tree Physiology*, 21: 223–232.

Fatichi, S., Leuzinger, S. and Körner, C. (2014) Moving beyond photosynthesis: From carbon source to sink-driven vegetation modeling. *New Phytologist*, 201: 1086–1095.

Field, T.S., Lec. D.W. and Holbrook, M.M. (2001) Why leaves turn red in autumn; the role of anthocyanins in senescing leaves of red-osier dogwood. *Plant Physiology*, 127: 566–574.

Flexas, J., Gallé, A., Galmés, J., Ribas-Carbo, M. and Medrano, H. (2012a) The response of photosynthesis to soil water stress. In: Aroca, R. (ed.) *Plant Responses to Drought*. Springer, Berlin, Germany, pp. 129–144.

Flexas, J., Loreto, F. and Medrano, H. (2012b) *Terrestrial Photosynthesis in a Changing Environment: A Molecular, Physiological, and Ecological Approach*. Cambridge University Press, Cambridge, UK.

García-Plazaola, J.I. and Flexas, J. (2012) Special photosynthetic adaptations. In: Flexas, J., Loreto, F. and Medrano, H. (eds) *Terrestrial Photosynthesis in a Changing Environment: A Molecular, Physiological, and Ecological Approach*. Cambridge University Press, Cambridge, UK, pp. 85–97.

Gilson, A., Barthes, L., Delpierre, N., Dufrêne, É., Fresneau, C. and Bazot, S. (2014) Seasonal changes in carbon and nitrogen compound concentrations in a *Quercus petraea* chronosequence. *Tree Physiology*, 34: 716–729.

Hirons, A.D. (2012) Straightening out the Askenasy curve. *Arboriculture and Urban Forestry*, 38: 31–32.

Hoch, G., Richter, A. and Körner, C. (2003) Non-structural carbon compounds in temperate forest trees. *Plant Cell Environment*, 26: 1067–1081.

Koike, T., Kitao, M., Maruyama, Y., Mori, S. and Lei, T.T. (2001) Leaf morphology and photosynthetic adjustments among deciduous broad-leaved trees within the vertical canopy profile. *Tree Physiology*, 21: 951–958.

Körner, C. (2003) Carbon limitation in trees. *Journal of Ecology*, 91: 4–17.

Körner, C. (2005) An introduction to the functional diversity of temperate forest trees. In: Scherer-Lorenzen, M. Körner, C. and Schulze, E.-D. (eds) *Forest Diversity and Function: Temperate and Boreal Systems*. Springer, Berlin, Germany, pp. 13–37.

Körner, C. (2012) *Alpine Treelines: Functional Ecology of the Global High Elevation Tree Limits*. Springer, Berlin, Germany.

Körner, C. (2015) Paradigm shift in plant growth control. *Current Opinion in Plant Biology*, 25: 107–114.

Kozlowski, T.T. (1992) Carbohydrate sources and sinks in woody plants. *Botanical Review*, 58: 107–222.

Lambers, H., Stuart Chaplin, F.S.III and Pons, T.L. (2008) *Plant Physiological Ecology*, 2nd edition. Springer, Berlin, Germany.

Larcher, W. (2003) *Physiological Plant Physiology*, 4th edition. Springer, Berlin, Germany.

Machado, J.-L. and Reich, P.B. (2006) Dark respiration rate increases with plant size in saplings of three temperate tree species despite decreasing tissue nitrogen and nonstructural carbohydrates. *Tree Physiology*, 26: 915–923.

Marschner, P. (2012) *Mineral Nutrition of Higher Plants*, 3rd edition. Academic Press San Diego, CA, USA.

Mauseth, J.D. (2016) *Botany: An Introduction to Plant Biology*, 6th edition. Jones and Bartlett Learning, Burlington, VT, USA.

McDowell, N., Pockman, W.T., Allen, C.D., Breshears, D.D., Cobb, N., Kolb, T. *et al.* (2008) Mechanisms of plant survival and mortality during drought: Why do some plants survive while others succumb to drought? *New Phytologist*, 178: 719–739.

Miyazawa, Y. and Kikuzawa, K. (2005) Winter photosynthesis by saplings of evergreen broad- leaved trees in a deciduous temperate forest. *New Phytologist,* 165: 857–866.

Morales, F. and Warren, C.R. (2012) Photosynthetic responses to nutrient deprivation and toxicities. In: Flexas, J., Loreto, F. and Medrano, H. (eds) *Terrestrial Photosynthesis in a Changing Environment: A Molecular, Physiological, and Ecological Approach.* Cambridge University Press, Cambridge, UK, pp. 312–330.

Niinemets, Ü. (2007) Photosynthesis and resource distribution through plant canopies. *Plant Cell Environment,* 30: 1052–1071.

Niinemets, Ü. and Sack, L. (2006) Structural determinants of leaf light-harvesting capacity and photosynthetic potentials. In: Esser, K., Lüttge, U.E., Beyschlag, W. and Murata, J. (eds) *Progress in Botany,* Vol. 67. Springer, Berlin, Germany, pp. 385–419.

Pearcy, R.W. (1988) Photosynthetic utilization of sunflecks by understorey plants. *Australian Journal of Plant Physiology,* 15: 223–238.

Pearcy, R.W. (2007) Responses of plants to heterogeneous light environments. In: Pugnaire, F.I. and Valladares, F. (eds) *Functional Plant Ecology,* 2nd edition. CRC Press, Boca Raton, FL, USA, pp. 213–257.

Placová, L. and Jansen, S. (2015) The role of xylem parenchyma in the storage and utilization of non-structural carbohydrates. In: Hacke, U. (ed.) *Functional and Ecological Xylem Anatomy.* Springer, Berlin, Germany, pp. 209–234.

Richardson, A.D., Carbone, M.S., Hugget, B.A., Furze, M.E., Czimczik, C.L., Walker, J.C., *et al.* (2015) Distribution and mixing of old and new nonstructural carbon in two temperate trees. *New Phytologist,* 206: 590–597.

Richardson, A.D., Carbone, M.S., Keenan, T.F., Czimczik, C.I., Hollinger, D.Y., Murakami, P., *et al.* (2013) Seasonal dynamics and age of stemwood nonstructural carbohydrates in temperate forest trees. *New Phytologist,* 197: 850–861.

Sala, A., Fouts, W. and Hoch, G. (2011) Carbon storage in trees: Does relative carbon supply decrease with tree size. In: Meinzer, F.C., Lachenbruch, B. and Dawson, T.E. (eds) *Size- and Age-Related Changes in Tree Structure and Function.* Springer, Berlin, Germany, pp. 287–306.

Sala, A. Woodruff, D.R. and Meinzer, F.C. (2012) Carbon dynamics in trees: feast or famine? *Tree Physiology,* 32: 764–775.

Salguero-Gómez, R., Shefferson, R.P. and Hutchings, M.J. (2013) Plants do not count… or do they? New perspectives on the universality of senescence. *Journal of Ecology,* 101: 545–554.

Schaberg, P.G., Murakami, P.F., Turner, M.R., Heitz, H.K. and Hawley, G.J. (2008) Association of red coloration with senescence of sugar maple leaves in autumn. *Trees,* 22: 573–578.

Taiz, L., Zeiger, E., Møller, I.M. and Murphy, A. (2014) *Plant Physiology,* 6th edition. Sinauer Associates, Sunderland, MA, USA.

Thomas, H. (2013) Senescence, ageing and death of the whole plant. *New Phytologist,* 197: 696–711.

Thomas, S.C. and Martin, A.R. (2012) Carbon content of tree tissues: A synthesis. *Forests,* 3: 332–352.

Valladares, F., García-Plazaola, J.I., Morales, F. and Niinemets, Ü. (2012) Photosynthetic responses to radiation. In: Flexas, J., Loreto, F. and Medrano, H. (eds) *Terrestrial Photosynthesis in a Changing Environment: A Molecular, Physiological, and Ecological Approach.* Cambridge University Press, Cambridge, UK, pp. 239–256.

Valladares, F. and Niinemets, Ü. (2008) Shade tolerance, a key plant feature of complex nature and consequences. *Annual Review of Ecology, Evolution and Systematics*, 39: 237–257.

Vargas, R., Trumbore, S.E. and Allen, M.F. (2009) Evidence of old carbon used to grow new fine roots in a tropical forest. *New Phytologist*, 182: 710–718.

Vogel, S. (2012) *The Life of the Leaf*. University of Chicago Press, Chicago, IL, USA.

Warren, C.R., Garcia-Plazaola, J.I. and Niinemets, Ü. (2012) Ecophysiology of photosynthesis in temperate forests. In: Flexas, J., Loreto, F. and Medrano, H. (eds) *Terrestrial Photosynthesis in a Changing Environment: A Molecular, Physiological, and Ecological Approach*. Cambridge University Press, Cambridge, UK, pp. 465–487.

Way, D.A. and Pearcy, R.W. (2012) Sunflecks in trees and forests: from photosynthetic physiology to global change biology. *Tree Physiology*, 32: 1066–1081.

Weston, D.J. and Bauerle, W.L. (2007) Inhibition and acclimation of C3 photosynthesis to moderate heat: a perspective from thermally contrasting genotypes of *Acer rubrum* (red maple). *Tree Physiology*, 27: 1083–1092.

White, T.C.R. (2009) Catching a red herring: Autumn colours and aphids. *Oikos*, 118: 1610–1612.

Wright, I.J., Reich, P.B., Westoby, M., Ackerly, D.D., Baruch, Z., Bongers, F., *et al.* (2004) The worldwide leaf economics spectrum. *Nature*, 428: 821–827.

Würth, M.K.R., Peláez-Riedl, S., Wright, S.J. and Körner, C. (2005) Non-structural carbohydrate pools in a tropical forest. *Oecologia*, 143: 11–24.

Xu, L. and Baldocchi, D.D. (2003) Seasonal trends in photosynthetic parameters and stomatal conductance of blue oak (*Quercus douglasii*) under prolonged summer drought and high temperature. *Tree Physiology*, 23: 865–877.

8

Tree Nutrition

In nature, trees are the dominant form of vegetation in all but the most extreme habitats. Without human interference, they successfully compete for space both above and below ground on flood-plains, rocky mountainsides and virtually every habitat in between. Relative success in these contrasting environments is driven by the ability of the root system to acquire water and mineral resources (nutrients) in order to fund the tree's growth, development and, ultimately, reproduction.

Nutrition is fundamental to tree performance in natural environments and managed landscapes. With the exception of water and light, nutrients are the environmental factor that most constrain tree growth. For that reason, the benefits (ecosystem services) offered by groups of trees and individuals will be affected by each tree's ability to acquire and use nutrients. Consequently, those managing trees should have some understanding of tree nutrition and how this can be managed.

Managing tree nutrition is highly complex as there are so many variables that can exert influence over both the capacity of the soil to supply nutrients and the ability of the tree to acquire them. Physical attributes of the soil, such as compaction, texture and water status (see Chapters 4 and 6), can influence the nutrient availability. Chemical factors, such as pH, salinity, mineral composition and the cation exchange capacity (CEC), impact nutrient supply. Many of the common nutrients, such as calcium and potassium (Table 8.1), have a positive charge when dissolved in water and are called cations; the CEC is the ability of soil to hold on to these before they are leached away, keeping them available to plants. An array of microorganisms and other soil fauna may also compete for or contribute to soil nutrients. Some microorganisms, such as mycorrhizal fungi (see Chapter 9), even form symbiotic relationships with tree roots that aid nutrient uptake. Temperature, precipitation and evaporation have a marked effect on biological and chemical processes, and climate also has a large impact on the capacity of soils to supply nutrients for tree development. The rooting morphology of different tree species will affect how they interact with soil biota and soil chemistry. The cumulative effect of all these variables (and more) interacting in complex ways can make it difficult to decide which factors are most limiting the availability of nutrients, as these are likely to change from site to site. Nevertheless, these factors are the architects of what we might refer to as 'soil fertility'.

From a management perspective, it is important to determine how fertile any particular site is and to what extent site management or tree management could manipulate it. What opportunities exist to improve tree performance? Just as importantly, what are

Applied Tree Biology, First Edition. Andrew D. Hirons and Peter A. Thomas.
© 2018 John Wiley & Sons Ltd. Published 2018 by John Wiley & Sons Ltd.

Table 8.1 Essential nutrients, the available forms and their functions within the tree. When dissolved in water, these include nutrients that have a positive charge (cations) and those with a negative charge (anions). Forms and functions are included that are not discussed elsewhere in the text: they are included for completeness and the interested reader can readily find information on these elsewhere. Concentrations found in parentheses under each element indicate approximate concentrations per gram of dry plant shoot matter sufficient for adequate growth, based on Epstein and Bloom (2005). Other data are from Marschner (2012) and Jones (2012).

Nutrient	Typical available forms	Functions
Macronutrients		
Nitrogen (N) ($\sim$1000 μmol g^{-1})	Nitrate (NO_3^-) Ammonium (NH_4^+)	Proteins, phospholipids, nucleic acids, chlorophyll, co-enzymes, phytohormones, secondary metabolites
Potassium (K) ($\sim$250 μmol g^{-1})	K^+ cation	Enzyme activation, proteins, regulation of stomatal aperture, phloem transport (loading), stress resistance
Calcium (Ca) ($\sim$125 μmol g^{-1})	Ca^+ cation Calcium carbonate ($CaCO_3$) Calcium sulfate ($CaSO_4$)	Cell wall stabilisation, cell extension, secretory processes, membrane stabilisation, osmoregulation
Magnesium (Mg) ($\sim$80 μmol g^{-1})	Mg^+ cation	Chlorophyll, enzyme activation, phosphorylation
Phosphorous (P) ($\sim$60 μmol g^{-1})	Dihydrogen phosphate ($H_2PO_4^-$) Monohydrogen phosphate (HPO_4^{2-}) Al, Fe, and Ca phosphates	ATP (energy transfer), nucleic acids, phospholipids, co-enzymes, starch, sugars
Sulfur (S) ($\sim$30 μmol g^{-1})	Sulfate (SO_4^{2-})	Amino acids, proteins, co-enzymes, secondary metabolites, cellular resistance to dehydration and frost damage
Micronutrients		
Chlorine (Cl) ($\sim$3 μmol g^{-1})	Cl^- anion	Role in photosynthetic oxygen production, osmoregulation
Boron (Bo) ($\sim$2 μmol g^{-1})	Hydrogen borate (H_3BO_3) Borate (BO_3^{3-}) Undissociated boron ($B(OH)_3$)	Role in cell wall structure, membrane function, reproductive growth and development, role in root elongation and shoot growth
Iron (Fe) ($\sim$2 μmol g^{-1})	Ferric cations (Fe^{3+}) Ferrous cations (Fe^{2+})	Chlorophyll synthesis, proteins, enzymes
Manganese (Mn) ($\sim$1 μmol g^{-1})	Mn^{2+}, Mn^{3+}, Mn^{4+} cations	Enzymes, co-factor to enzymes, photosynthetic oxygen production
Zinc (Zn) ($\sim$0.3 μmol g^{-1})	Zn^+ cation	Component of enzymes, activation of enzymes, involved in protein synthesis, involved in carbohydrate metabolism
Copper (Cu) ($\sim$0.1 μmol g^{-1})	Low molecular weight humic and fulvic acids. Cupric ion (Cu^{2+})	Proteins, important for lignification, role in pollen formation and fertilisation
Nickel (Ni) ($\sim$0.001 μmol g^{-1})	Ni^{2+} cation	Component of enzymes, role in nitrogen metabolism
Molybdenum (Mo) ($\sim$0.001 μmol g^{-1})	Molybdate anion (MoO_4^{2-})	Enzyme for nitrogen fixation, component of enzymes and enzyme co-factors

the risks of doing so? Further, will the outcome be worth the expense of management? This latter question is much easier to answer in agriculture, production horticulture and commercial forestry, where plant performance can easily be related to the economic value of the crop, than it is in landscape trees where the cost–benefit ratio of management can be more difficult to calculate. Before exploring potential management strategies to enhance the nutrition of trees, it is important to develop an understanding of the role of nutrients and how they are acquired.

Essential Nutrients

There are currently 14 known *essential nutrients* found in plants (Table 8.1). To be classed as such, the plant must be unable to complete its life cycle in the absence of the nutrient; it must not be replaceable by another nutrient; and it must be directly involved in plant metabolism (Arnon and Stout 1939). Despite all being important, the relative concentrations of these nutrients within the plant differ by several orders of magnitude. Approximately 10 000 times more nitrogen is required than molybdenum, but both are essential. Such differences in the quantity of nutrients required have led to the division of nutrients into macro- and micronutrients (Table 8.1). Most micronutrients are predominantly used in enzymes so they tend to be required in small concentrations (0.001–3 μmol per gram dry weight). Macronutrients are found in much higher concentrations (30–1000 μmol per gram dry weight) as they are used in larger organic compounds, such as proteins and nucleic acids, or act to modify the osmotic potential of the cell (Kirkby 2012). Nitrogen, phosphorus and potassium are the macronutrients that most frequently limit plant growth so are often the focus of soil or plant analysis.

In addition to the 14 essential nutrients, some other elements, such as sodium, silicon, cobalt, selenium and aluminium, are considered *beneficial elements*. These tend to stimulate growth in some way, by alleviating the toxicity of another element (e.g. silicon helps with manganese toxicity) or by assisting a beneficial symbiotic organism (e.g. symbiotic nitrogen-fixing bacteria need cobalt), but they are not fundamental to basic plant metabolism (Broadley *et al.* 2012).

Nutrient Uptake

Most nutrients are taken into the plant by being dissolved in the water of the soil (the soil solution). Solid nutrients are simply no use for roots: they must be held in solution so that the roots can absorb them as *solutes*. This is one of the reasons why dry soils can never be fertile soils, even if they are potentially nutrient rich.

So how are nutrients stored and released from the soil? One of the most important components of the soil is the *colloidal particles*, comprised of clay-sized mineral particles[1] and organic humus material. These soil particles have a colossal *specific surface area* (surface area to volume ratio) and typically have an electrostatic charge (Ashman and Puri 2002). It is this electrostatic charge that attracts ions with an opposing charge and gives soil the ability to store nutrients, as well as toxins and

---

1 A clay particle is defined as being <2 μm, 0.002 mm or 0.000002 m.

hydrogen ions that are responsible for soil acidity (pH). Providing soils are not too acidic or alkaline, the charge most frequently associated with the colloidal particles is negative. Opposite charges attract, so any elements with a positive charge (cations) are drawn to the surface of the particle, forming the *cation exchange capacity* (CEC). Those cations closest to the clay particle are very strongly held but subsequent layers of cations are held progressively weakly as they settle further away from the colloidal particle. The electrostatic charge of the nutrient also has a bearing on the magnitude of attraction between soil colloidal particle and nutrient ion. For example, trivalent cations, such as Al^{3+}, bind more strongly than divalent cations, such as Ca^{2+}, which bind more strongly than monovalent cations, such as K^+ (White 2012). Importantly, cations can be *exchanged* between the colloidal surface, the layers of cations and the surrounding soil solution, making them available to roots. Physical, biological and chemical properties affect this dynamic process in which different ions jostle for position around the soil particle.

The CEC is a measure of the soil's ability to absorb cations and, by implication, supply the cation nutrients, such as K^+, Ca^{2+}, Mg^{2+}, NH_4^+ and Na^+, to the plant (Table 8.1). The CEC is dependent on colloidal particles, so soils with high proportions of clay and organic matter have high CECs and soils with a sandy texture and low amounts of organic matter have low CECs. Nutrients with negative charges (anions) are not attracted to the negatively charged colloidal surfaces. For this reason, nitrates such as NO_3^-, and phosphates such as $H_2PO_4^-$ and HPO_4^{2-}, are easily leached out of the soil. In soils polluted with heavy metals or that are very acidic, hydrogen ions can stick to the colloidal particles (the process of protonation) causing them to become positively charged. In such cases, anions can attach to the colloidal surfaces and be less prone to leaching. Anions may also be held more tightly in the soil when they form associations with other soil elements that are capable of latching on to the negatively charged soil particles.

Nutrients are brought into contact with the root in one of three ways: by roots directly touching soil particles as they grow; through the flow of the soil solution; or diffusion of nutrient ions down a concentration gradient. This provides a further problem for some nutrients, such as the phosphate in the form of PO_4^{3-}, in that they are very immobile because of their low solubility. As a result, their movement within the soil is restricted, so uptake relies on root (or mycorrhizal) growth into a nutrient-rich patch rather than the movement of the nutrient to the root. Roots can also actively release chemicals into the rhizosphere.[2] These either help release nutrients chemically held within the soil or help microbes make nutrients available to the tree. More detailed information on soil chemistry can be found in textbooks such as Ashman and Puri (2002), Brady and Weil (2016) and Binkley and Fisher (2013).

As a consequence of the chemical properties of nutrients, plants can exert a degree of selection about which nutrients, and how much of each, they take up into their roots. Certain nutrients are taken up preferentially whilst others can be largely excluded. It also means that nutrients can be shifted around the plant and held at higher concentrations inside the cell than in the external environment, or vice versa.

Uptake of nutrients occurs mostly through the fine roots <2 mm in diameter. Nutrients enter the root with water, as explained in Chapter 6, passing through the outer cortex of the root either between or through cells. Once they reach the endodermis (see Figure 6.5), nutrients have to be taken into the cells through a cell membrane.

_____

2 The rhizosphere is defined as the area of soil influenced by the root.

Electrochemical gradients generated by the small electrical charges associated with the different nutrient ions require that nutrients are transported across cell membranes by an *active* process that uses energy. Although this has a cost to the tree, it is worthwhile because it gives tight control of nutrient uptake and movement within the tree. However, this does mean that factors such as the supply of oxygen and carbohydrates to root cells also influence nutrient uptake (White 2012). For trees, this is particularly relevant in waterlogged or compact soils which can have very low oxygen levels, and in trees that have lost their lower branches through shading or pruning (because lower branches will typically export carbohydrates to their roots). Indeed, excessive crown raising and the removal of epicormic growth from low on the trunk can greatly reduce the quantity of carbohydrates exported to the roots and therefore constrain nutrient uptake.

The fine roots take up most nutrients, and only very minor amounts of nutrients are absorbed through the larger woody roots which are mainly used for the movement of sap, anchorage and storage. For this reason, biotic damage (such as root herbivory) or abiotic factors (such as flooding, soil compaction or mechanical injury) that disrupt fine root development and survival fundamentally affect nutrient acquisition. Trees that have had their fine roots removed during transplantation or that are growing in highly compacted soils (see Chapter 4) are thus vulnerable to nutrient deficiencies, regardless of the apparent nutritional status of the soil. However, as fine roots are also responsible for the uptake of water, trees experiencing substantive root loss often show symptoms of water deficit earlier than those of nutrient deficiency.

Symbiotic Relationships That Help Nutrient Acquisition

One of the most profound influences on the nutrient uptake in trees is their relationship with symbiotic microorganisms. In natural ecosystems, trees form symbiotic relationships with *mycorrhizal* fungi that colonise the fine roots. As described in Chapter 9, the symbiosis is based on the fungi taking carbohydrates (sugars) from the tree in return for using its vast infrastructure of hyphae (fungal threads) to assist in the acquisition of nutrients.

A relatively small but diverse group of trees has a symbiotic relationship with bacteria. Most trees are unable to take nitrogen (N_2) from the atmosphere, even though it makes up 78% of the air. However, some specialist bacteria have developed a mechanism to 'fix' atmospheric N_2 into ammonia (NH_3), a form of nitrogen that can be used by plants. Trees that have been able to form this symbiotic relationship with bacteria have a major advantage, particularly on low-fertility sites. The relationship is analogous to that of mycorrhizas in that the tree provides a source of carbon for bacteria and, in return, the tree obtains a share of the nitrogen fixed by the bacteria.

The largest group of plants that forms this relationship belong to the Fabaceae: the pea family. As well as including important agricultural crops such as peas, beans and lentils, a number of important woody genera[3] are also included in this group. In these, *Rhizobium* spp. (belonging to one of the big bacterial groups, the proteobacteria)

─────

3 Genera in the Fabaceae family that form associations with *Rhizobium* spp. to fix nitrogen from the atmosphere: *Acacia, Albizia, Bauhinia, Caesalpinia, Caragana, Cassia, Centrolobium, Ceratonia, Cercis, Cladrastis, Dalbergia, Delonix, Dialium, Enterolobium, Erythrina, Gleditsia, Guibourtia, Gymnocladus, Hymenaea, Inga, Laburnum, Leucaena, Lysiloma, Machaerium, Millettia, Mimosa, Myroxylon, Parkinsonia, Peltogyne, Piscidia, Pithecellobium, Platymiscium, Prosopis, Pterocarpus, Robinia, Schizolobium, Senegalia, Styphnolobium, Tamarindus, Tipuana, Vachellia* and *Vatairea*.

Figure 8.1 Actinorhizal nodules made up of short lateral roots on a section of alder *Alnus glutinosa* root. These contain the bacteria *Frankia alni* which fix nitrogen that then becomes available to the tree.

colonise the roots, forming characteristic root nodules (swellings) that become the sites where carbon and nitrogen is exchanged. As with their herbaceous cousins, the seedpods are often highly valued as fodder because they contain high levels of protein.

In addition to *Rhizobium* spp., another group of N-fixing soil bacteria, known as *Frankia* (an Actinomycete, a group that has features of both bacteria and fungi), also associates with about 220 species from a range of woody genera (Santi *et al.* 2013).[4] These *actinorhizal* plants form specialised structures that resemble clusters of short lateral roots that host the symbiotic bacteria (Figure 8.1). As with members of the Fabaceae, species with *Frankia* associations can survive in some very challenging environments. For example, the actinorhizal desert oak *Allocasuarina decaisneana* is one of the few trees that can grow in the low-nutrient, dry soils of central Australia (Figure 8.2).

Clearly, as the root systems of trees are tasked with nutrient acquisition, the co-evolution of symbiotic soil fungi (mycorrhiza) and bacteria alongside roots has been an important advance. However, this may be only part of the picture. Nitrogen-fixing bacteria have also been found in the stems of riverside (riparian) species, such as black cottonwood *Populus trichocarpa* and Sitka willow *Salix sitchensis* (Doty *et al.* 2005, 2009). This is highly significant given the ecology of these species. As with all riparian species, frequent inundation of floodwater regularly leads to the death of fine roots; by implication, any symbiotic associates of the roots would be lost. Keeping the symbionts in the stem thus protects this asset. Further, both the above species use stem fragments to propagate so carrying the symbiotic bacteria with them provides a valuable source of nitrogen as they establish on the nutrient-poor river gravels. Detailed studies using a 'heavy' form of nitrogen that is easy to detect (a ^{15}N isotope compared to normal nitrogen with an atomic weight of 14) suggest that in the poplar hybrid *Populus trichocarpa* × *P. deltoides*, nitrogen fixed by bacteria is largely responsible for the nitrogen found in leaves and stems, whilst mineral

---

4 Woody genera forming associations with *Frankia* include: *Allocasuarina, Alnus, Casuarina, Ceanothus, Cercocarpus, Chamaebatia, Coriaria, Elaeagnus, Gymnostoma, Hippophaë, Morella, Myrica, Purshia* and *Shepherdia*.

Figure 8.2 Desert oak *Allocasuarina decaisneana* can grow in very impoverished soils as a consequence of its relationship with *Frankia* bacteria. Note the characteristic narrow juvenile form of young trees seen in the foreground. These trees were photographed in Uluru-Kata Tjuta National Park, NT, Australia.

nitrogen from the soil remained in the roots (Knoth *et al.* 2014). These studies underline how little is known about the complex relationships between trees and the diverse biological communities that they live in, but they also provide fascinating insights into the strategies that trees use to overcome challenging environments.

How much nitrogen is fixed? This varies considerably, depending on species and environment. In temperate environments, both false acacia *Robinia pseudoacacia* (using *Rhizobium*) and red alder *Alnus rubra* (using *Frankia*) have been estimated to fix up to 300 kg of nitrogen per hectare, per year. However, estimates are often less than this, particularly on sub-optimal soils. Soil factors, such as acidity, salinity, temperature, moisture and physical characteristics, will affect bacterial infection and nodulation, and therefore nitrogen fixation (Cooper and Scherer 2012). Nevertheless, even modest amounts of fixed nitrogen can be a major advantage to a species by delivering a nutritional boost in otherwise impoverished soils. In fact, nitrogen-fixing species can also be a huge asset to nearby vegetation as they share nutrients via the leaf litter they produce. As shown in Table 8.2, the litter produced by grey alder *Alnus incana* has six times more nitrogen than that of Norway maple *Acer platanoides*, and over three times the average nitrogen content of non-nitrogen-fixing species (Berg and McClaugherty 2008). Thus, over time, leaf litter from nitrogen-fixing species can raise the fertility of the soil by releasing nitrogen into the soil. In turn, this allows other species to grow which further

Table 8.2 Macronutrients in leaf litter of some temperate tree species. Note that litter of the nitrogen-fixing grey alder *Alnus incana* has significantly higher levels of nitrogen.

Species	Concentration of nutrient (mg g^{-1})					
	Nitrogen	Phosphorus	Sulfur	Potassium	Calcium	Magnesium
Norway maple *Acer platanoides*	5.1	3.15	–	13.1	20.4	1.46
Grey alder *Alnus incana*	30.7	1.37	6.12	15.6	12.3	2.32
Silver birch *Betula pendula*	7.7	1.05	0.80	4.7	11.8	3.30
Beech *Fagus sylvatica*	9.5	1.40	1.30	2.3	7.4	1.20
European ash *Fraxinus excelsior*	8.6	1.96	–	15.3	33.2	2.28
Aspen *Populus tremula*	8.2	0.93	–	5.1	29.9	4.69
Pedunculate oak *Quercus robur*	15.9	0.73	–	0.75	7.2	0.68
Rowan/mountain ash *Sorbus aucuparia*	7.1	0.31	–	10.8	12.4	2.86
Average	**10.9**	**1.45**	**2.74**	**9.6**	**18.2**	**2.59**

Source: Berg and McClaugherty (2008). Reproduced with permission of Springer.

enriches the developing ecosystem. For this reason, it is possible to utilise nitrogen-fixing species to establish tree cover on nutritionally poor sites, and to aid soil fertility in agroforestry and amenity landscape plantings.

However, the advantage given to trees by their nitrogen-fixing bacterial friends can make them rather invasive. Indeed, in parts of Europe false acacia *Robinia pseudoacacia* is considered invasive (Cierjacks *et al.* 2013).

Other Factors That Influence Nutrient Availability –pH, Moisture, Aeration, Temperature

Soil pH has a direct effect on the availability of nutrients for root uptake (Figure 8.3). In alkaline soils (pH >6.5; see Box 4.2), the availability of phosphorus, iron, manganese, molybdenum, copper, boron and zinc is very low. In acid soils (pH <5.0), the availability of most nutrients declines but the increased toxicity of elements such as aluminium can also interfere with nutrient uptake. For this reason, the *target pH* of soils under management is typically in the range of pH 6.2–6.5, as all nutrients tend to be available in this range. However, the natural distribution of trees testifies to the fact that many trees can operate quite successfully outside this range. In boreal coniferous forests, the pH can be as low as 3, whilst soils formed on calcareous parent rock can have a pH of 8. As a general rule, unless the pH is rather extreme (<4 or >8), it seems to have only a minimal effect on tree growth and nutrition (Binkley and Fisher 2013). Therefore, adjusting the entire pH of the soil through the addition of amendments is rarely of value for most trees. However, pH has been implicated in nutrient deficiencies in some species that are widely planted in amenity landscapes; for example, pin oak *Quercus palustris* and various rhododendrons that require acidic soil (Figure 8.4). A number of herbaceous food crops have also been shown to be sensitive to soil pH, hence the prominence of the topic in many standard soil texts.

Soil pH

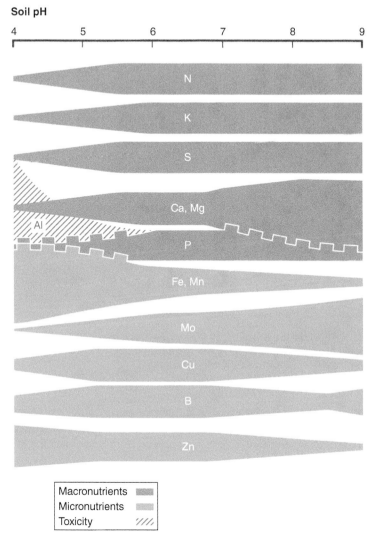

Figure 8.3 The influence of soil pH on nutrients. Bandwidth indicates the relative availability of the nutrients; blocked margins indicate that nutrients become immobilised and unavailable to the plant. Hatching describes the increase in potential aluminium toxicity as pH is reduced. A pH of 7 is considered chemically neutral; <7, acidic; >7, alkaline. However, ecologically, a 'neutral' soil has pH 5.0–6.5, and a pH of 6.2–6.5 is generally considered to be the range in which nutrients are most available. Full chemical names are given in Table 8.1. Based on data from Landon (1991), Ingram *et al.* (2008) and Brady and Weil (2016).

Nutrients need to be held in solution, so the availability of soil moisture is vital for the uptake of nutrients. However, too much water is generally not helpful as more mobile nutrients can get leached through the soil and away from the fine roots. Waterlogged or flooded soils also lead to oxygen deficiency that directly affects nutrient uptake because the process requires cellular energy. These conditions can also lead to root mortality and new roots will need to develop before normal root activity resumes. Soil temperature will affect the chemical and biological activity within the soil, so it will be relevant

Figure 8.4 Chlorosis, or yellowing of the leaves, in pin oak *Quercus palustris* is frequently seen in amenity landscapes as a result of a soil pH that is too high for this tree. This species requires acidic soils (pH <5) and at a higher pH will show nutrient deficiency symptoms.

to key processes of decomposition as well as the uptake of nutrients by roots and associated symbionts. Clearly, frozen soils present particular challenges for nutrient uptake, but these tend to occur when there is very low demand for nutrients by the tree so they tend not to be too problematic.

Nutrient Cycling

Nutrients are precious commodities to trees. It takes energy to grow roots, to take up the nutrients and to host symbiotic organisms. Consequently, once nutrients are safely captured, trees go to great lengths to preserve them. Strategies to achieve this can be seen at all sorts of scales. As described in Chapter 3, trees have evolved evergreen and deciduous leaves to conserve nutrients in response to how they cope with climatic factors. Leaves have also evolved in response to the nutrition available from soils. Evergreen leaves are held on the tree for more than one growing season which reduces the quantity of nutrients needed from the soil in any one year, but this does require leaves to be robust enough to cope with periods of substantial environmental stress caused by cold winters or periodic drought. In more fertile environments, trees can often afford to construct new leaves each year and it is cheaper to grow disposable leaves that are shed during a harsh winter or summer drought. However, even under these circumstances, conservation of nutrients is still prevalent; a tree that wastes nutrients is unlikely to succeed in the long-term.

Table 8.3 Mean (± standard deviation) nitrogen (N) and phosphorous (P) concentration in mature leaves (in milligrams per gram of leaf), and resorption efficiency. Different letters within columns indicate statistical difference.

	Concentration (mg g^{-1})		Resorption efficiency (%)	
	N	P	N	P
Evergreen trees and shrubs	13.7 (±5.2)[a]	1.02 (±0.56)[a]	46.7 (±16.4)[a]	51.4 (±21.7)[a]
Deciduous trees and shrubs	22.2(±7.4)[b]	1.60 (±0.92)[b]	54.0 (±15.9)[b]	50.4 (±19.7)[a]
Mean	**17.95**	**1.31**	**50.4**	**50.9**

Source: Adapted from Aerts (1996). Reproduced with permission of John Wiley and Sons.

On a smaller scale, the austere nature of both deciduous and evergreen trees means that they try to recuperate as many nutrients as possible from individual leaves prior to them being shed. This *resorption efficiency* can vary quite substantially across species, different sites and for different nutrients, but, generally, about 50% of key nutrients, such as nitrogen and phosphorus, are drawn back into the tree prior to leaf fall (Table 8.3). The remaining nutrients locked up in the cellular fabric of the leaves only become available to the tree after the leaf litter has decomposed.

Those nutrients that are lost in falling leaves, however, are not wasted because they are vital for soil organisms involved in decomposition. These provide a service to the tree by breaking down organic matter and releasing nutrients that can be re-used by the tree. Not all nutrients will return to the tree from whence they came, as individual plants in an area compete for their share of nutrients, but they will all contribute to a healthy ecosystem that is necessary for sustained tree growth. At least, that is how nature operates in the world's forests, where about 90% of nutrients come from the recycling of organic matter and only about 10% of the nutrients come from atmospheric deposition or weathering of parent rock (Chaplin 1991).

Nutrient recycling is therefore fundamentally important to tree and forest health (Figure 8.5). In essence, abscised leaves, fine roots and other parts of the tree contribute to the soil organic matter. Various microorganisms, aided by invertebrates, decompose this organic matter and release nutrients in forms that plants can take up (the process of *mineralisation*) into the soil solution. Many of these nutrients will interact with the anion and cation exchange sites within the soil; some may be immobilised; others will be leached below the reach of the roots; and still others may be released back to the atmosphere as gases in the process of *volatilisation*. Nutrients are absorbed from the soil solution by mycorrhizal fungi and fine roots and internally distributed to where they are needed throughout the tree. Some of the hard-won nutrients will be lost by leaching from the leaves by heavy rain, but strategically placed fine roots may well recapture some of these leached nutrients from the water that runs down the stem (*stemflow*) or drips through the canopy as *throughfall*, particularly around the dripline at the edge of the crown. Some nutrients will be lost from the system each year by being washed away through the soil or blown away as leaf fragments, but the input from rock below the soil, and especially from rain and dust above (*dry and wet deposition*) more than compensates for this loss: in the long-term, it is a self-sustaining system.

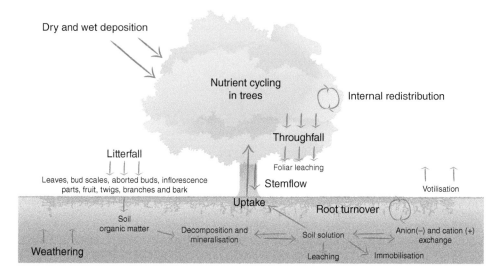

Figure 8.5 A simplified view of nutrient cycling in trees. *Source:* Hirons (2015). Reproduced with permission of the Royal Horticultural Society.

It is important to note that significant quantities of nutrients are held within the litter layer (Table 8.4) and only become available if they are recycled through the decomposition and mineralisation processes (Figure 8.5). Compared with the potential quantities of nutrients in the litter layer, the annual nutrient uptake rates of trees in temperate deciduous forests are fairly low (Table 8.5): 362 kg of nitrogen is held in the litter layer in 1 hectare at any one time compared to a need of 75 kg per hectare by the trees each year (Landsberg and Gower 1997). For the most important nutrients, the annual uptake rates are below the average values present in the litter layer (Tables 8.4 and 8.5). This, combined with the fact that most nutrients are made available to the tree by recycling, shows the importance of leaf litter as a source of nutrients for trees. In colder northern areas, decomposition is much slower and can form a bottleneck slowing tree growth

Table 8.4 Nutrient content of the litter layer in temperate deciduous forests in kilograms per hectare (kg ha^{-1}). This is based on data from 20 sites across Europe and North America. Values given are the mean and range, and are also translated to an area of 100 m^2 which is taken to represent the canopy area of a single mature tree.

	Nutrient			
	Nitrogen	Phosphorus	Potassium	Calcium
Range	65–1050	3–72	12–123	6–517
Mean	362	29	56	141
kg 100 m^{-2} (based on mean)	3.62	0.29	0.56	1.41

Source: Adapted from Khanna and Ulrich (1991).

Table 8.5 Annual nutrient uptake rates (kg ha^{-1} year^{-1}) in temperate deciduous forests. Values given are as in Table 8.4.

	Nutrient			
	Nitrogen	Phosphorus	Potassium	Calcium
Range	43–115	48–169	40–59	48–169
Mean	75	6	51	85
kg 100 m^{-2} (based on mean)	0.75	0.06	0.51	0.85

Source: Adapted from Landsberg and Gower (1997).

because most nutrients remain locked up in soil organic matter. In temperate areas, however, the litter layer can sufficiently supply the annual nutrient requirements of a tree, providing decomposition is encouraged. By implication therefore, the removal of litter material can have serious consequences for the long-term provision of nutrients for the tree.

Even this relatively simple analysis of nutrient dynamics in temperate trees and forests can help us make management decisions in our gardens, parks and streets. Left to their own devices, temperate tree ecosystems are more than capable of supplying their own nutritional requirements. However, it is abundantly clear that the disruption to normal recycling processes, by hard surfaces that prevent nutrients returning to the soil (Figure 8.6) or management practices such as leaf-litter removal, will strongly reduce the availability of nutrients to trees. Trees must then go on producing biomass every year with an ever-diminishing nutrient supply. For this reason, it is important to look for opportunities to mimic forest conditions in our managed landscapes.

(a)

(b)

Figure 8.6 (a) Temperate forests, such as this beech *Fagus sylvatica* woodland, are capable of meeting the nutritional requirements of the trees through the recycling of nutrients held within the litter layer. (b) In contrast, sealed surfaces in urban sites with no leaf litter significantly disrupt any nutrient cycling and can lead to nutrient deficiencies in trees. (b inset) Close-up of trunk base showing the limited space for nutrients to return to the soil.

Managing Tree Nutrition

The central aim of nutrition management should be to maintain the nutrient supply within an 'adequate' range without allowing nutrients to become either deficient or toxic (Figure 8.7). It would be possible to go through cycles of plant and/or soil analysis on each site, followed by prescribed nutritional amendments. However, this would likely be rather labour intensive and expensive. Therefore, unless there is evidence of serious nutritional problems it is better to simply mimic what happens in nature. Promoting the natural recycling of nutrients is one of the most effective ways of ensuring sustainable nutrition for trees. If at all possible, the leaf litter layer should be allowed to accumulate and break down over the root plate of the tree. Clearly, this is easier to achieve in parks and gardens than it is in more urban environments where other factors influence management decisions. Where leaves cannot be left for aesthetic or safety reasons, it may be preferable to collect the tree litter and replace it with an organic mulch of some kind (Figure 8.8). As discussed further in Chapter 5, this will provide the raw material for nutrient recycling and a healthy, biologically active soil, capable of supplying nutrients to the tree. If applied correctly to a depth of 5–10 cm over the root plate of the tree (but not burying the stem of the tree), mulching can have a host of beneficial effects (see Figure 5.23) without adversely affecting gas exchange between the soil and the atmosphere. Care should be taken with mulches of pure wood chips because in the short term they can reduce the amount of nitrogen available to the trees (as this is locked up inside the microorganisms decomposing this difficult substrate). By suppressing grass or other herbaceous plants, mulch also helps reduce competition for nutrients.

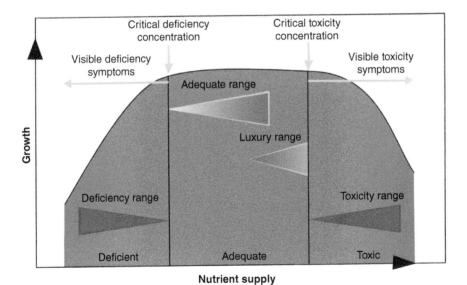

Figure 8.7 The relationship between nutrient supply and growth. Nutrient supply should be kept within an adequate range and not be allowed to limit growth through deficiency or toxicity. Adapted from Römheld (2012). *Source:* Marschner (2012). Reproduced with permission of Elsevier.

Figure 8.8 A well-maintained mulch ring over the root plate of white mulberry *Morus alba* in the Royal Botanic Gardens, Sydney, Australia. This type of mulching not only improves the conditions for root development and supplies nutrients, but also discourages people from walking over the root plate.

Whilst mulching should be high up on any tree-nutrition management plan, it is vital that the roots are not physically restricted by heavily compacted soils underneath the mulch (discussed in Chapter 4). If roots are unable to grow in the soil under the mulch, there is little value in applying the mulch in the first place. Equally, nutrients are taken up via the soil solution, so adequate soil moisture is essential. However, care should be taken not to over-irrigate trees as this can leach valuable nutrients away from the roots and lead to waterlogging.

In most cases, if leaf litter is left in place or replaced by mulching, and the soil is suitable for fine root development (i.e. aerated, moist and not compacted), nutrients will not be limited. However, as shown in Figure 8.9, visual symptoms, such as deformed leaves (deformations), yellowing (chlorosis) or dead parts (necrosis), may indicate a nutritional disorder. Where high value amenity trees or commercial tree crops are concerned, it can be worth the cost of confirming a suspected deficiency with analysis of soil or plant material. Even then, visual symptoms are useful in deciding what deficiencies to test for because a testing laboratory typically only tests for the nutrient(s) requested: there is no point wasting time (and money) evaluating unlikely candidates for deficiency. Once a nutritional problem is identified, it may be appropriate to add fertiliser. However, it is important that factors such as soil compaction are sorted out first, otherwise fertiliser application may prove ineffective. It is also important to make sure that soil pH is not interfering with the availability of nutrients. If this is the case, adding fertilisers is unlikely to resolve the underlying cause of any deficiency.

Despite this cautious advice, applying fertilisers is, in some cases, a valid management strategy. Fertilisers can be applied in a wide variety of forms and using a range of methods.

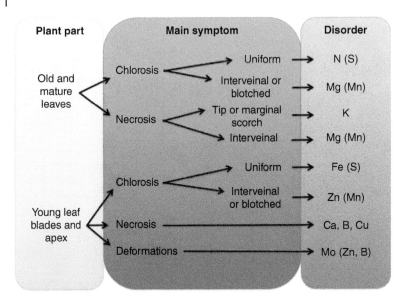

Figure 8.9 Visual signals that indicate nutritional deficiencies in plants. Full nutrient names are given in Table 8.1. See text for definitions of main symptoms. Adapted from Römheld (2012). *Source:* Marschner (2012). Reproduced with permission of Elsevier.

The fertiliser used should, of course, address any specific deficiency identified. However, the most appropriate application technique is likely to vary from site to site: fertilisers may be broadcast in granular form, applied by an irrigation system (*fertigation*), injected into the soil (as a liquid or granules) or injected into the stem. Granular fertiliser spread over the root-zone works well in some circumstances, although surrounding plants, such as turf-grass, may benefit the most. In such cases, the tree will gain little from the application. Liquid fertiliser injected into the soil can be used to good effect, particularly in soft landscapes such as parks and gardens. This approach delivers the soluble fertiliser into the root-zone of the tree at a greater depth than would be possible using a broadcast method and, to some extent, limits the interception by other plant roots (depending on their depth).

Stem injections can deliver very precise amounts of nutrients and clearly have no impact on non-target plants. They can be a good option in hard landscapes where other application methods are difficult. Stem injections are often criticised because they cause injury to the tree but, providing they are used judiciously to target specific problems, and alternative approaches have been ruled out, they should be considered a perfectly legitimate option. Care should be taken, however, to ensure delivery is into actively conducting sapwood (see Chapter 2) so that the material is taken up and distributed around the crown.

Fertilisers, particularly those high in nitrogen can increase the tree's vulnerability to both abiotic and biotic stress. The promotion of shoots over roots increases the leaf area and shoot:root ratio. This increases the water demand from the crown without increasing the ability of the tree to take up water (Brunetti and Fini 2017). Fertilisation may also lead to rapid growth of highly nutritious new shoots that often have reduced levels of defensive compounds. This has been shown to increase the tree's susceptibility to feeding insects (Herms 2002).

The timing of fertiliser application is critical. Highly soluble fertilisers should be applied when the roots are active, otherwise the nutrients may have leached beyond the reach of the roots by the time they are needed. Slow-release fertilisers will reduce the likelihood of excessive leaching and are preferred where possible. It is also best to avoid giving fertilisers late in the growing season as this can trigger new growth that is then vulnerable to cold injury.

References

Aerts, R. (1996) Nutrient resorption from senescing leaves of perennials: Are there general patterns? *Journal of Ecology*, 84: 597–608.

Arnon, D.I. and Stout, P.R. (1939) The essentiality of certain elements in minute quantity for plants with special reference to copper. *Plant Physiology*, 14: 371–375.

Ashman, M.R. and Puri, G. (2002) *Essential Soil Science: A Clear and Concise Introduction to Soil Science.* Blackwell, Oxford, UK.

Berg, B. and McClaugherty, C. (2008) *Plant Litter: Decomposition, Humus Formation, Carbon Sequestration*, 2nd edition. Springer, Berlin, Germany.

Binkley, D. and Fisher, R.F. (2013) *Ecology and Management of Forest Soils*, 4th edition. Wiley-Blackwell, Chichester, UK.

Brady, N.C. and Weil, R.R. (2016) *The Nature and Properties of Soils*, 15th edition. Pearson, London, UK.

Broadley, M., Brown, P., Cakmak, I., Ma, J.F., Rengel, Z. and Zhao, F. (2012) Beneficial elements. In: Marschner, P. (ed.) *Mineral Nutrition of Higher Plants*, 3rd edition. Academic Press, San Diego, CA, USA, pp. 249–269.

Brunetti, C. and Fini, A. (2017) Fertilization in urban landscapes. In: Ferrini, F., Van Den Bosch, C.C.K. and Fini, A. (eds.) *Routledge Handbook of Urban Forestry*. Taylor & Francis, Abingdon, UK, pp. 434–448.

Chaplin, F.S. III (1991) Effects of multiple environmental stresses on nutrient availability. In: Mooney, H.A., Winner, W.E. and Pell, E.J. (eds) *Response of Plants to Multiple Stresses*. Academic Press, San Diego, CA, USA, pp. 67–88.

Cierjacks, A., Kowarik, I., Joshi, J., Hempel, S., Ristow, M., von der Lippe, M., *et al.* (2013) Biological Flora of the British Isles: *Robinia pseudoacacia. Journal of Ecology*, 101: 1623–1640.

Cooper, J. and Scherer, H. (2012) Nitrogen fixation. In: Marschner, P. (ed.) *Mineral Nutrition of Higher Plants*, 3rd edition. Academic Press, London, UK, pp. 389–408.

Doty, S.L., Dosher, M.R., Singleton, G.L., Moore, A.L., Van Aken, B., Stettler, R.F., *et al.* (2005) Identification of an endophytic *Rhizobium* in stems of *Populus. Symbiosis*, 39: 27–35.

Doty, S.L., Oakley, B., Xin, G., Kang, J.W., Singleton, G., Khan, Z., *et al.* (2009) Diazotrophic endophytes of native black cottonwood and willow. *Symbiosis*, 47: 23–33.

Epstein, E. and Bloom, A.J. (2005) *Mineral Nutrition in Plants: Principles and Perspectives.* Sinauer, Sunderland, MA, USA.

Herms, D.A. (2002) Effects of fertilization on insect resistance of woody ornamental plants: Reassessing an entrenched paradigm. *Environmental Entomology*, 31: 923–933.

Hirons, A.D. and Percival, G.C. (2012) Fundamentals of tree establishment: A review. In: Johnston, M. and Percival, G. (eds) *Proceedings of the Urban Trees Research Conference, 'Trees, People and the Built Environment'*. Forestry Commission, Edinburgh, UK, pp. 51–62.

Ingram, D.S., Vince-Prue, D. and Gregory, P.J. (2008) *Science and the Garden*, 2nd edition. Wiley-Blackwell, Chichester, UK.

Jones, J.B. (2012) *Plant Nutrition and Soil Fertility Manual*, 2nd edition. CRC Press, Boca Raton, FL, USA.

Khanna, P.K. and Ulrich, B. (1991) Ecochemistry of temperate deciduous forests. In: Röhrig, E. and Ulrich, B. (eds) *Temperate Deciduous Forests*. Ecosystems of the World: 7, Elsevier, Amsterdam, The Netherlands, pp. 121–163.

Kirkby, E. (2012) Introduction, definition and classification of nutrients. In: Marschner, P. (ed.) *Mineral Nutrition of Higher Plants*, 3rd edition. Academic Press, San Diego, CA, USA, pp. 3–5.

Knoth, J.L., Kim, S.-H., Ettl, G.J. and Doty, S.L. (2014) Biological nitrogen fixation and biomass accumulation within poplar clones as a result of inoculations with diazotrophic endophyte consortia. *New Phytologist*, 201: 599–609.

Landsberg, J.J. and Gower, S.T. (1997) *Applications of Physiological Ecology to Forest Management*. Academic Press, San Diego, CA, USA.

Langdon, J.R. (1991) *Booker Tropical Soil Manual*. Longman Scientific and Technical, Harlow, UK.

Marschner, P. (2012) *Mineral Nutrition of Higher Plants*, 3rd edition. Academic Press, San Diego, CA, USA.

Römheld, V. (2012) Diagnosis of deficiency and toxicity of nutrients. In: Marschner, P. (ed.) *Mineral Nutrition of Higher Plants*, 3rd edition. Academic Press, San Diego, CA, USA, pp. 299–312.

Santi, C., Bogusz, D. and Franche, C. (2013) Biological nitrogen fixation in non-legume plants. *Annals of Botany*, 111: 743–767.

White, P. (2012) Ion uptake mechanisms of individual cells and roots: short distance transport. In: Marschner, P. (ed.) *Mineral Nutrition of Higher Plants*, 3rd edition. Academic Press, San Diego, CA, USA, pp. 7–47.

9

Interactions With Other Organisms

Trees as Habitats and Hosts

Outside of Africa with its large mammals, trees tend to be the biggest organisms and usually have the largest biomass per unit area of landscape. Thus, they are a very important part of any ecosystem in which they are found. They obviously change the microenvironment, which affects what can live in their shadow, and they also provide a huge number of habitats and niches. These range from providing nesting sites for birds through to providing dead wood for many insects, fungi and bacteria. They are also a food supply for a wide range of organisms, from small insects to large mammals that browse the shoots, twigs and leaves. The litter they produce can be a nightmare for those wanting to 'tidy up' urban areas and yet, in a more natural environment, the litter and detritus is a key component for a whole food-web of detritivores and decomposer organisms that release nutrients needed by the tree.

Some of these interactions are potentially harmful to trees in that they can lose part of their biomass, or can be killed outright. Others are fairly neutral in their effect on a tree, such as most epiphytes, or can even be positively beneficial, as in most mycorrhizal relationships. Understanding these interactions, and knowing which to control, is an important part of tree management.

Plants and Epiphytes

A large number of living things, including algae, lichens and plants, live on the outside of trees. They are classed as *epiphytes* because they take nothing from the tree and are just using the tree as a handy place on which to grow; as such, they are different from parasites which take things from the tree. Epiphytes include lichens and green algae (usually a *Pleurococcus* species and often the orange, powdery *Trentepohlia* spp.) on bark, but can also include tree seedlings and other flowering plants growing in pockets of humus trapped in branch forks or rooting into rotting wood (Figure 9.1). Other epiphytes are rooted in the ground but lean on the tree, saving them from having to invest heavily in their own self-supporting, woody skeleton to reach the light high up in the canopy – the European ivy *Hedera helix* is a good example in the temperate world but numerous species of lianas can be seen in tropical rainforests.

Applied Tree Biology, First Edition. Andrew D. Hirons and Peter A. Thomas.
© 2018 John Wiley & Sons Ltd. Published 2018 by John Wiley & Sons Ltd.

(a)   (b)

Figure 9.1 (a) Green algae *Pleurococcus* on the bark of a yew *Taxus baccata*; (b) epiphytes on Kauri *Agathis australis* in New Zealand.

Do these do any harm to the tree? The short, and rather surprising, answer is no. Even the luxuriant growth of mosses and lichens found in moist and unpolluted parts of the world do not damage the tree. It is sometimes suggested that these can block the lenticels – the corky breathing holes that allow oxygen and other gases to pass in and out of the tree – but the epiphytes produce such an open growth that this does not happen. Some might consider them aesthetically undesirable on trees that are grown for their ornamental bark but, short of using herbicides for the mosses and fungicides for the lichens, there is little that can be done to dissuade them from growing. Gentle scrubbing with a mild soapy solution can help but this is obviously only practicable on small trees. Humus pockets in the canopy will tend to get larger if they are hosts to epiphytes, such as ferns, as they accumulate dead parts of the epiphytes. Tropical trees, including mangroves, often produce roots from the branches that invade these pockets, giving them an extra source of nutrition. Such roots can be found in temperate trees, such as beech *Fagus sylvatica* and limes *Tilia* spp. but, unlike their tropical cousins, it is unlikely that they gain much extra nutrition.

Surely ivy damages trees and should be removed (Figure 9.2)? Ivy is rooted in the ground and does little except use the tree for support. As such, it does no appreciable damage to healthy trees and its presence is mainly an aesthetic issue. Its flowers provide nectar late in the year, and its dense growth is habitat to many birds and insects, so it should be left, unless unsightly, and not routinely removed from vigorous trees. However, ivy can become a problem on old, weak trees because the ivy adds weight to the crown and can present a large 'sail' area for wind, particularly in the winter when winds are generally strongest and deciduous trees are otherwise leafless, making the trees more prone to windthrow. Similarly, on slow-growing old trees the vigorous flowering shoots can smother the leaves, outcompeting them for light, particularly on new epicormic

(a)

(b)

Figure 9.2 (a) Ivy *Hedera helix* growing on an oak, creating a large mass of evergreen foliage; and (b) ivy cut through at the base to kill the plant above; this is not normally necessary to help the tree. *Source:* Thomas (2014). Reproduced with permission of Cambridge University Press.

shoots (see Expert Box 9.2 later in this chapter). In these cases, the removal of ivy can be justifiable. When inspecting trees, ivy may also obscure fungi, cavities or other structural defects that may be pertinent to a tree risk evaluation.

Some plants are parasites and so are a little more demanding of their hosts. European mistletoe *Viscum album* and the broadleaf mistletoes of the Americas (*Phoradendron* spp.) are green and take only water from their hosts; they do little damage (strictly called a *hemiparasite*). Others are complete *parasites*, stealing not just water and nutrients, but also sugars from the host. This includes the small, often yellow, dwarf mistletoes (such as *Arceuthobium* spp.) found in North and Central America, Asia and Africa. These can be very damaging to conifers in western North America and Asia, causing distorted and twisted growth of branches (which can be pruned out), and heavy infestations can lead to reduced vigour and even death.

Trees may also support parasitic plants on their roots (Figure 9.3). A notable example in the UK is toothwort *Lathraea squamaria*, a chlorophyll-less white plant that only appears above ground to flower. A number of other similar-looking plants, such as bird's-nest orchid *Neottia nidus-avis* and yellow bird's-nest *Monotropa* spp., are not parasitic on trees but on fungi in the soil; these used to be called saprophytes but are really *myco-heterotrophs*. None do appreciable damage to the trees, and in most cases are quite rare – something to be cherished rather than a cause of concern.

(a)   (b)

Figure 9.3 (a) Toothwort *Lathraea squamaria*, a parasitic plant on the roots of various trees, particularly hazel *Corylus* and elm *Ulmus*. The flower head is up to 30 cm tall, with no chlorophyll, and is the only part of the plant seen (as the rest is below ground living on the tree roots). (b) *Rafflesia* is a genus of some 28 species, parasitic on woody vines in south-east Asia (Mursidawati *et al.* 2015). Larger flowers can be up to 1 m in diameter. They smell and look like rotting meat to encourage pollinating flies, hence the various common names that translate to 'corpse flower'.

Microorganisms

Trees, like humans, are covered in bacteria, yeasts and fungi. Some are epiphytes but most are *heterotrophs*, deriving a living from their hosts. On leaves, the *phylloplane* community (growing on the leaf surface) feed mostly on plant exudates and animal products (including the sugary honeydew of aphids), or on dead tissues and do little harm. Other bacteria and fungi are parasitic on leaves, penetrating into the leaf and exploiting the living tissue, causing areas of necrosis. In some cases, such damage is more of an aesthetic than a health problem. For example, the tar spot fungus *Rhytisma acerinum*, creating the characteristic black spots on leaves of sycamore *Acer pseudo-platanus*, does little to harm the long-term growth of the tree, despite huge epidemics in years with humid weather.

Symbiotic Fungi

Fungi tend to have a bad press in causing weakness and death in trees but many fungi form invaluable *mycorrhizal* relationships: an association between the roots of a plant and one or more fungi to the benefit of both (symbiosis). The fungus benefits by receiving sugars and other compounds from the tree while, in turn, supplying water and nutrients, particularly phosphorus, but also nitrogen and a few others for which competition is intense. The tree may also receive some protection from toxic levels of salinity and heavy metals. Moreover, exposure to the mycorrhizal fungus, which acts as a very mild form of disease, can 'immunise' the tree against this disease in the process of *systemic acquired resistance* (SAR). Mycorrhizas are found in over 80% of the world's vascular plants and in many cases, such as beech *Fagus* spp., oak *Quercus* spp. and pines *Pinus* spp., are essential. Yet maples *Acer* spp. and birches *Betula* spp. can grow happily without them, and others, notably members of the Proteaceae (such as *Protea, Banksia* and *Grevillea* in the southern hemisphere), rarely if ever form mycorrhizas. In turn, some fungi are only found in mycorrhizal relationships (termed *obligate symbionts*) while others can be free-living (*facultative symbionts*) and even aggressive pathogens, given the right conditions.

Mycorrhizas found on trees exist in two main forms: *ectomycorrhizas* (ECM) and *arbuscular mycorrhizas* (AM; also called VAM, AMF or endomycorrhizas). In ECMs, the fungus (usually Basidiomycetes but sometimes Ascomycetes) forms a glove-like sheath over the root tips with fungal hyphae penetrating between the cells of the root. Such mycorrhizas are comparatively rare around the world – found in just 3% of flowering plants – but are very important in trees. These include 90% of trees of the northern hemisphere, including most conifers, some tropical families, such as the dipterocarps (Dipterocarpaceae), and southern hemisphere trees such as *Nothofagus* and *Eucalyptus.* In contrast, AMs are less easy to see in the field because the loose network of fungal hyphae (usually Glomales; formerly included in the Zygomycetes but now assigned to a new group, the Glomeromycetes) penetrates into the cells of the root without the glove-like covering, and so can be easily overlooked. These are found in a wide variety of families around the world and include common trees, such as maples *Acer*, elms *Ulmus*, ashes *Fraxinus* and poplars *Populus* spp.[1] Five other types of mycorrhizas have been identified across the plant kingdom (Smith and Read 2008); the most relevant of these to trees is the *ericoid* type which has many similarities to AMs and can be found in members of the Ericaceae family which includes most heathers.

AMs are more typically found on nutrient-rich soils where nitrogen is fairly abundant and phosphorus tends to be limiting. Here, the dead leaf litter rapidly decomposes and the mycorrhiza helps the tree, primarily by increasing the effective surface area of the roots as the hyphae spread out into the soil (sometimes called the *extramatrical mycelium*). Despite this enormous increase in effective soil exploration, it is quite an easy role for the fungus and so is fairly cheap for the tree: 2–15% of carbon fixed is given to the fungi.

By contrast, ECMs are more frequent at higher latitudes, on colder soils where nitrogen is in short supply. Here, the main role of the fungus is in directly breaking down the litter and passing the nutrients to the tree. This is a more costly relationship because it requires more fungus to be effective. In pines *Pinus* spp. there can be up to 1000–8000 cm of fungal hyphae for every centimetre of root (Lipson and Näsholm 2001) and typically the fungus uses up to 20% of carbon fixed by the tree, but on these soils it is sugar well-spent to acquire scarce nitrogen. It can get even better: the ECM fungus *Laccaria bicolor* has been found to trap and digest springtails and pass the nitrogen on to its tree hosts. Klironomos and Hart (2001) found that eastern white pine *Pinus strobus* could gain up to 25% of its nitrogen from these soil animals. Mycorrhizas can also

---

1 Families with arbuscular mycorrhizas include: Araucariaceae: monkey puzzle, kauri; Betulaceae: birches, alders; Cupressaceae: cypresses, junipers, redwoods, swamp cypress; Ginkgoaceae: maidenhair tree; Hamamelidaceae: sweet gums, witch hazels; Juglandaceae: walnuts, hickories; Magnoliaceae: magnolias, tulip trees; Oleaceae: olives, ashes; Salicaceae: willows, poplars; Sapindaceae: maples; Taxaceae: yews; Ulmaceae: elms.

aid in the establishment of seedlings. Horton *et al.* (1999) found that Douglas fir *Pseudotsuga menziesii* seedlings could only successfully establish near to bushes of *Arctostaphylos* spp. with which they shared a number of ECM fungi. As climate change increases, it is likely that ectomycorrhizal associations will become even more valuable to the host trees (Gehring *et al.* 2014).

In terms of tree care, the relationship between mycorrhizal fungi and root can appear to be a clear and simple 'good thing': ensure the mycorrhizas are working and everything will be fine. In many cases, this is largely true – mycorrhizas are extremely beneficial – but it is not always quite so clear-cut. The net benefit of the mycorrhiza to the tree may be strongly influenced by the interaction between the fungus, the species of tree and the microbial community sheathing the roots in the rhizosphere, with each having its own effect on nutrient availability (Prescott and Gayston 2013). This can make it less easy to predict the effect of mycorrhizas on tree growth. Trees on nutrient-rich soil will have less need of mycorrhizal fungi, particularly ECMs, because they become an unnecessary cost if sufficient nutrients are readily available. In some cases, seedlings need the mycorrhizal association to help them become established in a highly competitive environment, such as under dense shade, but would grow better later in life without the fungus. There is evidence from Scots pine *Pinus sylvestris* that when nitrogen is in short supply, the ectomycorrhizal fungi will increasingly hoard it (Näsholm *et al.* 2013). As the tree supplies more sugar to encourage nitrogen delivery, this results in yet more nitrogen being locked up in the fungus, creating a positive feedback loop to the detriment of the tree. In this case, it would be better for the tree to get rid of the mycorrhizal fungi but the tree can be trapped in a binding agreement, unable to sever the connection with the fungus, even though it would show better performance without it (Franklin *et al.* 2014). Moreover, in weak trees, the mycorrhizal fungi can become parasitic if the tree is not providing sufficient sugars, again to the further detriment of trees. It can become more complicated still because each ectomycorrhizal fungus may be connected to different trees and each tree is usually colonised by more than one fungal individual. In this case, it becomes a market with each player trying to maximise their own benefits. Rising amounts of aerial nitrogen pollution will also sway the cost–benefit analysis for each partner, adding yet another variable to the equation. The take-home message from this is that in natural forests, ECMs are not always cosy helpful associations.

Commercial Inoculants

A number of commercial mycorrhizal inoculants are available either applied as a powder or in water to planted seeds and seedlings, or applied as liquid suspensions to the roots of mature trees, usually through holes drilled into the soil 0.5 m apart around the drip-line of the tree. The inoculant usually contains a mixture of generalist AM and ECM fungi so that one inoculant fits many potential host species. A study in Chile has shown that a non-indigenous pine could establish with just one ECM fungal species, suggesting that artificial inoculation of even a limited range of fungal species should allow most planted trees to establish mycorrhizal links (Hayward *et al.* 2014).

In natural forests, mycorrhizas, particularly ECMs, may not always be beneficial to the tree, but in urban soils, with their myriad problems, including compaction and/or pollution and which may also be lacking natural sources of mycorrhizal fungi, inoculations may be beneficial. This is particularly true if non-native trees are planted because they would be expected to find a smaller range of compatible fungi in the soil than would native species (Lothamer *et al.* 2014); but even native trees can have very variable colonisation by mycorrhizas when they leave the nursery. This is especially the case as a number of studies have shown that the mycorrhizas in nursery soils do not necessarily survive in urban soils. However, the results of studies using inoculated mycorrhizas have been very mixed, with a few showing benefits to some tree species under certain conditions (e.g. Wiseman *et al.* 2009; Ferrini and Fini 2011), while others have shown no benefits (e.g. Appleton *et al.* 2003; Wiseman and Wells 2009), or benefits only after several years. Some of the perceived benefits may in fact come from other additives to the inoculant mixture, such as clay granules, fertiliser or rhizosphere bacteria. In most cases, the success of inoculations has been evaluated in terms of height and stem diameter growth, but Garbaye and Churin (1996) found that in 7–9 cm diameter silver limes *Tilia tomentosa* planted in Paris, leaf yellowing in autumn was delayed after inoculation. However, most importantly, none of the studies have shown any negative effects. In terms of practical advice, it can be concluded that in *urban* soils the use of commercial mycorrhizal inoculations are likely to benefit only a limited number of trees, but as they do not appear to do any harm they can be used where it is economically feasible. In all cases, recently produced inoculum containing a range of mycorrhizal species should be used to maximise the proportion of live spores and the chances of fungal compatibility with the target trees.

Pathogenic Fungi

As well as the symbiotic mycorrhizas, some fungi are also parasites (living on their hosts but not killing them), while others are saprotrophs (decomposing the dead wood) and yet others are pathogens (causing disease and potentially killing their hosts). It is also possible that the same fungus can work in different ways at different times. For example, a fungus can invade and parasitise a tree without causing great harm, but if the tree is weakened by, say, a drought, the fungus becomes a pathogen that may overwhelm the host, and then live off the dead wood (acting as a saprotroph) while looking for another host tree. Saprotrophic fungi in wood decompose cellulose, hemicellulose and lignin in various proportions, depending on the specific host–fungi relationship. This can clearly have consequences for the tree's stability as these are the compounds that provide wood with its mechanical properties (for further details see Decay in Stems).

Pathogenic fungi can be truly formidable, exemplified by Dutch elm disease caused by *Ophiostoma novo-ulmi* that killed 15 million elms in the UK alone in the 1970s, and chestnut blight caused by *Cryphonectria parasitica* that killed 99.9% of American chestnuts *Castanea dentata* (about 3.5 billion trees) in eastern North America in less than a century. It has also been said, probably with a little exaggeration, that the growth of every tree in the UK is suppressed by honey fungus *Armillaria* spp., even if it is not killed by it. Even here, it is not always straightforward. Of the different honey fungus species, *Armillaria mellea* is the most harmful pathogen, especially in the lowlands, but is usually less virulent in woodlands because the presence of other fungi diminishes its effectiveness as a pathogen. Moreover, the pathogenicity of *Armillaria* is affected by the condition of the host. If the host is stressed by recent drought or waterlogging, then *Armillaria* already in the tree as a parasite is likely to become more virulent and lethal (Popoola and Fox 2003).

(a) (b)

Figure 9.4 The fungus *Sirococcus tsugae* on Atlas cedar *Cedrus atlantica*, causing severe shoot blight and defoliation that is currently an aesthetic problem rather than being fatal to the tree. The disease was first reported in Britain in 2013. (a) The infected cedar in the foreground is losing many of its needles and becoming very unsightly. Deodars *Cedrus deodara* are currently unaffected in Britain, although they are susceptible in North America. The fungus also attacks hemlocks *Tsuga* spp. (hence the species name of the fungus), and infected trees have been found in the south-west UK. (b) The affected needles turn a characteristic pink before going brown and producing the fungal fruiting bodies.

Other fungi may not be voracious killers, but they can disfigure ornamental trees and are more likely to lead to the tree being removed as it becomes too unsightly or in danger of shedding dead wood before the disease can kill it. The current fungal infection of Atlas cedar *Cedrus atlantica* (Glauca group) is a good example (Figure 9.4).

Various *Phytophthora* species are also causing problems around the world, from the USA to New Zealand – these are an Oomycete, a fungus-like organism. One of the most potent of these appears to be *Phytophthora ramorum*, the cause of sudden oak death in California and southern Oregon, killing hundreds of thousands of tanoak *Notholithocarpus densiflorus* and Californian black oak *Quercus kelloggii*. It has also now been found in at least another 100 species. A different strain of the same pathogen is now established in Europe. First found in the UK in 2003, it is now found in *Camellia, Larix, Magnolia, Pieris, Viburnum* and particularly *Rhododendron*, and poses a threat not only to oaks, but also to *Acer, Castanea, Fagus*, and a variety of shrubs. Fortunately, the two oaks native to the UK (pedunculate oak *Quercus robur* and sessile oak *Q. petraea*) are members of the white oak group which are likely to be less vulnerable to the disease. Detecting *Phytophthora* is not straight-forward, and to be certain of the species' presence requires DNA testing or microscopy. The problem is made more difficult because Europe has been suffering oak decline for decades, ascribed to a variety of causes, such as root-infecting fungi, other *Phytophthora* species, drought and various insect attacks. *Phytophthora* is likely to keep on spreading because the spores are readily moved in water and mud transported on shoes or vehicle tyres. Although this is still a challenging group of pathogens to manage, improving site drainage and soil decompaction can slow disease progression. Phosphite fungicide applications delivered via stem injection in combination with a bark spray and surfactant have been found to give some control of *Phytophthora ramorum* (Garbelotto *et al.* 2007). Other studies show the use of biochar (Zwart and Kim 2012) and pure mulches (Percival 2013) also show promise as treatments to reduce the severity of disease caused by *Phytophthora* spp. Management of pests and diseases is discussed further in Expert Box 9.1.

Expert Box 9.1 Tree Health Care Technology: Pests and Diseases
Glynn C. Percival

Until recently, pest and disease management has played a minor part for professionals involved in urban tree management. However, the dramatic increase in pests and disease outbreaks within urban forests over the past 10 years now means management strategies are of fundamental and economic importance. Pertinent examples for the UK alone include *Pseudomonas* bleeding canker *Pseudomonas syringae* pv. *aesculi*, sudden oak death *Phytophthora ramorum*, horse chestnut leaf miner *Cameraria ohridella*, oak processionary moth *Thaumetopoea processionea*, Massaria canker of London plane *Splanchnonema platani*, sweet chestnut blight *Cryphonectria parasitica*, ash die-back *Hymenoscyphus fraxineus, Chalara fraxinea* and Asian long-horned beetle *Anoplophora glabripennis*. Uncontrolled, history has shown that these types of pests and diseases can prove devastating to urban treescapes.

Pests and Disease Management: A Suggested Protocol

1) *Identification and diagnostics*

   Correct identification of a pest or disease at the early stages of attack is critical. The saying 'prevention is better than cure' is highly relevant. Successful management of a pest or disease when severity or infestation is high is extremely difficult, requiring a significant input of labour, finance and resources. Management at the early stages of attack is more likely to succeed with a fraction of the resources required. This can be achieved by regular plant inspections. Types of information that should be collected during inspections include:

   - Site information (i.e. climatic conditions); existing and previous landscape practices; relevant building and/or construction work; soil conditions, to include structure (clay, loam, sand) and elemental composition (macro/micronutrients, pH, organic matter, salinity, heavy metals).
   - Tree information: condition, developmental and phenological stage.
   - Pest information to be collected includes identification, population level, and life cycle stage. Note, the presence of an organism on a tree does not imply that it is the causal agent. Improper identification of an organism can lead to misdiagnosis of the problem or improper prescription of a management tactic.

2) *Aims of the management strategy*

   Prior to commencing a pest and disease management strategy, the first question to ask is 'What is the aim of the management strategy?'

   Is it prevention? Many diseases and insect pests can be effectively controlled by applying plant protection products (fungicides, bactericides, insecticides) before their populations have the opportunity to build up.

   Alternatively, it might be decided to try to eradicate a pest or disease. In this case, it is worth bearing in mind that total eradication is extremely difficult to achieve. However, eradication strategies may be necessary when dealing with highly noxious pests, such as oak processionary moth, that presents a public health hazard.

   The 'norm' in most situations, however, is suppression. In this instance, the intent is to reduce pest and disease populations and associated landscape damage to a tolerable level.

3) *Development of a management strategy*

   Due to recent developments in tree protection technology, several pest and disease management options exist that include the use of chemical, bio-control, nutritional,

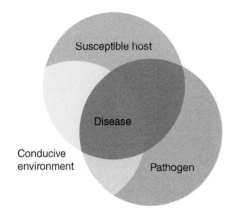

Figure EB9.1 Classic pest and disease triangle. Environment=Total of conditions influencing pest and disease development. Pathogen=Virulence and abundance of pest and disease. Host=Total of conditions influencing host development. Disease=Population and severity of pest and/or disease attack.

cultural and physical tactics. Ideally, all management tactics should be integrated into a single management strategy for maximum efficiency and effectiveness.

Integrated approaches to pest and disease management need to ensure each tactic focuses on one side of the classic pest and disease triangle (host, pathogen and environment; Figure EB9.1). Pests and diseases require each of these parameters to be suitable in order to infect a host. Integrated control measures should focus on adapting one or more of these parameters in favour of tree vitality, to create conditions that detrimentally disrupt pest and disease life cycles.

4) *Management tactics for pest and disease control*

Management tactics should be selected that will have the least detrimental impact on the landscape ecosystem. Options include the following:

- Species selection: Practitioners should be aware of the information that exists linking tree species to a specific pest and/or disease susceptibility or resistance. Selecting high-quality nursery stock also minimises the risk of introducing pests from the nursery into the landscape.
- Plant diversity: Urban landscapes are, typically, contrived, with limited tree species and genera diversity, yet a high diversity of species and genera has been proposed as a solution to a healthy and sustainable urban tree population. The consensus of these studies is that the maximum proportion of any one species should be 5–10% of the population.
- Cultural practices: Cultural control is based on the premise that conditions, suboptimal for tree growth, tend to be favourable to pest and disease attack. Creating a growing environment optimal for plant growth in turn creates one unfavourable for pest and disease attack. This can be achieved by manipulation of the growing landscape environment by ensuring optimal soil nutrition (fertilisers, mulches), soils are not compacted and are well aerated (use of air excavation technology) and root–mycorrhizal associations are encouraged.
- Plant protection products: One of the most effective but controversial means of managing pest and disease outbreaks is through the use of chemical plant protection products (i.e. insecticides and fungicides). Internationally, countries have access to a broad range of chemical plant protection products that are registered as foliar sprays, soil injections and/or trunk injections. Within the UK, however, tight legislative restrictions regarding the use and application of insecticides and

fungicides means only a limited number of products are commercially available. Readers are advised to obtain copies of the latest Pesticide Safety Directorate, published by the British Crop Protection Council, CABI Publishing, Oxon, UK, which details the range of products currently registered for use within the UK amenity environment. Analogous guidance from relevant authorities should be sought in other countries.

To Spray, or Not to Spray?

Currently, management tactics rely heavily on repeat spray technology of a registered fungicide or insecticide. Within UK urban landscapes, sprays tend to be applied by either high or ultra-low/low volume technology (i.e. electrostatically). Both have their advantages and disadvantages. High volume sprays rapidly apply a product, ensuring thorough coverage of the leaf surface; however, drift can be problematic. Ultra-low and low volume sprays apply far less product but substantially reduce drift. Current consensus within the UK is a preference for low over high volume spray technology. Irrespective of application technology, for effective, responsible pest and disease management, the following procedure should be adopted:

- Select an appropriate officially registered product.
- Use the correct dosage.
- Apply the product at the correct life cycle stage and with appropriate frequency.
- Ensure the individual employed to apply any form of plant protection product has been appropriately trained and qualified.

Plant protection products are classified as either contact or systemic. Contact products kill pests and diseases through direct physical contact with the material or its residue following application. Systemic pesticides are applied to the root-zone, stem or foliage of the tree, and translocated in the vascular system to affected tissues. The use of an appropriate product can have a marked impact on controlling a specific pest, as seen in the Case Scenarios.

Case Scenario 1: Royal Botanic Gardens, Kew, London

Several years ago, the Royal Botanic Gardens, Kew experienced an outbreak of the oak processionary moth (OPM), a defoliating insect of English, sessile and Turkey oaks *Quercus robur, Q. petraea* and *Q. cerris*. OPM is also a risk to human health. The older caterpillars become covered in irritating hairs that contain a toxin. Contact with these hairs can result in skin irritation, such as dermatitis, and allergic reactions, such as rashes and conjunctivitis. If hairs are inhaled, respiratory distress such as asthma or anaphylaxis can occur. These problems are significant because the Royal Botanic Gardens receives up to 10 000 visitors a day.

For this reason, it was decided to spray all susceptible oak trees within the gardens: no mean feat when over 700 mature trees had to be sprayed, with most trees being over 30 m high. In year 1, it was decided to use a bio-control product known as DiPel®. DiPel is comprised of live bacteria (*Bacillus thuringiensis*) which produce a toxin that controls caterpillars when ingested by the insect. DiPel also has the advantage in that it can be used near water, and collateral damage (impacts on non-target insects) is low. Unfortunately, results were variable, influenced by weather conditions at the time of treatment, and it was estimated that only 60% control of OPM was achieved. Results also demonstrated that the older the OPM caterpillars became, the less vulnerable they were to *Bacillus thuringiensis* and therefore the lower the population kill.

As a result of the poor control rates achieved in year 1, in year 2 it was decided to use a synthetic pyrethrum (deltamethrin) which combines rapid control with long residual activity on the foliage of treated trees. Effects were apparent within 2 hours of spraying, with a 100% kill of OPM recorded. However, deltamethrin is a broad-spectrum insecticide which also caused substantial collateral damage to non-target insects (e.g. aphids, leaf hoppers, leaf miners and flea beetles).

From year 3 onwards, OPM management relies heavily on an insect growth regulator, Dimilin® Flo. Basically, treated caterpillars do not reform an exoskeleton after moulting. Similar to DiPel, Dimilin Flo is most effective on young caterpillars but, importantly, only affects caterpillars and moths. Dimilin Flo has very little effect on other insects, such as ladybirds, beetles and honey bees. Consequently, there is little collateral damage from using this product. Data collected also shows that greater than 90% control can be achieved. This high kill: low collateral damage ratio makes Dimilin Flo the preferred OPM management option.

However, despite the high target kill rate and low collateral damage, use of sprays within urban landscapes is still, in the majority of cases, to be avoided because of possible spray drift contamination of surrounding vegetation, traffic and pedestrians.

A Way Forward for the Future?

Over the past few years there have been major developments in trunk injection technology. Systems widely used in the USA, Canada and Australia quickly inject an insecticide or fungicide directly into the tree's vascular system. Trunk injection technology has a number of advantages over spray technology:

1) Injecting directly into the sap stream of a tree uses the tree's own natural transport system. Consequently, no air, soil or groundwater contamination occurs and no product is wasted.
2) Once injected, products are moved quickly throughout a tree.
3) Once injected, up to 2–3 years' control can be achieved.

Case Scenario 2: Barnes Common, Richmond, London

English oaks located at Barnes Common, a 40+ ha area of common land within the borough of Richmond upon Thames, provides a vital green space for the capital. It is greatly utilised by a local population of almost 13 000 as well as many overseas visitors. However, following an outbreak of OPM within the common, 40 infested English oak trees were injected with an insecticide. The degree of OPM control was recorded by counting the number of OPM nests over the following 3 years. Results to date have been very impressive, with none of the injected trees having OPM nests while those trees that were not injected (or injected with water) having, on average, six nests per year. Importantly, trunk injection systems were discreet and, unlike spraying, could be performed irrespective of weather conditions, such as rain or high winds.

To Spray or inject?

Whether to spray or inject will be heavily influenced by tree population size and location. For example, in environmentally sensitive areas, like schools, parks, golf courses, playgrounds and near waterways, injection technology would be preferred. Likewise, injections can be scheduled during inclement weather such as heavy rain and high winds. However, trunk injections are more time-consuming and expensive than spray technology. Consequently, in areas of large tree populations, spray technology has the advantage that hundreds of trees can be sprayed in a day. Recent research from the USA and Canada has shown that injection of 20% of ash trees resulted in an 80% control of emerald ash borer populations. Such a result indicates that injection of an entire tree population may not be necessary when controlling insect pest outbreaks, and that strategic selection of specific trees may be sufficient to lower populations to tolerable levels.

Conclusions

The recent marked increase in pests and disease occurrence is already threatening significant numbers of urban trees. Some pests, such as OPM, also have important implications for human health. Well-planned, effective pest and disease management is therefore of fundamental importance. Pest and disease management programs should adopt a range of appropriate strategic practices:

1) Correct species selection, as resistance to a pest or disease can vary immensely within trees from the same family.
2) Evaluating and correcting soil nutrient deficiencies through appropriate fertilisation and remediation.
3) Mulching to promote root vigour as a means of increasing tree vitality.
4) Visually inspecting trees on a regular basis, allowing preventative rather than therapeutic remedial measures.
5) Use of innovative plant protection technologies and application methods.

If practitioners do not adapt current management systems to embrace new approaches and new ways of thinking, then history has shown that many dominant trees are likely to disappear from our landscapes.

Defence of Stems

Most trees can cope quite well living in complex ecosystems, providing their bodies remain intact and they stay healthy. However, to many microorganisms and insects, tree stems represent a colossal larder that could contribute to the success of their own kind – they just need to find a way in. Fungi can enter trees through breaks in the bark either above or below ground, such as where a branch breaks from the tree or a root naturally dies. Pathogenic fungi, such as *Armillaria*, can aggressively invade the living tissue of roots and so it is almost inevitable that a tree will become infected by pathogenic fungi at some point in its life. Whether this threatens the future of the tree very much depends on what cocktail of fungal species are involved, their virulence, and the vigour of the infected tree. Underlying this, a tree has a number of defences against the invaders.

Although forest pathologists have been observing the relationship between wood and microorganisms for well over 100 years (Merrill 1992), Alex Shigo and his colleagues at the US Forest Service transformed the way these ideas were communicated through their *Compartmentalisation of Decay in Trees* (CODIT) model (Shigo and Marx 1977; Shigo 1979, 1984). This model describes how trees seal off a portion of their xylem using four boundaries or 'walls'. *Wall 1* restricts decay through the stem (in an axial direction) by plugging tracheids and vessels; *wall 2* restricts decay towards the centre of the stem (radial) using the resilience of latewood; *wall 3* restricts decay around the stem (tangential) through chemical modification of rays; and *wall 4* represents the barrier zone that seeks to separate xylem present at the time of injury from new xylem grown afterwards. The 'walls' become progressively more effective as barriers to decay from 1 through to 4.

While this model remains popular, it has been criticised by a number of authors. Some mycologists took exception to the model because it is not solely defensive 'walls' that protect the wood from decay, but also unfavourable conditions for fungal growth (Boddy and Rayner 1983; Rayner and Boddy 1988). In particular, the low oxygen content of functional sapwood helps to restrict fungal spread. Only when the wood is damaged, allowing extensive aeration, is there enough oxygen to allow rapid fungal growth. A further criticism of the CODIT model is the choice of the word 'decay' in the acronym. As Dujesiefken *et al.* (1989) pointed out, many of the reactions that were within the CODIT model aimed to keep the xylem conducting water effectively, rather than preventing decay. Problems for the tree relating to wood decay are far less pressing than hydraulic dysfunction. It was therefore suggested that *dysfunction* would be a more appropriate 'D' word for CODIT. This certainly has merit. Pearce (1996) provides an excellent review that attempts to integrate the active microbial defences of the CODIT model with the passive environmental constraints on the growth of pathogens.

The strength of the CODIT model is that it recognises that the tree is, by nature, a highly compartmented structure. It therefore provides a useful framework for describing the development of dysfunction within the tree. However, the discussion that follows focuses on three major components of stem defence: the inherent resilience of the stem; the reactions of living cells; and the reaction of the cambia (Table 9.1). This approach has the advantage of considering the entire woody stem, including the bark, as a unit rather than just focusing on the wood (xylem).

The challenge for the tree is to keep out potentially pathogenic microorganisms or damaging insects without hindering those that may help the tree. The stem – the trunk and branches – and roots are crucial to the long-term survival of a tree in their role of providing biomechanical support and in transporting materials to where they are needed. Thankfully, although some of the details can differ between trees, there are a number of shared defences (Table 9.1).

To survive, trees have had to develop an inherent resilience against pests and diseases. Most insects and microorganisms simply do not have the right tools to get into the tree, let alone break down its complex carbon-based structures. Lignin, cellulose, suberin

Table 9.1 Summary of the major components of stem defence.

Component of stem defence	Key examples
Inherent resilience and passive defence	Bark, spines and thorns
	Lignified tissues
	Regions of high wood density
	Low oxygen in functional sapwood
	Heartwood with high levels of protective compounds
	Passive closure of bordered pits
Reaction of living cells	Anaplasia of parenchyma to form wound periderm or repair the vascular cambium
	Synthesis and distribution of resins, gels and other extractives to sites of vulnerability
	Blockage of vessel lumen via tyloses and gels
	Chemical modification of axial and radial parenchyma with compounds such as suberin, phenols, tannins and terpenes
Reactions of the cambia	Reinstatement of the periderm
	Production of a barrier zone with a modified xylem composition; typically this includes fewer conducting elements and higher proportions of fibres and/or parenchyma. For some species an increase in secretory canals is also important
	Development of wound callus

and the host of other compounds found in wood and bark are not readily digestible: it takes highly specialised enzymes and, in the case of some insects, mouthparts before the tree's body becomes a viable food source.

The outermost defence is the periderm and, where present, the rhytidome (see Chapter 2). These encapsulating tissues keep potentially harmful agents out, maintain the water status of the sapwood and control gas exchange into and out of the stem. In many ways, it is analogous to our skin: infection is far more likely without it. Very thick barks, such as in cork oak *Quercus suber* and giant sequoia *Sequoiadendron giganteum*, are even capable of protecting the tree from fire. In cold environments, thick bark also provides useful protection from frost injury.

Outer bark, being largely made up of dead tissue, provides a *passive defence*. In other words, the defensive capabilities are put in place while the tissue is growing and, crucially, do not require any additional energy to make them work once in place. In addition to providing a physical barrier, the lignin, suberin and waxes of the bark have antimicrobial and hydrophobic properties that keep water in the stem, as well as reducing the likelihood of colonisation by microorganisms. Protective substances, such as tannins and phenols, may also be deposited in the bark to reduce its palatability further. In a number of species, specialist cells known as *idioblasts* produce calcium oxalate crystals to form 'stone cells' that physically deter browsers and wood-boring insects from trying to chew their way to the nutritious phloem beneath (Hudgins *et al.* 2004). These factors, along with the low nutritional value of bark, make it an excellent barrier.

Inner bark (secondary phloem) also provides resilience by using thickened scleren-chyma (phloem fibres) and calcium oxalate crystals. It also has living cells that provide *active defence*; that is, cells that remain alive and use energy to modify or create new cells in response to injury or invasive organisms (Pearce 1996). Some parenchyma cells in the phloem are specialised to produce and accumulate phenolic compounds. In coni-fers, these cells, known as *polyphenolic parenchyma cells* (PP cells), are often produced in rings around the tree and may remain alive for decades (Krekling *et al.* 2000). They are also capable of increasing their toxicity in reaction to specific invading organisms (Franceschi *et al.* 2005). Such a tailored response makes these bands of specialised parenchyma cells key structures in stem defence, where present.

Effect of Wounding to the Bark

So what happens if the bark is breached? Many species use the living cells within the inner bark to produce substances that hinder the development of microorganisms and insects: resins in many conifers; gels, mucilages, gums, resins, oils and latex in dicotyle-donous trees; and kino (a bright red, resin-like material) in *Eucalyptus* and a number of other trees. These substances are either produced directly by the parenchyma around the wound or ooze out of canals (resin) or veins (kino) from specialised reservoirs. Even in species where resin canals are not normally found, *pathological* or *traumatic canals* can be induced through mechanical wounding, microorganism or insect activity.[2] These substances work by either being toxic to pathogens or by physically clogging up the mouthparts of insects. However, some highly specialised microorganisms can use these complex substances as a food source.

In many cases, these defensive substances can be economically and socially impor-tant. For example, myrrh *Commiphora myrrha* and frankincense *Boswellia* spp. are derived from tree resins. Interestingly, it is likely that these were the reason for the first (*c.* 1550–1292 BC) long-distance tree transplanting that occurred from the Horn of Africa, present day Somalia, to Egypt (Dixon 1969). Most ancient societies valued resins and gums so highly that the expense of transplanting was worthwhile. Today, one of the most widely used defensive substances commercially harvested is the latex produced upon superficial wounding of the rubber tree *Hevea brasiliensis* (Figure 9.5).

The initial protection of wounds by chemicals is rapidly followed by repair. Minor injury to the surface of the bark can be repaired by the initiation of *wound periderms* (Biggs 1992). These usually occur from the *anaplasia* of parenchyma cells within the cortex of the inner bark.[3] Initially, a new phellogen or cork cambium is formed, and before long this creates new corky cells (see Chapter 2) that can eventually link up with the adjacent uninjured parts of the existing bark (Fink 1999). The wound periderm has the role of restoring a continuous physical barrier between the sapwood and the out-side world, and can easily be spotted, as it looks different from the normal bark. It is for

---

2 Pathological or traumatic resin canals (= ducts) are formed in genera such as *Abies, Cedrus, Sequoia* and *Tsuga* in response to injury. Resin canals are a normal feature of the xylem of genera such as *Larix, Picea, Pinus* and *Pseudotsuga,* but traumatic resin canals may also be formed.

3 Anaplasia is defined as the dedifferentiation of already differentiated mature cells back into meristematic cells (Fink 1999). This is a particularly important process in the repair of wounds in trees. Intact stem parenchyma adjacent to the wound can become meristematic again in order to produce new protective tissues such as the wound periderm.

Figure 9.5 Wounding the stem of the rubber tree *Hevea brasiliensis* causes the tree to produce latex, seen here as a white liquid running down the wounded stem. Once collected, the latex is turned into rubber. Although originally from Brazil, rubber trees are now economically important trees in many tropical countries; this plantation is in Thailand. Note the partially processed latex hanging in the barn behind this recently wounded tree.

this reason that surface injuries, such as those caused by scratching (e.g. by bears sharpening their claws or by romantics carving their initials into a tree), can be seen long after the wound has been sealed by new bark. The formation of wound periderms involves energy to create new tissues, so it is a form of active defence. As such, wound periderms can only form during the growing season while conditions are suitable for tissues to grow.

Where bark has been removed from large areas of the stem, a wound periderm is unlikely to be able to replace all the lost bark. However, parenchyma cells on the wound surface occasionally become modified by suberisation and lignification of their walls (*metacutisation*) to become cork cells (Fink 1999). How well this works in creating a barrier is highly dependent on the number and distribution of parenchyma cells on the exposed surface. In practice, metacutisation can only ever create a partial boundary but it may help to prevent desiccation of the underlying tissues. Eventual closure of the wound surface will depend on *wound callus* created around the edge of the wound by the vascular cambium, potentially in combination with new meristematic cells formed by anaplasia (Blanchette 1992). New wood and bark starting from the callus grows inwards from the margins of the injury (Figure 9.6) and, over time, may completely cover the exposed surface of the wound. On larger wounds, this can take many years or

Figure 9.6 Occluding tissues forming around a stem injury to European ash *Fraxinus excelsior*.

may never occur. This new xylem tissue tends to have fewer vessels than normal xylem as it does not need to be highly conductive.

Wounds tend to be most easily colonised by fungi and bacteria immediately after the bark is damaged, before the tree can respond. This is helped by the loss of bark and drying of any damaged wood letting in enough oxygen for the microbes to grow. So the outcome of an injury depends on how rapidly microorganisms can colonise the injury, compared with the ability of the tree to establish wound responses that can reduce the oxygen content of the wood (Boddy 1992).

Effect of Wounding to the Sapwood

Inevitably, the removal of the bark or mechanical damage to the stem exposes sapwood to desiccation and embolism. Limiting the hydraulic dysfunction of wounded xylem is of primary importance as it can threaten the immediate survival of the tree. As described in Chapter 2 (see Sapwood and Heartwood), the tree has various mechanisms to stop air being pulled into the xylem. In tracheids this involves the aspiration of bordered pits (a passive defence that requires no energy expenditure). The fact that this event does not require energy means that most conifers do not require extensive axial parenchyma for active responses in their xylem. In vessels, however, parenchyma cells exude gels or push tyloses into the vessel lumen via shared pits, within a few hours of wounding. These features, possibly in combination with the encrustation of vessel perforations, block the vessel lumen. Gels and tyloses not only physically help to prevent any further entry of gas ingress, but can become chemically toxic to microorganisms and so slow their spread up through the xylem. Blocking vessels requires the synthesis of new materials and so is considered an active defence mechanism.

The parenchyma cells within the xylem that produce toxic gels and tyloses may themselves also contain substantial quantities of antimicrobial substances, such as suberin, pectin, terpenes, tannins and other phenolic compounds. Living cells between the xylem tubes (*axial parenchyma*) are joined to those in the rays (*ray parenchyma*) to create a three-dimensional lattice of living tissue throughout the sapwood (Figure 9.7). The precise construction of this lattice is species specific. Species that have bands of axial parenchyma throughout their xylem are very good at resisting microbial growth in a radial direction (bark to pith or the other way round).[4] These bands of parenchyma seem particularly important for defence in many tropical species that lack growth rings created by seasonal changes in the make up of the xylem (Morris *et al.* 2016). Those species without complete bands of axial parenchyma have to rely more on the passive resistance of the xylem to decay, and cannot use active defence mechanisms to the same extent. This is largely brought about by the highly lignified cell walls of fibres or tracheids that increase wood density in the latewood. Rays, particularly the large multiseriate rays possessed by some species, provide a barrier to colonisation in a tangential direction around the tree (parallel to the vascular cambium).[5] As rays are living, these can chemically modify and provide an active response to pathogens.

The structural and chemical changes produced by the living parenchyma lead to the formation of a narrow zone of discoloration between the wound-affected sapwood and the surrounding functional sapwood, called *wound wood*. This region is often referred to as the *reaction zone* because the change in colour is ostensibly brought about by a reaction of the tree to injury (Shain 1995). The reaction zone is broadly equivalent to walls 1, 2 and 3 in the CODIT model. Similar processes are also involved in the production of regular heartwood, so wound wood is similar to some forms of heartwood. In trees such as *Fagus* or *Acer* species, which do not normally form a distinct heartwood, the reaction of the parenchyma cells to injury or other trauma can sometimes convert sapwood into 'false heartwood' (see Chapter 2), which has defensive properties. The major difference between wound wood and regular heartwood is that the former is an active defence mechanism produced in response to external cues (the wound), whereas the latter is produced in response to internal cues, as part of normal xylem development (Spicer 2005). The ability of wound wood to resist the invasion of fungi and bacteria depends on the extent of the parenchyma lattice and the potency of the defensive chemicals produced.

In the absence of chemical or physical changes to the xylem produced by the living parenchyma, defence against microorganisms is down to the structure of the wood. The lumina of vessels and tracheids make good corridors for fungal growth as soon as they

---

4 Genera from northern temperate regions known to have banded axial parenchyma include: *Alianthus, Alnus, Amelanchier, Betula, Carpinus, Castanea, Corylus, Fagus, Juglans, Koelreuteria, Morus, Nyssa, Ostrya, Pistacia, Platanus, Prunus* and *Styrax*. Based on information from the InsideWood database (2004–onwards; Wheeler 2011).

5 Genera from northern temperate regions with large multiseriate rays include: *Acer, Ailanthus, Casuarina, Celtis, Cladrastis, Cornus, Fagus, Fraxinus, Gymnocladus, Juglans, Morus, Pistacia, Platanus, Prunus, Quercus, Sambucus, Staphylea, Styrax, Symplocos* and *Ulmus*. Based on information from the InsideWood database (2004–onwards; Wheeler 2011).

(a)   (b)

(c)   (d)   (e)

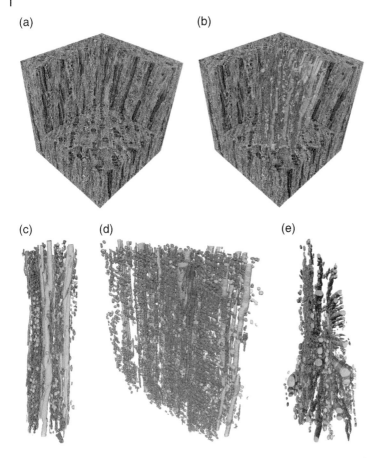

Figure 9.7 Three-dimensional views of a cube of wood from a *Ziziphus obtusifolia* stem. The wood structure is shown in (a) and various tissues, shown in different colours, have been added in (b). In *Ziziphus*, the rays (red) are arranged in a planar orientation (radiating out from the stem centre) and create discrete sectors between vessel rows, as visualised from different perspectives: (c) in tangential, (d) in oblique radial and (e) in transverse. The axial parenchyma (purple) surrounds the vessels (blue). Calcium oxalate crystals can be found distributed throughout the ray parenchyma (green) and act as a chemical defence mechanism. Scale varies with perspective. Edges of the grey wood cube: 650 μm. *Source: Morris et al.* (2016), http://journal.frontiersin.org/article/10.3389/fpls.2016.01665/full. Licensed under CC BY 4.0.

become partly gas-filled. This helps to explain why trees have evolved mechanisms to block vessels, and (in gymnosperms) the pits of tracheids, as soon as gas gains entry.

Tracheids, vessels and fibres (usually) are dead when mature so cannot react to the presence of microorganisms. However, the lignin in woody cell walls is highly resistant to decay. Thus, if the cell walls are relatively lignin-rich or if they are thick in proportion to cell diameter, the wood will be more resistant to decay (and denser). Branch junctions tend to have regions of dense wood and contain cells with rather tortuous pathways. Both of these factors will also bestow resilience to microorganism colonisation and will delay them growing from a branch into the stem.

A further line of defence to the invasion of fungi, bacteria and insects happens when the tree continues growing after it has been wounded. When the vascular cambium near the wound resumes growth, the new xylem is modified too so that it becomes more difficult for microorganisms to move from the wound wood into wood that is subsequently grown. This *barrier zone*, represented by wall 4 in the CODIT model (Shigo 1984), protects new xylem from microorganisms that may have already colonised the wound. To achieve this, the composition of the barrier zone xylem is quite different from xylem that is more typical of the species. For example, in sugar maple *Acer saccharum*, the xylem usually consists of about 15% vessels, 15% parenchyma and 70% fibres, whereas the xylem formed after wounding has been found to consist of 5% vessels, 95% parenchyma and no fibres (Rademacher *et al.* 1984). In European ash *Fraxinus excelsior*, earlywood vessels quadrupled in number but decreased their lumen size by around three-quarters within the barrier zone. In addition, rays increased in size and frequency. These changes reduced conductivity but improved the hydraulic safety of the wound-induced xylem to embolism (Arbellay *et al.* 2012). In balsam poplar *Populus balsamifera*, fibres were found to be the dominant cell type of the barrier zone (Rioux and Ouellette 1991). Barrier zones dominated by parenchyma presumably have the ability to respond actively to ongoing threats; although these are energetically expensive, they provide better defence than barrier zones dominated by fibres. Traumatic resin, kino and gel canals may also be found in the barrier zone to give further protection; these are found in *Eucalyptus, Liquidambar* and *Prunus* species, among others.

In smaller diameter stems, the barrier zone may be continuous around the full circumference of the stem. In larger stems, however, the barrier zone is often *discontinuous* and only present around a portion of the circumference (typically around 30° either side of the wound) and a short way above and below the wound. Usually, within a few years of the injury, the xylem reverts back to its normal pattern of development, but a tree may expand the reaction and barrier zones if microorganisms breach the original boundaries. These internal changes remain encapsulated in the wood and provide a record of injury useful to those seeking to date past disturbances (Schweingruber 2007; Stoffel *et al.* 2010).

Pruning and Wounding

When a tree is pruned, a range of protective mechanisms ensure the integrity of the stem, maintaining its mechanical structure and its ability to conduct water and sugars. Pruning can cause immediate retraction of water columns from the cut surface, leading to at least some embolism below the cut. In gymnosperms, this will cause the aspiration of bordered pits and an influx of resin to the wounded region. In angiosperms, entry of air into vessels is likely to be initially halted by the perforation plates, although if the tree is short of water leading to a high tension within the water column, air may be pulled farther into the stem. For this reason, trees under drought stress can be left with high levels of xylem dysfunction after pruning. Within a few days, the dysfunctional vessels are likely to be blocked, typically by tyloses or gels (for more details see Chapter 2). Parenchyma cells on the cut surface that do not die from desiccation will often become corky (through impregnation with suberin), which provides some protection from subsequent drying and microbes. Wound callus then forms from the margins of the cut and over time will give rise to new wood and bark, which may cover the exposed xylem, although this might take many years. Passive defences from anatomical structures and preformed chemicals will give some resilience to colonisation from microorganisms in the heartwood but the active defence processes described above are critical to the protection of sapwood.

In temperate trees, these active processes are only likely to occur during the growth season as they tend to be limited by low temperatures. For example, tyloses tend not to be able to form below 5°C (Schwarze *et al.* 2000b), and presumably formation of all protective substances is similarly limited by temperature. Further, active responses can only occur in sapwood because living parenchyma is only found in this region. So, any pruning that exposes heartwood must rely on the resilience of the age-altered wood, rather than any active response, to restrict colonisation by microorganisms. For this reason, the nature of the heartwood has a profound effect on the likelihood of microbial colonisation within the tree's central core. Species with regular heartwood are much more durable than many irregular heartwood species such as those that only form ripewood (see Chapter 2). This explains why more smaller wounds are always more favourable than fewer large wounds: less heartwood is exposed in this way. A particularly dangerous scenario is when pruning creates a series of wounds in the same region of the stem (Lonsdale 1999). Here, zones of dysfunction associated with wounding can coalesce to form large volumes of dysfunctional wood. This can aid microbial colonisation and potentially lead to substantial cavities if decay fungi become established. Ultimately, this threatens the mechanical integrity of the entire stem and can lead to tree failure.

Stems provide a theatre in which different actors strive to gain advantage. It is a highly complex arena in which the interaction of numerous species can be relevant. Will the tree contain or 'compartmentalise' damage using strategies honed over millennia, or will the larder be unlocked by the ingenuity of microorganisms and insects? The answer is, of course, that both outcomes occur somewhat simultaneously: trees are capable of resisting infection or infestation, but under certain circumstances there is a shift in favour of the fungi, bacteria or insect. That is how nature works. Processes are dynamic, they are complex; they involve a theatre with many stages and, importantly, a diverse cast.

Decay in Stems

Despite the host of defences that trees possess, many fungi have evolved ways to colonise and decay wood (indeed, this is essential for the recycling of woody material). To use wood as a food, they have developed specialised enzymes to break down the complex carbon structures of cellulose, hemicellulose and lignin into simple sugars that they can use in their own metabolism (Stokland *et al.* 2012).

Traditionally, decay fungi have been categorised into various types of rot (Figure 9.8). *Brown rots* preferentially decay cellulose and hemicellulose but degradation of lignin is limited. *White rots* are capable of removing cellulose, hemicellulose and lignin but differ in the way they go about this. *Selective delignification white rots* preferentially decay lignin and hemicellulose and only move on to cellulose at quite an advanced stage. *Simultaneous white rots* are capable of degrading lignin and cellulose and hemicellulose from the early stages of decay: a particularly acute problem for the tree if the decay is very extensive or in a mechanically critical zone. Cellulose and lignin confer different mechanical properties to the xylem, so their degradation is also associated with different

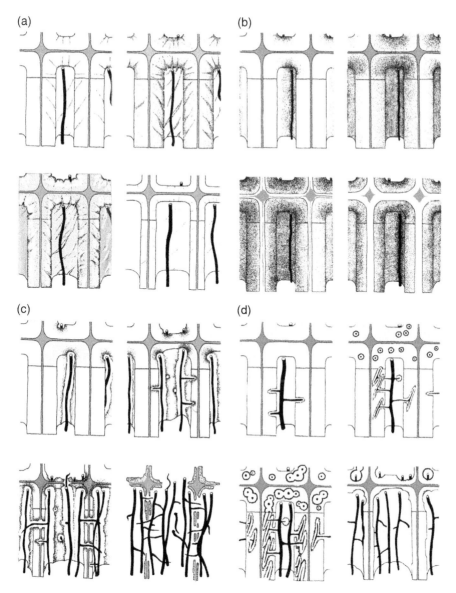

Figure 9.8 The features of different decay types. (a) Brown rot by *Fomitopsis pinicola*. At an early stage of decay, enzymes are secreted by fungal strands (hyphae) growing on the S_3 layer that diffuse into the cell wall. At a more advanced stage, the enzymes have penetrated into the entire secondary wall, causing extensive breakdown of hemicellulose and cellulose. Wood shrinkage leads to the formation of numerous cracks and clefts within the secondary wall. Even at this advanced stage the S_3 remains intact, supported by a matrix of modified lignin. (b) Selective delignification white rot by *Heterobasidion annosum*. At an early stage of decay, enzymes (shown as dots) diffuse into the secondary wall from hyphae. These begin the degradation of hemicellulose and lignin within the secondary wall, extending to the middle lamella. At advanced stages, the preferential degradation of pectin (which holds the cells together) and lignin results in the separation of individual cells from one another. Initially, cellulose remains intact. (c) Simultaneous white rot by *Fomes fomentarius*. At an early stage, degradation occurs around the abundant hyphae growing within the lumen. The cell wall is progressively degraded from the lumen outwards. Individual hyphae penetrate into the cell wall at right angles. The cell wall becomes increasingly thinner and numerous boreholes pass through adjacent cells. At a more advanced stage, degradation is hampered by the strongly lignified compound middle lamella. (d) Soft rot by *Kretzschmaria deusta*. At an early stage of decay, hyphae diffuse into the secondary wall, branch and grow along the cellulose microfibrils in the S_2 layer. This leads to the formation of cavities with conically shaped ends. At an advanced stage of decay, the secondary wall is nearly completely broken down, whereas the guaiacyl-rich compound middle lamella persists. *Source:* Adapted from Schwarze and Baum (2000). Reproduced with permission of Springer.

kinds of loss in mechanical performance. Removal of cellulose reduces the tensile strength of wood; removal of lignin reduces the compressive strength. *Soft rots* are characterised by their fungal strands (hyphae) that tunnel through the cell walls and preferentially decompose cellulose and hemicellulose, although they are capable of slightly modifying lignin (Schwarze *et al.* 2000b).

To enable fungi to overcome the defences of wood, they have evolved some very specific adaptations. Some decay specialists, such as *Ganoderma adspersum*, have developed the capacity to digest the phenolic compounds deposited in tylosis and present in gels that block vessels (Schwarze and Baum 2000). Other fungi show an ability to break down the chemically altered parenchyma cells. Indeed, many fungi show a remarkable versatility that challenges our attempts at classifying them. The 'white rot' fungus *Inonotus hispidus* is able to penetrate the reaction zones of London plane *Platanus × acerifolia* by tunnelling through the cell walls like a 'soft rot' fungus, thus circumventing the phenolic deposits blocking the vessels (Schwarze and Fink 1997). As a result, the toxic compounds produced by the tree are avoided. Once the fungus passes blockages in the vessels, it continues in 'white rot' mode. A similar plasticity has been seen in *Meripilus giganteus*, on large-leaved lime *Tilia platyphyllos* and beech *Fagus sylvatica* (Schwarze and Fink 1998). Moreover, the 'brown rot' fungi *Fistulina hepatica* uses a soft rot mode during the initial decay of the tannin-rich heartwood of oaks *Quercus* spp. (Schwarze *et al.* 2000a). Fungi are incredibly inventive in their ways of overcoming their host's defences and, although the categories of brown, white and soft rot are useful, many fungi do not fall easily into one type, or they may change their mode of decay under certain circumstances (Eaton 2000; Schwarze 2007).

Perhaps it should not be surprising that the hugely diverse fungal world does not fit neatly into categories. For decay fungi, trees simply represent a substrate for growth, an opportunity to expand their territory and a food source to fund their own reproduction. They are extremely important, not only in the cycling of dead wood, but also as creators of habitat for a wide range of other species (see Expert Box 9.3 later in this chapter). Some fungi are highly specialised and may even be restricted to a single species of tree; others are more generalist and can be found on a number of different tree species. Some may only be able to colonise dead wood; others are expert pathogens that can invade living trees. Some are root specialists whilst others are only associated with dead wood above ground. Decay fungi are part of the complex and competitive fungal communities that are highly responsive to their biotic and abiotic environments (Boddy 2001).

Bacteria

Some bacteria are decidedly beneficial to trees. As discussed in Chapter 8, members of the pea family (Fabaceae), such as *Acacia, Laburnum, Robinia, Styphnolobium* and *Vachellia*, have nodules on the roots that contain bacteria (*Rhizobium* spp.) that convert atmospheric nitrogen into a form that is usable by plants. This is a symbiotic relationship similar to mycorrhizas with a cost to the tree, so where soil nitrogen levels are higher, and the tree has less need of the bacteria, there is less nodulation of the roots. A wide range of other trees also have a symbiotic relationship with Actinomycetes, usually *Frankia* spp., a type of filamentous bacteria. These are mostly temperate hardwood trees, including *Alnus, Ceanothus, Elaeagnus* and *Myrica*. These nodules are often a supplement to the normal mycorrhizal relationships, used as an extra source of nutrition.

(a) (b)

Figure 9.9 Chestnut bleeding canker on horse chestnut *Aesculus hippocastanum*, caused by the bacterium *Pseudomonas syringae* pathovar *aesculi*. (a) An early stage with the exudate oozing from the initial infection site. *Source:* Courtesy of Duncan Slater; (b) a more serious infection with the remains of the black ooze at the top, and extensive areas of dead bark caused by the infection.

Rhizobacterial and other nitrogen-fixing bacteria, along with phosphate-solubilising bacteria (e.g. *Agrobacterium radiobacter,* a free-living bacterium that helps make phosphorus more available in the soil), have been applied to soils, primarily in agriculture, to boost the nutrition of plants.

On the negative side, bacteria, particularly species of *Pseudomonas, Enterobacter*, and *Clostridium*, are responsible for small-scale volumes of *wetwood* in trees, and sometimes growth abnormalities (Scott 1984). Areas of bacterial wetwood become obvious as a result of the brown exudate (or 'flux') that oozes from the wound from fermentation pressure. The flux may even smell of alcohol, but usually the liquid is rapidly invaded by other bacteria and yeasts, causing it to have a very unpleasant smell. *Pseudomonas syringae* pathovars, in European ash *Fraxinus excelsior* and *Prunus* spp., go further in producing raised bacterial cankers.[6] In *Prunus*, the bacteria also infect the leaves, killing small areas that fall out, leaving red-ringed holes, known as *shot-holes* because it looks as if the plant has been attacked with a shotgun. More seriously, *P. syringae* pv. *aesculi* in horse chestnut *Aesculus hippocastanum* causes the potentially fatal *chestnut bleeding canker*, seen as a black, bleeding ooze leading to extensive areas of dead bark as the phloem dies (Figure 9.9).

Fortunately, most of the wetwood infections usually do little damage to the tree and do not seriously weaken the structure. Moreover, because the wood around the bacterial

---

6 Pathovar or pv. is a contraction of pathogenic variety.

infection is so wet (and hence low in oxygen) it tends to inhibit the development of wood-decaying fungi. Although unsightly, the best treatment is to leave wetwood infections alone because most attempts to drain or remove the rot act to worsen or spread the infection.

Insects

Pollinators and Defenders

Most major forest trees, certainly in higher latitudes, are wind-pollinated. These tend to form forests dominated by one species, so the high number of individuals of the same species makes wind a very efficient mover of pollen between trees. It is also true that the huge numbers of flowers they produce would require more pollinators than can possibly become available. Conversely, many small trees that grow in small groups or singularly over a wide area are insect pollinated, including *Sorbus*, *Crataegus* and *Prunus* species. The value of this should not be underestimated: in parts of China, apple production is only maintained by laborious hand-pollination of flowers after the decimation of bee populations (Partap *et al.* 2001).

A number of trees around the world use insects, particularly ants, as a form of defence. These are found mainly in tropical trees but also a number of temperate trees. The ants defend the trees, killing or dislodging insects and even birds, and in return are given lodgings and/or food. In the North American black cherry *Prunus serotina*, the red-headed ant *Formica obscuripes* is attracted from its nearby nest to the extrafloral nectaries on the marginal teeth of the tree leaves in early spring. While in the tree, it also helps itself to the young grubs of the eastern tent caterpillar *Malacosoma americanum*, a major defoliator of the cherry (Tilman 1978): both the ant and the tree benefit.

Temperate trees also use much smaller insects to defend themselves. The leaves of many tree species have small tufts of hair or pockets of tissue called *domatia* on the underside of the leaf where the veins join together (Grostal and O'Dowd 1994). These are home to various predatory insects and microbivores (i.e. eating microbes) which patrol the leaf, eating their fill and at the same time helping to protect the leaves from herbivorous damage or disease.

Sap Suckers and Defoliators

A large number of aphids and scale insects steal the sugary phloem sap through piercing individual phloem bundles in vulnerable trees, such as limes *Tilia* spp. and some maples (such as sycamore *Acer pseudoplatanus*). They particularly seek the small amount of nitrogen in the sap, and so large quantities of excess sugar are excreted – this is the stuff that sticks to car windscreens and other surfaces, including bark and lower leaves. The sugary liquid is readily invaded by sooty moulds, creating an unsightly black mess that is more an aesthetic nuisance than harmful to the tree. Some aphids can be more damaging. A number of woolly aphids (aphids covered in a waxy wool, such as *Phyllaphis fagi*, found on beech *Fagus sylvatica*; *Eriosoma lanigerum*, found on apple *Malus* spp., and *Adelges piceae*, found on firs *Abies* spp.) can cause curling of leaves, shoot swellings and dieback. Moreover, the damage they cause can let in fungal pathogens. For example, infestation by woolly aphid *Eriosoma lanigerum* can lead to apple canker caused by *Nectria* species. In northeastern North America, beech bark disease (caused by two *Neonectria* species) affects American beech *Fagus grandifolia* trees following infestations of the beech scale *Cryptococcus fagisuga*, an invasive non-native insect (Cale *et al.* 2015).

Insect larvae can be very effective at defoliating trees. In North America, insects and pathogens cause five times more economic damage than fire (Dale *et al.* 2001). Severe defoliation can be caused by a wide range of insects but particularly by moths, including 'looper' caterpillars (Geometridae, including the winter moth *Operophtera brumata*, and the mottled umber *Erannis defoliara*), and the green oak roller *Tortrix viridana*. In the UK, these are being joined by new pests, such as the oak processionary moth *Thaumetopoea processionea*, native to central and southern Europe, a pest that has already caused huge infestations in the Netherlands and nearby Belgium. It was first discovered in the UK in 2005. In addition to stripping leaves from a variety of trees (the native pedunculate *Quercus robur* and sessile *Q. petraea* oaks, as well as Turkey oak *Q. cerris* and *Carpinus*, *Castanea*, *Fagus*, *Corylus* and *Betula* in mainland Europe), the caterpillars have *urticating* hairs that can cause skin and respiratory problems (see Expert Box 9.1). With climate change and milder winters, a greater number of insect species will be able to survive winter and build up effective populations. This, with the increasing number of pests likely to find their way into new countries by slipping past *phytosanitation* (biosecurity) checks in the growing international plant trade, means we can expect more defoliation in the future.

As a rule of thumb, removal of up to 30% of the leaves by defoliating insects will reduce growth by very little in any one year and, biologically, even losing 50% of leaves will usually have little impact, as long as it is not associated with any other damage to the tree. Moreover, most trees will recover from complete defoliation by regrowing a new set of leaves later in the summer, once the insects have finished. Defoliation is one of the causes of this *lammas growth* (named after Lammas day, the first day of August). Those trees that cannot or do not have time for new growth that year will regrow a new set of leaves as usual the following spring, but the effect of just one year of defoliation can be seen as reduced growth for up to 5 years. The biggest problem comes from repeated defoliations year on year because removal of the leaves greatly reduces the production of sugars and starch, and each new set of leaves is a further drain on the dwindling reserves stored inside the tree. This can also reduce investment in chemical defences, leading to greater susceptibility to biotic as well as abiotic stressors. Death can result.

It is worth saying at this point that many insects that attack leaves do not pose a significant threat to the tree; but some do. A harmful example is the horse chestnut leaf miner *Cameraria ohridella*. The small caterpillar of this moth lives within the leaf blade, eating out the internal tissue and leaving a brown trail behind it. In the UK, it was first seen in London in 2002, but it is now spreading and can occur in such high densities that all the leaves across the canopy turn brown. There is evidence that loss of green leaf area can lead to a 35% reduction in photosynthesis. More important are the effects on reproduction because the leaf miner has been seen to reduce seed weight by 47%, and in some cases germination is up to 32% lower (Takos *et al.* 2008; Percival *et al.* 2011). Moreover, the leaf miner makes the tree more susceptible to horse chestnut bleeding canker (caused by the bacterium *Pseudomonas syringae* pv. *aesculi*) by suppressing key defensive enzymes (Percival and Banks 2014). More extreme is the pine beauty moth *Panolis flammea*, a native to the UK, found on the needles of Scots pine *Pinus sylvestris*. In 1976 it was found on lodgepole pine *Pinus contorta* on wet peats soils of northern Scotland where its caterpillars have resulted in tens of thousands of hectares of plantation to be sprayed with pesticide to keep the trees alive.

Physical removal of insects by brushing or washing is possible on small trees but the most effective methods of control are chemical, particularly during the winter or early spring when the insects are most vulnerable and before they cause the next wave of damage. These chemicals obviously have environmental consequences and their application is often impracticable for large infestations. In such cases, the best management is to ensure the tree is otherwise healthy to withstand the defoliation.

Climate change was once thought to help trees because caterpillars of insects such as the winter moth *Operophtera brumata* were thought to be more responsive to early spring temperatures than the oak they lived on: they were emerging before the buds had opened and subsequently starving to death. However, nature being ever resourceful, the insects have now adapted. Nevertheless, some individual trees may benefit by variation in their spring behaviour. Trees that now flush even earlier, before the insects have grown sufficiently to eat much, may survive, because leaves become tougher with age because of the build-up of greater concentrations of chemical deterrents, such as tannins. Moreover, very late flushing trees may also survive largely unscathed because the young vulnerable leaves appear after the young insect larvae have pupated. Thus, the effects of climate change can be complex and difficult to predict.

Wood and Bark Borers

Insects can cause considerable damage by burrowing into the wood. A wide range of insects use wood as home and food. In warmer parts of the world, termites (Isoptera) make up a most impressive group of 2000 species or so. Many are fairly benign but there are around 100 species capable of causing significant damage to dead timber and buildings. In temperate areas, most wood borers are beetles (Coleoptera) and moths (Lepidoptera), with a few sawflies and wood wasps (Hymenoptera). The majority of beetles are restricted to dead or stressed trees and branches and are not a problem but some can be serious pests of healthy trees. In some of these, the adult causes damage to wood but usually it is the larvae that do most of the boring, spending a large part of their life cycle safely hidden in the wood. It is important to note that many insects can live in wood too dry for fungal decay. The common furniture beetle *Anobium punctatum*, the cause of most woodworm damage, can survive in wood down to 12% moisture, but usually does best in wetter wood.

Some insect borers can do considerable damage to trees, just by the larvae chewing through the wood. The longhorn beetles come into this category, including the Asian longhorn beetle *Anoplophora glabripennis* and citrus longhorn beetle *A. chinensis* (Figure 9.10). Both are native to Asia and have been introduced to the USA, Canada and a number of European countries, and have been found in the UK sporadically since 1994 (in each case being exterminated before it could spread). The damage is caused by the sheer size of the larvae (up to 5 cm long) and the large number of wide tunnels they carve through the wood. The exit holes they leave through the bark are up to 1 cm in diameter so the internal damage can be substantial. They have been found in a wide

Figure 9.10 Citrus longhorn beetle *Anoplophora chinensis. Source:* https://en.wikipedia.org/wiki/ Citrus_long-horned_beetle#/media/ File:Anoplophora_malasiaca.jpg. Licensed under CC BY 2.5.

range of trees, including *Acer, Aesculus, Betula, Salix, Populus* and *Ulmus* (and to a lesser extent in *Alnus, Fagus, Fraxinus, Malus, Platanus, Prunus* and *Pyrus*), so the effect on the landscape where they become established could be huge, especially in areas with poor species diversity (Sjöman *et al.* 2014).

Some smaller borers have an even larger effect if their host is ring-porous (see Chapter 2). In these, the outermost ring of wood is vital for water conduction, and so the tree is very vulnerable to beetles that invade the phloem–xylem junction and damage that outermost ring. Of growing importance here is the emerald ash borer *Agrilus planipennis*, a vivid green beetle native to Asia and eastern Russia, first found in North America in 2002 (probably introduced in wooding packing material) and considered by some to be the most destructive forest pest ever seen in North America (Thomas 2016). It is now found in >13 000 km² of the USA and Canada, and has killed tens of millions of ash trees. In wider Europe, it is causing particular problems in Moscow, having killed >1 million trees, and is moving west and south at the rate of 30–40 km per year (seemingly hitching rides on vehicles along main highways; Straw *et al.* 2013). Once infected, trees usually die within 2–3 years. Some insects do even more rapid and extensive damage by bringing along a pathogen. The most classic example of this is the elm bark beetles (mainly *Scolytus scolytus* and *S. multistriatus* in Europe and *Hylurgopinus rufipes* in North America) which carry the Dutch elm disease fungus *Ophiostoma novo-ulmi* that is responsible for killing over 20 million elm trees in Europe and 40 million in North America. The disease is particularly devastating because elms *Ulmus* spp. are, of course, ring-porous.

In more general terms, wood-boring insects do not usually physically weaken the tree so that it is in danger of snapping, although they can greatly reduce the quality of the timber, but many otherwise innocuous beetles can damage trees by allowing entry points for pathogens and other pests. However, it is not always a straightforward case that all boring beetles should be removed where possible. The two-spotted oak buprestid *Agrilus biguttatus* is considered a harmful pest in mainland Europe linked to oak decline, but is a rare British Red Data Book species only associated with over-mature or veteran oaks. Efforts are even being made to conserve it.

Trees are, of course, not passive in letting the beetles in. A whole variety of toxic resins, gels and latexes are produced by the tree, often under a slight positive pressure, which will physically wash the beetle or other potential pest out of the bark as well as kill it. However, the arms race between beetle and tree is not all one-sided; a number of beetles have evolved counter defences. The eight-toothed spruce bark beetle *Ips typographus*, which burrows into white spruce *Picea abies*, brings with it the blue-stain fungus *Ceratocystis polonica*. This fungus acts to reduce resin production, allowing more beetles easier access to the trunk. The beetle normally invades weakened or diseased trees because it is easier for them to overcome the resin production; a much larger number of beetles is needed to invade healthy trees because more of the fungus is needed to slow resin production sufficiently enough to allow the beetle to invade. A similar process may happen with the various *Dendroctonus* beetles. Mountain pine beetle *D. ponderosae* has killed millions of pines over a large area of western North America and carries with it a fungus, *Grosmannia clavigera*, which may also slow down resin production. The fungus, however, may also be involved in helping the beetle larvae feed by breaking down the wood, making it more digestible.

When considering the management of pest wood-boring and bark beetles, prevention is undoubtedly the best method and is achieved primarily by maintaining tree vitality: the right tree being planted in the right place. Old, slow-growing trees, those with high competition for light, water or nutrients, and newly transplanted trees are likely to be the most susceptible. It is also possible to spray the bark of particularly valuable trees with insecticides but once the insect is inside the tree, insecticides will be largely ineffective. Pheromone traps have been used for a number of high-profile beetles, either using female scents to attract and catch males, leading to lower reproduction, or using repellent scents to push beetles away from valuable trees. Once trees are infested with the larvae of pest beetles there is little that can be done other than prune away affected limbs, or, in extreme cases, fell and ideally burn the whole tree or at least the bark. Of course, native beetles in dead wood that do little or no harm should be tolerated and even encouraged.

Underground damage by insects is comparatively rare, because moving through soil uses a lot of energy so most subterranean animals need to be carnivorous in order to gain enough energy (earthworms are a non-threatening exception). However, there are a few that do damage plants. The most important of these in Europe is the cockchafer *Melolontha melolontha*. The large adult beetle feeds on the leaves of oaks, where it does little damage. However, the eggs laid in the soil result in larvae (rookworms) that reach up to 30–45 mm long and feed for 2 years on roots in the soil. They can cause extensive problems on sports fields (where the turf can be rolled up like cloth because of the lack of intact roots) and in tree nurseries. The problem can be made worse by the digging of birds, badgers and foxes, trying to eat the succulent larvae. The easiest way to control these is by insecticides but biological control is available using parasitic nematodes and bacteria.

Synergy of Pests, Diseases and Environmental Stress

In the real world, pests and pathogens do not attack trees in isolation. It is often the case that a tree weakened by stresses or by one pathogen will be an easy target for others, and it can be difficult to identify the main cause of death. An example of this is seen in oak decline in Europe. Chronic oak decline was first recognised in the early 1900s and is characterised by a progressive dieback of the canopy, starting at the twig ends and leading to the death of large branches and, subsequently, the trunk. Some trees recover, surviving as stag-headed individuals. The condition appears to have no one cause but rather is the combined effect of a whole range of causes, potentially including honey fungus *Armillaria* spp. and other root pathogens (such as *Phytophthora* spp.), boring beetles (including *Agrilus biguttatus*), defoliating insects, powdery mildews and stressors such as drought or high temperatures. By comparison, acute oak decline, which can kill within 2–4 years, is primarily attributable to bacterial infections, creating dark weeping patches or cracks in the bark. Even here, a tree weakened by other agents is undoubtedly more prone to bacterial infection. Of course, neither of these is to be confused with sudden oak death, caused by *Phytophthora ramorum* and discussed at the start of this chapter.

In the above examples, the main causes of tree decline are pests and diseases, but equally the primary cause can be non-living or abiotic factors. Drought is the main contender, putting the trees in a difficult situation. As discussed in Chapter 6, most trees will aim to maintain their hydraulic conductivity by closing their stomata tightly to reduce water loss and prevent cavitation in the xylem. This limits gas exchange, and so photosynthesis rapidly declines for want of carbon dioxide which in turn may mean that the tree runs out of carbohydrates and starves to death. It is debatable whether trees will actually starve in this way (see Chapter 7), but declining carbohydrate reserves means less is available for plant defences, and the weakened tree is then much more susceptible to insects and pathogens.

An important corollary of this is that the main cause of a tree's demise may not be obvious from the most prominent symptoms. For example, seeing a tree suffering from defoliation by insects may stop us noticing an underlying problem of water stress caused by soil compaction that has weakened the tree, making it more susceptible to insects. Treating the insect infestation would not be as important as dealing with the major underlying problem of soil compaction. The moral of the story is to look beyond the obvious.

If a tree suffering from one or more stressors is heading towards death, it would be useful if we were able to detect the signs before it reached the point of no return (sometimes called the *irreversible tipping point*): if we could, we would have time to try and put things right. The good news is that some early warning signals have been found (Camarero *et al.* 2015). The less good news is that they seem to be specific to different species. In some cases, growth (as seen in tree ring width) becomes much more variable or declines as a tree goes into decline (as seen in European silver fir *Abies alba*), or the trees may show a rapid decline in stored food as detected by the change in carbohydrates in sapwood (as seen in Scots pine *Pinus sylvestris* and Aleppo pine *P. halepensis*). This is still an area of active research and it is to be hoped that more universal signs will be found.

Mammals and Birds

Seed Dispersers

In temperate forests and urban areas, birds are the most effective animals in moving seeds from the parent tree. It is well known that many of these seeds end up below perches, such as other trees, where the birds sit and eat or defecate. So manipulating the number of perches of the size that resident birds normally use can greatly change the number of *volunteer seedlings* – seedlings that 'volunteer' themselves and are not planted.

Injury by Birds and Mammals

Birds, especially large ones such as pigeons, can cause a huge amount of damage to trees by tearing off twigs and branches for nesting material. In woodland situations, this is of little concern as it is shared amongst many trees, but for lone or widely spaced trees in urban areas this can be a significant loss.

In more rural settings, deer can also be troublesome through browsing of low branches – even killing young saplings – and by debarking trees when rubbing the velvet off their young antlers. Fortunately, deer tend to use the same tree for rubbing so the damage is usually localised. Similarly, smaller animals, such as rabbits, voles and mice, can do considerable damage to bark, especially on young trees. Perhaps the most significant cause of damage is from grey squirrels *Sciurus carolinensis* introduced into Britain from North America in the 1880s. This rodent can strip large quantities of bark from hardwood trees in spring, partly to get at sweet phloem, but also as aggressive posturing to rival squirrels. Well-tended, vigorous trees tend to be favoured targets rather than self-sown trees in a dense woodland (which have a thicker bark and lower sap content), and so it is often specimen trees that are targeted. Some trees, such as species of maple *Acer* and beech *Fagus*, are more susceptible than ash *Fraxinus*, lime *Tilia* and cherry *Prunus*. Control of animal damage can be straightforward, as with deer and rabbits, when it is a matter of appropriate fencing; squirrels are obviously much more difficult, although a range of humane traps are available. Care needs to be taken to use these traps, or more direct methods of culling, with discretion and sensitivity.

Managing Trees as Habitats

While the emphasis of this book is on the tree itself, trees are also very important for the biodiversity that they hold. We have an ethical responsibility to preserve biodiversity, if for no other reason than that, based on the precautionary principle, we never know when other species might be useful – the basis of ecosystem services. Conserving biodiversity can also be an important reason for planting or keeping trees in our landscape. Some biodiversity will rapidly appear whether we intend it to or not, including epiphytic lichens and mosses, and the insects already discussed, but biodiversity can usually be supplemented by suitable management, such as leaving natural cavities as roosts or nesting areas for bats and birds. Fruit-bearing trees can be planted as a food source for insects, birds and mammals, where these fruits will not create a nuisance on paths and roads. Management can also be more extreme, such as planting groups of trees in the same hole. This *bundle planting* was much used in the Middle Ages, planting 2–7 saplings together that would coalesce to produce multi-stemmed trees with large spreading canopies and often abundant fruit production.

Older trees generally provide more habitats for conservation by virtue of being bigger and accumulating more deadwood in the canopy, providing more niches for wildlife. As such, ancient and veteran trees (Expert Box 9.2) are valuable in the range of habitats they provide, as well as being valuable in their own right. A step further is to keep and encourage dead wood in tree crowns.

Expert Box 9.2 Ancient and Veteran Trees: Their Importance and Management
David Lonsdale

What are Ancient and Veteran Trees?

An ancient tree has been defined as one 'that has passed beyond maturity and is old, or aged, in comparison with other trees of the same species' (Anon. 2008). This definition includes the words the 'same species' in order to allow for the tendency for some species to live much longer than others. For example, a 700-year-old yew *Taxus* sp. might have many more centuries of life, while a 70-year-old birch *Betula* sp. might be nearing the end of its life. Such differences occur because species vary in their capacity to cope with cumulative events and processes that can eventually become life-limiting. Although some species tend to be relatively short-lived, trees generally (other than palms) do not have an inherently limited lifespan. On the contrary, they have a theoretically unlimited capacity to continue growing by producing new shoots and new increments of wood and bark.

If we happen to know the age of a tree, we can judge whether it is ancient according to the above definition. If its age is unknown, its stage of development can be recognised only according to its physical characteristics. Ancient trees typically have several of the following characteristics:

- Large stem girth for the species concerned (Figure EB9.2).
- Gnarled appearance, associated with zones of growth that are irregular and/or associated with old wounds.
- Extensive heart-rot, usually including hollowing (perhaps with fungal fruit bodies visible).

Figure EB9.2 Ancient *Quercus robur*, showing large girth, gnarled appearance and extensive heart-rot with hollowing. Some of the lower branches are probably of secondary origin, having developed in response to retrenchment of the crown above. *Source:* © David Lonsdale; used with permission.

Figure EB9.3 Ancient *Quercus robur*, showing advanced retrenchment of the crown. The crown is healthy but is now very small in comparison with the girth of the stem. *Source:* © David Lonsdale; used with permission.

- Characteristic 'crown architecture', compared with that of a younger tree (Raimbault 2006).
- Signs of a progressive or episodic reduction in post-mature crown size, often known as 'retrenchment' (Figure EB9.3).
- Relatively rich and/or abundant covering of mosses, lichens or other epiphytes.

Also, the annual increments currently being formed in the stem of an ancient tree may be considerably narrower than the increments that were laid down at the peak of maturity. However, the width of a particular increment could vary greatly around the stem circumference.

The term 'veteran' can be applied to a tree of any age that has several of the above characteristics, provided that it is in good general health (i.e. its crown has been retrenching but is not irrecoverably dying back). The human analogy is that a relatively young soldier could be a war-scarred veteran. It follows that all (or nearly all) ancient trees are veterans but not all veteran trees are ancient. For brevity, they are all sometimes described as 'veterans', but the description 'ancient and other veteran trees' is sometimes used in order to emphasise the special importance of truly ancient trees for their special historic, cultural or ecological value.

Ancient trees are relatively rare, partly because they represent the relatively small proportion of trees that humans have allowed to live to a great age. However, there are certain places (e.g. old deer parks and traditional wood pastures) where concentrations of ancient trees can still be found, thanks to centuries of relatively sympathetic management.

Ancient trees deserve to be conserved not only because they are rare, but also because of their particular qualities. We value them not only for their sheer size and visual character, but also for their historic and cultural associations. Perhaps even more importantly, they have a wealth of dependent organisms, including fungi, lichens and invertebrates, many of which exist only in localities where such trees have been continuously present for at least several centuries.

In order to develop strategies for conserving ancient trees and their dependent species, we need, as far as possible, to emulate the conditions that would have favoured great longevity under natural conditions. The primeval landscape of the British Isles is now widely believed to have been a mosaic of open-grown trees, grassland, heathland and closed-canopy forest. The 'wildwood' used to be portrayed as consisting almost entirely of closed-canopy forest, except where glades briefly existed as a result of events such as wildfires and landslips. However, glades were probably numerous and extensive, being maintained for centuries by the grazing and browsing of large herbivores such as bison and the now-extinct aurochs (Vera 2000). It would otherwise be difficult to explain why a high proportion of our plants, animals and fungi require open habitats. These include species entirely dependent on today's ancient trees.

In Europe, the great majority of ancient trees occur in open-grown conditions and they typically have low branches, which would have become shaded out in dense woodland. In many cases, people helped to establish a framework of lower branches by pollarding trees as a source of fuelwood and/or fodder for livestock. The pollard cuts were created above the browsing height in order to allow the new growth to develop without being eaten. In other situations, trees were often coppiced in order to encourage the growth of multiple stems that could be used for many purposes, including the production of fencing materials. In many regions, pollarding and coppicing no longer take place on an economic basis but these practices have left a legacy of ancient trees, which in some respects represent continuity with the wildwood. However, such trees are often prone to life-shortening tree failure because they have heavy branches or stems that have not been cut for many decades.

The wildlife associated with ancient and other veteran trees represents another very important reason for conserving them. For the reasons explained in Expert Box 9.3, many rare invertebrates, fungi and lichens are found only in areas where ancient trees have been continuously present for at least a few centuries.

The overriding principle for managing ancient and other veteran trees, and/or their potential successors, is to avoid their unnecessary loss. The aim should therefore be to provide favourable growing conditions for such trees and to protect them from inappropriate removal or damage.

Protective Management of Ancient Trees, Including Lapsed Pollard and Coppice

The following objectives and practices can help to protect old trees, both from adverse processes and from activities (in the following list, the term 'veteran' includes chronologically ancient trees as well as younger veterans):

1) *Retention of veteran trees*
   - Retain veteran trees wherever they are found, not only in areas of key importance, such as old deer parks and wood pasture, but also in the wider rural landscape and in areas that have become, or are scheduled to become, urbanised.
   - When managing lapsed coppice stools, retain decaying wood habitat and old stems with sap-runs as much as possible.
2) *Protection from harmful influences*
   - Provide adequate root protection areas, generally erring on the side of caution rather than adopting the minimum indicated in published guidance. Thus, for a veteran tree, aim to protect a radius of 15 times the stem diameter at breast height or 5 m beyond the canopy perimeter, whichever is the greater. Given that root systems are often very asymmetric, the use of ground-penetrating radar should be considered for the mapping of root systems of important trees. Root protection should be implemented wherever possible in:
     o forestry (e.g. during timber extraction)
     o agriculture
     o sport or other recreational use of the surrounding land
     o construction work.
   - Carefully ameliorate adverse conditions (e.g. with mulching and/or mechanical decompaction) that have been caused by previous failure to provide adequate root protection.
   - Protect trees and their roots from fire, including badly sited bonfires.
   - Protect open-grown trees from excessive shading, whether from natural regeneration or from tree planting. If excessive shade is already present, reduce it gradually by selective removal or shortening of trees that are causing it ('haloing'; Lonsdale 2013a).
   - In order to maintain open ground between trees, preferably use grazing by suitable breeds of livestock but keep stocking densities low enough to avoid excessive compaction. Where necessary, protect individual trees from livestock by the use of deterrent barriers (e.g. thorny prunings).
   - Protect epicormic shoots from shading by ivy, given that they may have the potential to grow into new branches when old ones are lost, but take account of the habitat value of ivy.
   - Operate biosecurity in situations where advised in order to help mitigate the threat posed by alien pests and pathogens.
   - Support the introduction of effective measures to help prevent the establishment of alien pests and pathogens in the UK, recognising that some of these could do immense damage to our veteran trees, which cannot be 'replaced'. Some diseases, such as ash dieback, might eventually be managed by planting genetically resistant trees, but our existing ancient trees are irreplaceable.

3) *Tree succession*
- Assess the age-structure of the tree population and the 'death-rate' of the old trees in order to estimate the requirements for an unbroken succession of trees with veteran characteristics.
- Ensure a succession of trees in formal plantings, such as avenues, in order to meet both landscape/cultural and habitat requirements.
- Create new pollards from young trees as successors to existing old pollards.
- Where there is a need to fill a gap in the age structure, and where there are plenty of early-mature trees, perhaps 'veteranise' some of these (see Expert Box 9.3 for an explanation of the need for unbroken continuity of habitat).

Life Stages of a Tree as a Guide to Management

While a tree is young, it will tend to grow until its crown and its root system have reached a peak (mature) size, which is determined by its genetic characteristics and by the growing conditions. During this mature phase, the stem, branches and major roots of the tree continue to increase in girth by laying down successive increments of wood (except in the case of a palm tree, which does not grow by secondary thickening). From one year to the next, roughly the same cross-sectional area of wood is laid down, as long as the tree retains a full crown. The widths of successive increments, however, will slowly decrease because they are being spread around an increasing girth.

In late maturity, a tree's increase in stem-girth tends to become increasingly concentrated in particular sectors of the stem, which are well-connected to the more vigorous branches. These are also connected to well-developed parts of the root system. Each of these strips, with its foliage and its roots, increasingly becomes a semi-autonomous functional unit (Lonsdale 2013b): almost a tree within a tree. As the tree passes into its ancient phase of life, its stem tends to become increasingly fluted because of the relatively slow growth rate of the strips of tissue in between the well-developed 'functional units' (Figure EB9.4). The intervening slow-growing strips may eventually die. Meanwhile, the slower growing parts of the crown tend to die back or grow very little after being broken in storms. The crown, now in 'retrenchment', still retains plenty of healthy branches. Meanwhile, retrenchment is probably taking place in the root system but, for obvious reasons, we do not know much about processes occurring below ground.

The 'architecture' of a tree's crown changes as it passes through early growth, maturity and its ancient phase. These changes are partly the result of a decrease in shoot extension growth as the branches elongate. Also, depending on the tree species, there may be a tendency to develop a more pendulous twig pattern. If part of the crown is lost by pruning or breakage, the new shoots (if any) that grow in response may show a different, more juvenile, pattern of growth for several years until a mature pattern ensues. It has been suggested that the crown architecture of trees at different phases of life should be used as a guide to pruning. A model depicting 10 stages in the life of a tree has been proposed by Raimbault (1995).

Management to Help Prevent Life-Shortening Tree Failure

In lapsed pollards and in open-grown trees, the lower branches can form a residual crown after the upper branches have eventually died back or have broken in the process of retrenchment. Thus, the tree attains a more squat and stable form while maintaining the

Figure EB9.4 Veteran *Fraxinus excelsior,* which has undergone failure of its main stem. The tree now consists of a number of 'functional units' (see text), each consisting of a vigorous branch, a functional sector of the remaining main stem and a vigorous sector of the root system (out of sight). The branches are of different ages, as can be seen from their bark texture. *Source:* © David Lonsdale; used with permission.

leaf area required for further long-term survival. Low branches sometimes serve as natural props, owing to their capacity to bend down until they rest on the ground, rather than breaking. They can also root into the soil by the process of natural layering. Even if a low branch is shed, it might be succeeded by new ones developing from epicormic shoots, provided that such shoots have not been shaded out.

Natural propping can be encouraged in order to help prevent catastrophic tree failure, or to avoid a need to cut branches that are too large and too old to tolerate such severe treatment. Thus, if low, heavy branches have bent down towards the ground, the soil could be mounded (using well-aerated material) in order to provide just enough support to help prevent branch failure.

If, in order to help prevent life-shortening tree failure, an old tree needs to be pruned, one the key considerations is its capacity to produce new shoots afterwards. This depends very much on the tree species concerned, but also on the individual tree and its growing conditions past, present and future. Several factors need to be taken into account, including shade tolerance and the longevity of dormant buds (Lonsdale 2013a).

As a tree increases in girth, so does the proportion of its woody cross-section that consists of old, non-conductive wood. In some species, the living cells of the sapwood die in small numbers over many years until they are all dead. The term *ripewood* is sometimes used to describe the old, central wood of these species. Other species have a distinct boundary between their sapwood and their central non-living wood (i.e. heartwood). Depending on the tree species, heartwood varies a great deal in its durability, owing to the presence or absence of various kinds of natural preservatives (known in the timber industry as extractives).

In species that produce ripewood or non-durable heartwood (e.g. beech *Fagus sylvatica* or hornbeam *Carpinus betulus*), the exposure of such wood by breakage or pruning can be followed by rapid decay: perhaps so rapid that the mechanical strength of the tree, and sometimes also the viability of its sapwood, is compromised. It is necessary to bear in mind that the sapwood of an ancient tree could be very narrow and perhaps present only in certain sectors of the stem (see above, regarding 'functional units'). The development of decay is usually much slower in species that have durable heartwood (e.g. oak *Quercus robur* or yew *Taxus baccata*). It is perhaps partly for this reason that durable heartwood-forming species tend to live longer.

From an economic standpoint, the stem of a tree might be regarded as 'degraded' at a relatively early stage of decay. From a biological standpoint, however, decay is a normal and sustainable part of a tree's development as it passes from maturity into its ancient phase of life. Decay starts in discrete columns within the woody cross-section, where the natural dieback or breakage of branches and roots has allowed the ingress of oxygen and/or decay fungi. However, the latter are often already latently present in sound wood.

Provided that a decaying tree is still laying down increments of new wood on the outside, it can continue to stand and survive for an indefinite time, perhaps many centuries or even millennia. Such a tree can have immense value for the many species that require continuity of habitat, especially a decaying-wood habitat. It is interesting to realise that an ancient tree may be living in harmony with ancient fungi that colonised it, perhaps centuries ago. However, if a large proportion of the previously intact woody cross-section becomes exposed, for example by storm damage or excessive pruning, the rate of decay can overtake the tree's capacity to survive by keeping pace with it.

Decay sometimes extends into living sapwood, depending partly on the species of the tree and the fungus. In most tree species, however, the sapwood has natural defences, which can be very effective, providing that the sapwood is connected to abundant healthy foliage. It is partly for this reason that anyone who plans to prune a tree, especially an ancient one, takes account of its capacity to produce healthy shoots in response to cutting. More particularly, the likely response of each recognisable 'functional unit' of an old tree (Lonsdale 2013b) ought to be assessed before any pruning is carried out.

Given that unsuitable pruning can be very harmful to old trees for the above reasons, the following measures can sometimes help to minimise adverse effects:

- Phased pruning where appropriate (perhaps over many years of subsequent monitoring), to simulate natural retrenchment.
- Retention of stubs where needed for encouraging production of new shoots.
- Retention of epicormic shoots as potential new branches;
- Possible use of simulated natural fracture, which is thought sometimes to stimulate growth of new shoots.
- Making allowance for the shade tolerance/light demand of the species concerned, in order to help avoid excessive shading or, conversely, excessive insolation of new shoots when they develop.
- Individual assessment of lapsed pollards or coppiced trees for requirements (if any) for tree work – generally not to be cut back to the bolling (nor to ground level in the case of coppice).
- Special care to avoid pruning when trees are under physiological stress (e.g. from drought).

Deadwood

In the past, and certainly in the 1900s, deadwood inside trees and untidy broken branch stubs were a sign of neglect and lack of 'proper' management. Now, there is a greater appreciation of the conservation value of deadwood, and a trend to retain and even deliberately encourage it (Expert Box 9.3).

Expert Box 9.3 Improving Biodiversity in Trees
David Lonsdale

As noted in Expert Box 9.2, ancient trees are believed to have been of key ecological importance in the 'wildwood'. Their present-day successors occur mainly in the relatively few places (e.g. old deer parks and traditional wood pasture) that have not been developed for modern agriculture or urbanisation. They provide continuity of habitat, especially for invertebrates, fungi, lichens and other species, some of which appear to have very limited capacity to recolonise sites after continuity has been lost. The occurrence of such species is used in the assessment of the habitat quality of sites (Fowles *et al*. 1999; Alexander 2004).

In order to maintain continuity of the habitats that are uniquely associated with old trees, it is clearly necessary to avoid the deliberate destruction of such trees as far as possible; however, this is not sufficient. There is an equal need to protect such trees from damaging kinds of site management and from being shaded out. There is also a need for younger trees eventually to succeed them, and thus to provide habitats for the same vulnerable species when they become old enough to do so. Some of the measures for achieving these aims are summarised above.

Certain habitat features of old trees have the potential to be hazardous to people. Attached dead branches are likely to fall eventually. Nevertheless, living branches, with their greater leverage, sometimes fail more readily than dead ones, which can last for many years if they contain durable heartwood, as in stag-headed crowns of native oaks *Quercus robur* and *Q. petraea*. However, if dead branches are thought to be posing an unacceptable risk of harm to people, it may be necessary to take remedial action, preferably by 'moving the target' by such means as diverting paths, the use of warning signs or temporary site closures during severe storms. If dead branches need to be cut off or shortened, they should be retained nearby afterwards. In general, the 'arisings' from any pruning of old trees should be retained on-site as far as possible.

In areas where continuity of habitat is at risk of eventually being broken because of gaps in the succession of younger trees, such gaps might perhaps be bridged by deliberately damaging ('veteranising') younger trees. Such trees are usually not yet old enough to be providing much decaying wood habitat, but they are perhaps large enough to do so if the decay process can be accelerated. On the other hand, the bark of such trees might be too young to support rare lichens, with associated invertebrates, which live on the surfaces of very old trees. Also, even if an age gap can be bridged, some of the most vulnerable species might fail to disperse between suitable trees if these are too far apart.

Generally, it can be hoped that a range of tree species with a correspondingly wide range of habitats would benefit from veteranisation, but there may be a need to try to produce a certain kind of habitat for particularly vulnerable species. Examples of saproxylic species that are being conserved by such means include the violet click beetle *Limoniscus violaceus*

(Green 1995), the aspen hoverfly *Hammerschmidtia ferruginea* (Rotheray *et al.* 2015) and the pine hoverfly *Blera fallax* (Rotheray and McGowan 2000).

In order to decide whether 'veteranisation' is appropriate for a particular site, the age structure of the tree population needs to be studied in detail. In particular, it is necessary to estimate the mortality rate of the older trees in order to find whether there is likely to be a gap in habitat continuity. Given that veteranisation is likely to shorten the lives of the relatively young chosen trees, there is obviously a need to decide whether there are enough of them to be sacrificed by veteranisation, while others are retained as potential long-term successors of the ancient cohort.

Various methods for veteranisation have been suggested and, in some places, put into experimental practice (Lonsdale 2013a). Methods include severe pruning, bruising the bases of stems with a sledgehammer and, in more extreme cases, ring barking. The last method is especially useful in areas where younger trees are shading out their ancient neighbours. If so, veteranisation can be combined with the practice of 'haloing', whereby excessive shade, cast by surrounding trees, is reduced.

Veteranisation that involves damaging the main stem of a tree could help to provide continuity of heart-rot habitats, which are an especially important feature of old trees. There is also a need to ensure continuity of habitats associated with other features, such as attached dead branches. These could be provided by pruning, in order to retain long stubs which are likely to die back to some extent.

Deadwood – The Afterlife of Trees

Calendars and greetings cards often include photographs of beautiful woodland scenes, often carpeted with bluebells, but they tend to lack one vital (literally vital) ingredient: decaying wood. The importance of decaying wood habitats was acclaimed by the pioneering ecologist Charles Elton in the mid-twentieth century, but it has taken several decades to win a place for them in mainstream conservation. Elton suggested that the absence of decaying wood habitat could deprive an area of woodland of perhaps one-fifth of its fauna (Elton 1966); but this is now thought perhaps to be an underestimate.

Despite the image of a beautiful but over-tidy bluebell wood, many woodlands contain an apparent abundance of fallen deadwood, which provides habitats for a range of fungi and invertebrates (saproxylic species). Much of this wood is of relatively small diameter and does not provide the range of conditions and associated habitats that can exist in large-diameter material. If larger diameter material is abundant, a wider range of saproxylic species is likely to be present but most of these will probably be relatively mobile species that might be able to recolonise the site after an interruption in habitat continuity. The rarest and most vulnerable species are those that live in the decaying wood of living ancient trees (Alexander 2004). This provides a succession of habitat that can continue for centuries, while the tree continues to lay down new increments of wood.

A wide range of fungi, bacteria, actinomycetes and other microorganisms play a part in wood decay, but the process would be very slow without the presence of specialised wood decay fungi, which can penetrate deeply into large volumes of wood. Many of these are familiar by virtue of their bracket-like or toadstool-like fruit bodies. Without these fungi, deadwood would accumulate in vast quantities. By degrading wood, they have a key role in the cycling of carbon and of the minerals that are locked up during wood formation (Figure EB9.5).

Figure EB9.5 Ancient *Quercus robur*, with heart-rot caused by the fungus *Laetiporus sulphureus* (chicken of the woods). Both the heart-rot and the fruit bodies of this fungus are of particular importance as habitat for rare invertebrates that occur only where habitat continuity has existed, at least for several centuries. *Source:* © David Lonsdale; used with permission.

Invertebrates also play an important part in wood degradation, ranging from those that can feed on non-decayed wood to those that require their wood to be thoroughly pre-digested by a decay fungus. There are also some that feed on fungi in wood, rather than the wood itself. There is an almost infinite range of composition and texture of wood in an ancient tree, owing to the gradation from sound wood to almost completely decayed wood, together with the different kinds of decay induced by various fungi and a variation in moisture and mineral content. Also, under the general description of saproxylic inverte-brates, there are many that prey upon those that feed on wood or fungi.

The temperature of the wood is believed to be another important factor that can determine which species will thrive in a particular area. Many of the beetle species that are associated with large, open-grown trees appear to require relatively warm conditions (Dr K.N.A. Alexander, personal communication). The same relationship has been found in parts of continental Europe where summers are warmer than in the UK (Horák *et al.* 2010). These findings indicate that open-grown conditions are important, not only because they enable trees to survive to a great age, but also because such trees are warmed by the sun. However, there is also a need for decaying wood in cooler, moister conditions, which are favoured by many of the saproxylic two-winged flies (Diptera).

Insects make up the majority of saproxylic invertebrates and many of these have at least two lifestyles: one as larvae in deadwood or bark; the other as adults in the outside world. The food requirements of the adult stages vary between species. A large proportion require sources of nectar and/or pollen, which can be in short supply in the forest during the sum-mer. The relative abundance of flowering plants in the sunny conditions of deer parks or wood pasture is yet another reason why these areas are so important for saproxylic insects.

Sunny conditions are required by some of the rare lichens that live on the bark of old trees but there are some species of lichen that require shadier conditions. If there are rare lichens present, it is therefore helpful to seek expert advice about the management of the area before doing anything that will reduce or increase shading.

Deadwood, whether whole branches or the decaying wood inside living trees, is an important habitat for flora and fauna, with thousands of associated species in all sorts of microhabitats within the deadwood (Stokland *et al.* 2012). In the UK, more than 1700 different invertebrate species depend upon ancient trees, many using the deadwood (Butler *et al.* 2002). A good number of these deadwood species are listed as threatened or near threatened in national Red lists. At the same time, the amount of deadwood in our landscapes is usually far below natural levels (typically, one-third of biomass is dead in natural forests), and it tends to be clustered in natural reserves and other protected areas. Thus, there is a great need for supplies of deadwood, spread across the landscape. This does not mean that we should abandon any management of crowns, but the trend is to maintain deadwood in crowns where it is safe for people and for the long-term health of the tree, and within aesthetic limits.

Where dead stems and branches do need to be removed, there are strong advocates for moving away from the surgical, flat chainsaw cut to something more natural look-ing. These *natural fracture pruning* techniques, leaving something more jagged, create a whole range of microhabitats for deadwood organisms. Moreover, regrowth from adventitious buds is often stronger than after a clean saw cut. Some of the earliest trials took place at Windsor Great Park in southern England (Finch 1993) in an attempt to rejuvenate old veteran pollards by drilling holes or making saw cuts to weaken branches, before tearing them off by winch, or even by blowing them off with explosives (imagine the health and safety issues!). More recently, *coronet cutting* (Figure 9.11) has been used to mimic natural branch breakage. This form of cutting was originally used to make dead branches safe by shortening them, going as far as to remove most branches from a standing dead tree so that the dead standing trunk (referred to as a *monolith*) can be safely left upright as deadwood habitat. This technique is also now used for cutting back live branches, with the express aim of creating deadwood in the crown; a process of

Figure 9.11 Coronet cut on beech *Fagus sylvatica* at Myerscough College, Lancashire. The aim is to simulate natural breakage by skilful use of a saw leaving a jagged stub. The name obviously comes from the shape of the cuts. *Source:* Courtesy of Duncan Slater.

veteranisation. Some have gone further by drilling downward-sloping holes in healthy trees to encourage water retention and rot. For most tree managers this is a little extreme because we are aiming to prolong the life of trees, but it does show that deadwood can be tolerated or encouraged within the holistic management of a tree.

References

Appleton, B., Koci, J., French, S., Lestyan, M. and Harris, R. (2003) Mycorrhizal fungal inoculation of established street trees. *Journal of Arboriculture*, 29: 107–110.

Alexander, K.N.A. (2004) Revision of the Index of Ecological Continuity as used for saproxylic beetles. *English Nature Research Report No. 574*. English Nature, Peterborough, UK.

Anon. (2008) *Ancient Tree Guide No.4*. Ancient Tree Forum, c/o The Woodland Trust, Grantham, UK.

Arbellay, E., Fonti, P. and Stoffel, M. (2012) Duration and extension of anatomical changes in wood structure after cambial injury. *Journal of Experimental Botany*, 63: 3271–3277.

Biggs, A.R. (1992) Anatomical and physiological responses of bark tissues to mechanical injury. In: Blanchette, R.A. and Biggs, A.R. (eds) *Defense Mechanisms of Woody Plants Against Fungi*. Springer, Berlin, Germany, pp. 13–40.

Blanchette, R.A. (1992) Anatomical responses of xylem to injury and invasion by fungi. In: Blanchette, R.A. and Biggs, A.R. (eds) *Defense Mechanisms of Woody Plants Against Fungi*. Springer, Berlin, Germany, pp. 76–95.

Boddy, L. (1992) Microenvironmental aspects of xylem defences to wood decay fungi.In: Blanchette, R.A. and Biggs, A.R. (eds) *Defense Mechanisms of Woody Plants Against Fungi*. Springer, Berlin, Germany, pp. 96–132.

Boddy, L. (2001) Fungal community ecology and wood decomposition processes in angiosperms: From standing tree to complete decay of coarse woody debris. *Ecological Bulletins*, 49: 43–56.

Boddy, L. and Rayner, A.D.M. (1983) Origins of decay in living deciduous trees: The role of moisture content and a reappraisal of the expanded concept of tree decay. *New Phytologist*, 94: 623–641.

Butler, J., Alexander, K.N.A. and Green, T. (2002). *Decaying Wood: An Overview of its Status and Ecology in the United Kingdom and Continental Europe*. United States Department of Agriculture, Forest Service, General Technical Report PSW-GTR-181, Albany, CA, USA.

Cale, J.A., Teale, S.A., Johnston, M.T., Boyer, G.L., Perri, K.A. and Castello, J.D. (2015) New ecological and physiological dimensions of beech bark disease development in aftermath forests. *Forest Ecology and Management*, 336: 99–108.

Camarero, J.J., Gazol, A., Sangüesa-Barreda, G., Oliva, J. and Vicente-Serrano, S.M. (2015) To die or not to die: Early warnings of tree dieback in response to a severe drought. *Journal of Ecology*, 103: 44–57.

Dale, V.H., Joyce, L.A., McNulty, S., Neilson, R.P., Ayres, M.P., Flannigan, M.D., *et al.* (2001) Climate change and forest disturbance. *BioScience*, 51: 723–734.

Dixon, D.M. (1969) The transplantation of punt incense trees in Egypt. *Journal of Egyptian Archaeology*, 55: 55–65.

Dujesiefken, D., Ebenritter, S. and Liese, W. (1989) Wandreaktionen im Holzgewebe bei birke, buche und linde. *European Journal of Wood and Wood Products*, 47: 495–500.

Eaton, R. (2000) A breakthrough for wood decay fungi. *New Phytologist*, 146: 3–4.

Elton, C. (1966) *The Pattern of Animal Communities*. Chapman and Hall, London, UK.

Ferrini, F. and Fini, A. (2011) Results of a long-term project using controlled mycorrhization with specific fungal strains on different urban trees. In: Johnston, M. and Percival, G. (eds) *Trees, People and the Built Environment*. Forestry Commission, Edinburgh, UK, pp. 39–50.

Finch, R. (1993) An alternative method of crown reduction for ancient pollards and dead trees. In: Read, H. (ed.) *Pollard and Veteran Tree Management II*. Corporation of London, London, UK, pp. 98–99.

Fink, S. (1999) *Pathological and Regenerative Plant Anatomy*. Gebruder Borntraeger Berlin, Germany.

Fowles, A.P., Alexander, K.N.A. and Key, R.S. (1999) The Saproxylic Quality Index: Evaluating wooded habitats for the conservation of dead-wood Coleoptera. *The Coleopterist*, 8: 121–141.

Franceschi, V.R., Krokene, P., Christiansen, E. and Krekling, T. (2005) Anatomical and chemical defenses of conifer bark against bark beetles and other pests, Tansley Review. *New Phytologist*, 167: 353–376.

Franklin, O., Näsholm, T., Högberg, P. and Högberg, M.N. (2014) Forests trapped in nitrogen limitation: An ecological market perspective on ectomycorrhizal symbiosis. *New Phytologist*, 203: 657–666.

Garbaye, J. and Churin, J.L. (1996) Effect of ectomycorrhizal inoculation at planting on growth and foliage quality of *Tilia tomentosa*. *Journal of Arboriculture*, 22: 29–34.

Garbelotto, M., Schmidt, D.J. and Harnik, T.Y. (2007) Phosphite injections and bark application of phosphite+Pentrabark™ control sudden oak death in coast live oak. *Arboriculture and Urban Forestry*, 33: 309–317.

Gehring, C.A., Mueller, R.C., Haskins, K.E., Rubow, T.K. and Whitham, T.G. (2014) Convergence in mycorrhizal fungal communities due to drought, plant competition, parasitism, and susceptibility to herbivory: Consequences for fungi and host plants. *Frontiers in Microbiology*, 5: article 306.

Green, T. (1995) Creating decaying trees. *British Wildlife*, 6: 310.

Grostal, P. and O'Dowd, D.J. (1994) Plants, mites and mutualism: leaf domatia and the abundance and reproduction of mites on *Viburnum tinus* (Caprifoliaceae). *Oecologia*, 97: 308–315.

Hayward, J., Horton, T.R., Pauchard, A. and Nuñez, M.A. (2014) A single ectomycorrhizal fungal species can enable a *Pinus* invasion. *Ecology*, 96: 1438–1444.

Horák, J., Rébl, K., Vávrová, E., Horáková, J., Neradilová, I. and Loskotová, T. (2010) Beetles of Mr. President: Saproxylic beetles and sun-exposure gradient in pasture woodland in the Lány Game Park. In: 6th European Symposium and Workshop on the Conservation of Saproxylic Beetles, June 15–17 2010, Ljubljana, Slovenia, p. 30.

Horton, T.R., Bruns, T.D. and Parker, V.T. (1999) Ectomycorrhizal fungi associated with *Arctostaphylos* contribute to *Pseudotsuga menziesii* establishment. *Canadian Journal of Botany*, 77: 93–102.

Hudgins, J.W., Christiansen, E. and Franceschi, V.R. (2004) Induction of anatomically based defense responses in stems of diverse conifers by methyl jasmonate: A phylogenetic perspective. *Tree Physiology*, 24: 251–264.

Klironomos, J.N. and Hart, M.M. (2001) Food-web dynamics: Animal nitrogen swap for plant carbon. *Nature*, 410: 651–652.

Krekling, T., Franceschi, V.R., Berryman, A.A. and Christiansen, E. (2000) The structure and development of polyphenolic parenchyma cells in Norway spruce (*Picea abies*) bark. *Flora*, 195: 354–369.

Lipson, D. and Näsholm, T. (2001) The unexpected versatility of plants: Organic nitrogen use and availability in terrestrial ecosystems. *Oecologia*, 128: 305–316.

Lonsdale, D. (1999) Principles of tree hazard assessment and management. *Research for Amenity Trees No. 7*, The Stationery Office, London, UK.

Lonsdale, D. (2013a) *Ancient and Other Veteran Trees: Further Guidance on Management*. The Tree Council, London, UK.

Lonsdale, D. (2013b) The recognition of functional units as an aid to tree management, with particular reference to veteran trees. *Arboricultural Journal*, 35: 188–201.

Lothamer, K., Brown, S.P., Mattox, J.D. and Jumpponen, A. (2014) Comparison of root-associated communities of native and non-native ectomycorrhizal hosts in an urban landscape. *Mycorrhiza*, 24: 267–280.

Merrill, W. (1992) Mechanisms of resistance to fungi in woody plants: A historical perspective. In: Blanchette, R.A. and Biggs, A.R. (eds) *Defense Mechanisms of Woody Plants Against Fungi*. Springer, Berlin, Germany, pp. 1–12.

Morris, H., Brodersen, C., Schwarze, F.W. and Jansen, S. (2016) The parenchyma of secondary xylem and its critical role in tree defense against fungal decay in relation to the CODIT Model. *Frontiers in Plant Science*, 7: 1665.

Mursidawati, S., Ngatari, N., Irawati, I., Cardinal, S. and Kusumawati, R. (2015) *Ex situ* conservation of *Rafflesia patma* Blume (Rafflesiaceae): An endangered emblematic parasitic species from Indonesia. *Sibbaldia*, 13: 99–110.

Näsholm, T., Högberg, P., Franklin, O., Metcalfe, D., Keel, S.G., Campbell, C., *et al.* (2013) Are ectomycorrhizal fungi alleviating or aggravating nitrogen limitation of tree growth in boreal forests? *New Phytologist*, 198: 214–221.

Partap, U.M.A., Partap, T.E.J. and Yonghua, H.E. (2001) Pollination failure in apple crop and farmers' management strategies in Hengduan Mountains, China. *Acta Horticulturae*, 561: 225–230.

Pearce, R.B. (1996) Antimicrobial defences in the wood of living trees, Tansley Review. *New Phytologist*, 132: 203–233.

Percival, G.C. (2013) Influence of pure mulches on suppressing *Phytophthora* root rot pathogens. *Journal of Environmental Horticulture*. 31: 221–226.

Percival, G.C. and Banks, J.M. (2014) Studies of the interaction between horse chestnut leaf miner (*Cameraria ohridella*) and bacterial bleeding canker (*Pseudomonas syringae* pv. *aesculi*). *Urban Forestry and Urban Greening*, 13: 403–409.

Percival, G.C., Barrow, I., Noviss, K., Keary, I. and Pennington, P. (2011) The impact of horse chestnut leaf miner (*Cameraria ohridella* Deschka and Dimic; HCLM) on vitality, growth and reproduction of *Aesculus hippocastanum* L. *Urban Forestry and Urban Greening*, 10: 11–17.

Popoola, T.O.S. and Fox, R.T.V. (2003) Effect of water stress on infection by species of honey fungus (*Armillaria mellea* and *A. gallica*). *Arboricultural Journal*, 27: 139–154.

Prescott, C.E. and Grayston, S.J. (2013) Tree species influence on microbial communities in litter and soil: Current knowledge and research needs. *Forest Ecology and Management*, 309: 19–27.

Rademacher, P., Bauch, J. and Shigo, A.L. (1984) Characteristics of xylem formed after wounding in *Acer, Betula* and *Fagus. IAWA Bulletin*, 5: 141–151.

Raimbault, P. (1995) Physiological diagnosis. In: *The Tree in its Various States: Diagnosis and Architectural Training.* Proceedings of the Second European Congress of Arboriculture, Versailles, France, pp. 52–55.

Raimbault, P.F. (2006) A basis for morpho-physiological tree assessment. In: *Proceedings of the Seminar, Arboricultural Association.* Treework Environmental Practice, Ashton Court, Bristol, UK, 23 and 24 March 2006.

Rayner, A.D.M. and Boddy, L. (1988) *Fungal Decomposition of Wood: Its Biology and Ecology.* Wiley, New York, USA.

Rioux, D. and Ouellette, G.B. (1991) Barrier zone formation in host and non-host trees inoculated with *Ophiostoma ulmi.* I. *Anatomy and histochemistry, Canadian Journal of Botany,* 69: 2055–2073.

Rotheray, G.E. and MacGowan, I. (2000) Status and breeding sites of three presumed endangered Scottish saproxylic syrphids (Diptera, Syrphidae). *Journal of Insect Conservation,* 4: 215–223.

Rotheray, G.E., MacGowan, I., Rotheray, E., Sears, J. and Elliott, A. (2015) Conserving the aspen hoverfly. *British Wildlife,* 27: 35–40.

Schwarze, F.W.M.R. (2007) Wood decay under the microscope. *Fungal Biology Reviews,* 21: 133–170.

Schwarze, F.W.M.R. and Baum, S. (2000) Mechanisms of reaction zone penetration by decay fungi in beech wood. *New Phytologist,* 146: 129–140.

Schwarze, F.W.M.R. and Fink, S. (1997) Reaction zone penetration and prolonged persistence of xylem rays in London plane wood degraded by the basidiomycete *Inonotus hispidus. Mycological Research,* 101: 1201–1214.

Schwarze, F.W.M.R. and Fink, S. (1998) Host and cell type affect the mode of degradation by *Meripilus giganteus. New Phytologist,* 139: 721–731.

Schwarze, F.W.M.R., Baum, S. and Fink, S. (2000a) Dual modes of degradation by *Fistulina hepatica* in xylem cell walls of *Quercus robur. Mycological Research,* 104: 846–852.

Schwarze, F.W.M.R., Mattheck, C. and Engels, J. (2000b) *Fungal Strategies of Wood Decay in Trees.* Springer, Berlin, Germany.

Schweingruber, F.H. (2007) *Wood Structure and Environment.* Springer, Berlin, Germany.

Scott, E.S. (1984) Populations of bacteria in poplar stems. *European Journal of Forest Pathology,* 14: 103–112.

Shain, L. (1995) Stem defense against pathogens. In: Gartner, B.L. (ed.) *Plant Stems: Physiology and Functional Morphology.* Academic Press, San Diego, CA, USA, pp. 383–406.

Shigo, A.L. (1979) *Tree Decay: An Expanded Concept.* United States Department of Agriculture, Forest Service, Agriculture Information Bulletin 419, Washington, DC, USA.

Shigo, A.L. (1984) Compartmentalization: A conceptual framework for understanding how trees grow and defend themselves. *Annual Review of Phytopathology,* 22: 189–214.

Shigo, A.L. and Marx, H.G. (1977) *Compartmentalization of Decay in Trees.* United States Department of Agriculture, Forest Service, Agriculture Information Bulletin 405, Washington, DC, USA.

Sjöman, H., Östberg, J. and Nilsson, J. (2014) Review of host trees for the wood-boring pests *Anoplophora glabripennis* and *Anoplophora chinensis*: An urban forest perspective. *Arboriculture and Urban Forestry,* 40: 143–164.

Smith, S.E. and Read, D.J. (2008) *Mycorrhizal Symbiosis*, 3rd edition. Academic Press, Elsevier, Amsterdam, The Netherlands.

Spicer, R. (2005) Senescence in secondary xylem: heartwood formation and an active developmental program. In: Holbrook, N.M. and Zwieniecki, M.A. (eds) *Vascular Transport in Plants*. Elsevier, Amsterdam, The Netherlands, pp. 457–475.

Stoffel, M., Bollschweiler, M., Butler, D.R. and Luckman, B.H. (2010) *Tree Rings and Natural Hazards: A State of the Art*. Advances in Global Change Research, Volume 41. Springer, Berlin, Germany.

Stokland, J.N., Siitonen, J. and Jonsson, B.G. (2012) *Biodiversity in Dead Wood*. Cambridge University Press, Cambridge, UK.

Straw, N.A., Williams, D.T., Kulinich, O. and Gninenko, Y.I. (2013) Distribution, impact and rate of spread of emerald ash borer *Agrilus planipennis* (Coleoptera: Buprestidae) in the Moscow region of Russia. *Forestry*, 86: 515–522.

Takos, I., Varsamis, G., Avtzis, D., Galatsidas, Sp., Merou, Th. and Avtzis, N. (2008) The effect of defoliation by *Cameraria ohridella* Deschka and Dimic (Lepidoptera: Gracillariidae) on seed germination and seedling vitality in *Aesculus hippocastanum* L. *Forest Ecology and Management*, 255: 830–835.

Thomas, P.A. (2014) *Trees: Their Natural History*, 2nd edition. Cambridge University Press, Cambridge, UK.

Thomas, P.A. (2016) Biological Flora of the British Isles: *Fraxinus excelsior*. *Journal of Ecology*, 104: 1158–1209.

Tilman, D. (1978) Cherries, ants and tent caterpillars: Timing of nectar production in relation to susceptibility of caterpillars to ant predation. *Ecology*, 59: 686–692.

Vera, F.W.M. (2000) *Grazing Ecology and Forest History*. CABI, Wallingford, UK.

Wheeler, E.A. (2011) InsideWood: A web resource for hardwood anatomy. *IAWA Journal*, 32: 199–211.

Wiseman, P.E., Colvin, K.H. and Wells, C.E. (2009) Performance of mycorrhizal products marketed for woody landscape plants. *Journal of Environmental Horticulture*, 27: 41–41.

Wiseman, P.E. and Wells, C.E. (2009) Arbuscular mycorrhizal inoculation affects root development of *Acer* and *Magnolia* species. *Journal of Environmental Horticulture*, 27: 70–79.

Zwart, D.C. and Kim, S.H. (2012) Biochar amendment increases resistance to stem lesions caused by *Phytophthora* spp. in tree seedlings. *HortScience*, 47: 1736–1740.

10

Environmental Challenges for Trees

Wherever trees grow, a variety of environmental challenges act to reduce their growth or chances of survival. Some of these challenges are quite transient, as with some forms of flooding, whilst in other cases the local climate imposes challenging conditions lasting months. The greatest challenges in large regions of the world are limitations in water availability or extremes of temperature. Tolerance to salt is also considered in this chapter. However, because availability of light is so intrinsically linked to photosynthesis, it is dealt with in Chapter 7.

Each tree species will typically have a range of environmental conditions that they can grow within. With very specialist species, the limits of these conditions can be quite narrow, or at least they will not compete very effectively in different conditions. For example, many boreal trees can be grown in temperate climates but tend to be out-competed by more rapidly growing species that are better suited to the warmer climate. Similarly, many trees well-suited to warm and dry conditions are capable of growing in regions with higher levels of soil moisture, but they may not be able to compete with trees that have evolved in those conditions. Successful trees will therefore have a number of adaptations to cope with various environmental challenges, whilst remaining competitive against other plants and resilient to pests and diseases.

By understanding the traits and strategies that help trees cope with their natural environment, it is possible to apply this knowledge to the management of trees in our landscapes. Acknowledging the ecological heritage of the tree improves tree selection and the chances of successful tree establishment, which are critical if the provision of ecosystem services from amenity trees are to be secured for future generations. Basic knowledge of plant stress is also vital to understanding potential causes of ill-health in trees and design interventions that lessen the long-term impact of stress.

Avoidance and Tolerance of Plant Stress

Before dealing with some of the most important environmental challenges to tree growth, it is important to clarify the meaning of some terms relating to plant stress.

Stress is defined as an environmental factor that reduces the rate of a physiological process, such as growth or photosynthesis, below the maximum possible rate (Lambers *et al.* 2008). Therefore, one of the immediate effects of stress is a reduction in performance. Fortunately, plants have come up with various mechanisms to reduce the impact on plant performance and confer some degree of stress resistance.

Applied Tree Biology, First Edition. Andrew D. Hirons and Peter A. Thomas.

The precise nature of the mechanisms varies widely depending on the type of stress, its duration and the species concerned. However, it is useful to think of two overarching strategies: stress *avoidance* and stress *tolerance* (Levitt 1980; Kozlowski and Pallardy 2002). (Ephemeral plants may also *escape* periods of environmental stress but, as trees are perennial, this strategy is not possible for the whole tree.) Stress *avoidance* strategies act to, insofar as is possible, prevent the stress from affecting the plant tissues. For example, deep-rooting trees that can tap into the groundwater are able to avoid water deficits caused by prolonged dry periods. Stress *tolerance* allows the tree to survive in the presence of the stress. For example, species that can cope with a low leaf water content are stress tolerant as they can persist throughout a period of water deficit without needing to source water from elsewhere. Of course, every species will only be tolerant of different stresses within a certain range; equally, the thresholds for damage that lead to death are very different between species.

The use of avoidance and tolerance has two important implications for those managing trees. First, as avoidance is possible, it is difficult to be sure if a tree is inherently tolerant of particular conditions just by observing it growing within a certain environment. Therefore, a tree seen growing in a dry area in the middle of summer does not mean that the tree is tolerant of the conditions, as it may be avoiding them by rooting deeply. Secondly, species that grow together may adopt different strategies, so alternative solutions to the same problem may be found by trees growing in the same area. This becomes relevant if the performance of a tree on a particular site is closely linked to tolerance of a stress rather than its avoidance. For example, constrained rooting environments make it difficult for species to avoid water deficits by developing deep roots. Consequently, species that are tolerant of water deficits would be more likely to perform well on these sites.

Many regions, particularly those with seasonal climates, have several months of unfavourable conditions that may be too cold or too dry for tree growth. Survival will depend on the tree's threshold to tolerate or withstand these. This relates to the ability to cope with low winter temperature, limited water availability or a combination of both: temperate trees need to cope with freezing temperatures as well as (typically) some degree of water scarcity during summer. Depending on the habitat, it may be that other forms of stress are important. Riparian (flood-plain) trees will need to cope with periods of flooding, for example.

Acclimation and Adaptation

Once a *stress factor* (the stress stimulus) is present in the environment, a *stress response* results in an immediate decline in some aspect of tree performance. In some circumstances, *acclimation*, involving structural or physiological adjustments, helps compensate for the initial decline in plant performance. For example, previously shaded leaves can acclimate to higher light levels to reduce the impact of photoinhibition (see Chapter 3). Importantly, acclimation occurs within the lifetime of an individual and is usually initiated within days of the onset of stress.

In the longer term, *adaptation* is an evolutionary response, resulting in genetic changes in the population that compensate for the decline in performance associated with the stress (Lambers *et al.* 2008). Over generations, these adaptations help the plant cope better with the most limiting stress factors. The main stress factors within a particular type of climate are similar in many parts of the world, so trees have evolved similar adaptations in these areas. In other words, they have undergone *convergent evolution* and arrived at similar strategies for coping with similar climate regimes. For example, trees from Mediterranean climates are often broadleaved evergreen species with deep rooting capacity whether they are growing in the Mediterranean basin, Californian chaparral, Chilean matorral, South African fynbos or Western Australian mallee (Archibald 1995).

Specialisation for environmental stress enables survival in challenging environments but it can incur the cost of forgone opportunities. By adapting to one (or a series) of these stress factors, species often lose the ability to exploit and compete in more favourable environments (Keddy 2007). Advantages accrued through adaptation to particular environmental conditions, and the associated reduction in competitive ability in other environments, is largely responsible for determining the species composition within a given habitat.

Although it is not possible here to detail all the complexities of tree responses and adaptations to environmental stress, a broad overview of the most significant environmental challenges is presented.

Cold-Hardiness

Low temperature can be a major factor in limiting the growth and survival of trees. Resistance to damage caused by low temperatures (*cold-hardiness*) helps determine the natural distribution of species (Barnes *et al.* 1998) and is a major factor in determining where horticultural plants can be grown (Thomashow 1999). Trees native to warm regions do not develop enough cold-hardiness to establish in cold regions. Low temperatures kill trees that are not hardy enough for the area, are unable to acquire hardiness quickly enough during autumn or lose hardiness too quickly in spring. It is therefore critical to match the cold-hardiness of a tree to its planting location if the tree is to be grown outside of its natural range. For this reason, cold-hardiness is a crucial factor in selecting trees for planting and, in recognition of this, a great deal of effort has gone into characterising the hardiness of different species.

Acquiring Cold-Hardiness

A factor that makes this characterisation more difficult is that even the most cold-hardy of species varies considerably in its tolerance to cold throughout the year. This is because trees prepare for, or *acclimate* to, low temperatures before winter arrives, gradually becoming more cold-tolerant. After growth stops for the year, shortening day length and, to a certain extent, the decline of temperatures to slightly above freezing, provide the environmental cues for the acclimation process to begin. In this *pre-hardening* phase, various protective substances and carbohydrates accumulate in the cells in readiness for freezing temperatures. The vacuole (a fluid-filled sack at the centre of each cell) also splits up into many smaller vacuoles to reduce the potential for freezing damage. Once temperatures fall below zero, cell membranes and proteins are modified so that they can tolerate the loss of water caused by ice formation (Larcher 2003).

Acquiring cold-hardiness is therefore a gradual process in which successive phases occur in preparation for winter temperatures. To develop fully, hardening involves exposure to both environmental cues and sub-lethal low temperatures, in order to provide resistance to otherwise lethal temperatures.

Periods of warm weather during winter and in spring induce *dehardening*, where tissues become progressively less capable of surviving frost injury. As a consequence, the level of hardiness can vary throughout the winter, particularly in regions with mild winters that fluctuate between freezing temperatures and warmer periods. Species that have a modest chilling requirement (see Leaf Phenology in Chapter 3) can easily be lulled into a false sense of security by warm late-winter temperatures. If this results in early flushing of leaves, these unhardened tissues are vulnerable to injury from late frosts. When planting trees from different regions, it is therefore important to anticipate the potential impact the change in climate may have on the timing of tree development and susceptibility to frost injury.

Cold-Hardiness Maps

The range of minimum temperatures that trees have to cope with around the world is shown in Figure 10.1. Using data such as this, plant cold-hardiness maps have been developed for many countries (Figure 10.2). Widely used plant material will typically have details of its hardiness published in standard references, although these should be interpreted with some caution as hardiness ratings are not always reliable. These can be compared with plant hardiness maps to help in deciding where a given species may be planted without suffering frost injury. Probably because it encompasses such a wide variety of climates, the most widely adopted hardiness scheme is that produced by the

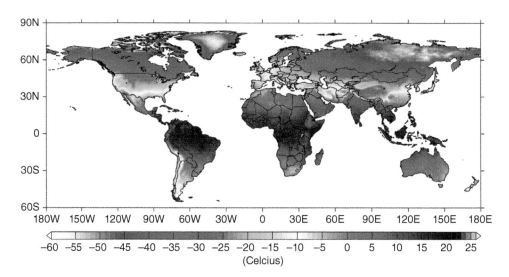

Figure 10.1 Average annual minimum temperature (°C) at 2 m above ground, using data from 1986–2015 from ERA-Interim Reanalysis. *Source:* Data from Dee *et al.* (2011). Plotted by Linda Hirons (National Centre for Atmospheric Sciences, University of Reading).

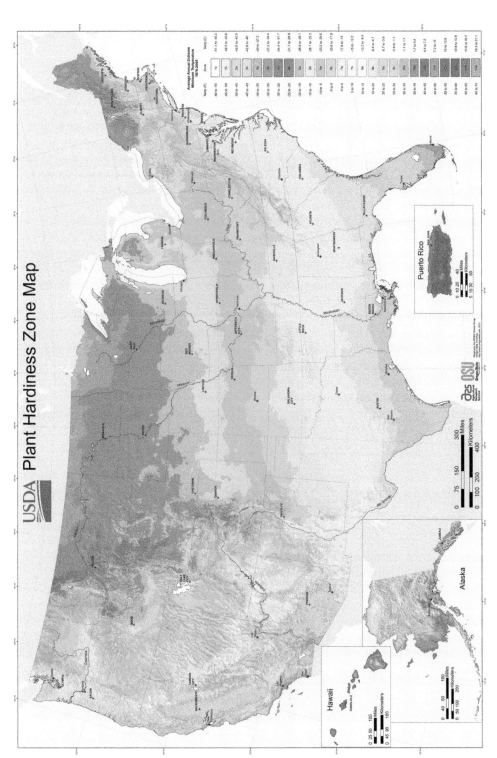

Figure 10.2 Plant hardiness zone map produced by the US Department of Agriculture. Reproduced courtesy of the United States Department of Agriculture.

US Department of Agriculture (USDA), shown in Figure 10.2. Of the 26 zones, the coldest zone (1a) has minimum annual temperatures of −51.1 to −48.3 °C (−60 to −55 °F) and the warmest hardiness zone (13b) only experiences minimum annual temperatures of 18.3–21.1 °C (65–70 °F).

Cold-hardiness maps are available for most countries and always involve some measure of annual minimum temperature. However, some countries, such as Canada, use additional climate variables (such as precipitation and snow depth) to produce a more complex hardiness index (Brady 2008). These maps are clearly helpful in determining if a species is likely to be vulnerable to frost injury but there are two issues that need to be considered. First, the minimum temperature that is fatal varies, depending on the age of the tree, time of year, climatic history and tissue type – leaves, flowers or woody parts (Sakai and Larcher 1987). This makes the absolute lowest temperature limit of a tree challenging to predict. Secondly, regions that have similar winter temperatures may have very different summer temperatures and rainfall, which can affect how the native trees cope with winter temperatures. For example, parts of the eastern USA share similar minimum winter temperatures with the UK but whilst the US regions may have well over 50 days above 30 °C during summer, the UK only has a handful of days at these temperatures. Across the year, this gives a very different growing environment, despite similarities in winter temperatures. This means that hardiness maps and zone ratings are useful but should not be used in isolation to determine the suitability of trees to new areas. Where possible, other information on a tree's native climate and habitat conditions within its natural range should be considered before introducing a species to a new area. In addition to this, careful experimentation over years is the only way of finding out the actual suitability of a species for a new area. At least some of this experimentation may well have already been done by local botanical collections and nurseries so, where possible, make use of local expertise before making a decision.

Cold Injury to Trees

Plant cells are damaged when the temperature drops below a critical value. Where this value is above freezing, damage is referred to as *chilling* injury, while damage caused by sub-zero (°C) temperatures is referred to as *frost* or *freezing* injury. Chilling injury is not really a problem in temperate or boreal species as cold winters have long since excluded species that are not capable of surviving at least some periods of freezing temperatures. However, trees of tropical and sub-tropical origin may experience chilling injury between about 10 and 0 °C.

With chilling, some cellular functions are impaired or stopped by the low temperatures, but if chilling continues for more than a few days damage typically becomes permanent. The primary effect of chilling is to disrupt the function of cell membranes. This affects the working of cells and may lead to their death (Larcher 2003). Whole trees can be sensitive to low temperatures, or the chilling injury may be confined to only the flowers or fruit. As with all types of low-temperature injury, differences in sensitivity occur at different times of the year and at different ages (i.e. seedling, sapling, mature tree).

It is possible to grow trees sensitive to chilling in temperate environments because summer temperatures are usually adequate. However, they will need protection as soon as air temperatures start to cool. In many cases, this puts it into the realm of technically possible but financially untenable. Such was the demand for 'exotic' *Citrus* species across northern Europe in the sixteenth and seventeenth centuries, that the societal elite containerised these trees so they could be moved into specially built orangeries for protection against low temperatures during winter. Quite rightly, botanical collections in cold climates often have some form of protected growing environment for cold-sensitive species, but this level of investment is rarely possible outside of specialist institutions.

Over half of the Earth's land area is exposed to frost (<0 °C) during the course of the year (Figure 10.1), so an ability to cope with low temperatures is crucial for the survival of many trees. This includes trees in Mediterranean, temperate, boreal and high-altitude regions.

Avoiding Freezing in Below-Zero Temperatures

No plant cell can survive ice crystals in the internal tissues (the protoplast), so surviving temperatures below freezing involves the prevention of ice formation within the cell (Körner 2012). In winter-deciduous trees, the prospect of freezing damage to leaves is removed by their being shed. All remaining tissues (including foliage on evergreen species) must survive freezing conditions either by avoiding freezing or by tolerating ice formation outside the protoplast.

Freezing can be avoided by accumulating substances (solutes), such as sugars (sucrose, raffinose and stachyose), that can *depress the freezing point* within cells to between about −1 and −5 °C. This is a fairly reliable method of avoiding gentle freezing but it is certainly not sufficient to protect trees from intense cold.

Remarkably, trees can also depress the freezing point of their cells without the use of solutes. This *supercooling* maintains water in a gel-like state at temperatures below freezing without the water becoming solid. Although our understanding of this process is somewhat incomplete, it seems that anti-nucleation agents, such as flavonoid glucosides (Kasuga *et al.* 2008), prevent the formation of ice crystals in temperatures well below freezing. In foliage, this prevents damage down to around −12 °C. In xylem parenchyma cells, however, a similar process of *deep supercooling* can prevent ice formation in cells at temperatures down to −10 °C in tropical trees, −40 °C in temperate trees and almost −70 °C in boreal trees (Fujikawa *et al.* 2009). The lowest temperature this can protect against will inevitably vary between species and the level of cold acclimation, but, once this limit is passed, water freezes inside the cells and lethal damage occurs.

Ice Formation Outside of the Cell Protoplast

Clearly, it is impossible to prevent all water within the tree from freezing in very cold temperatures. Plants avoid the damage that this would cause by neatly moving the water outside of the protoplast and putting it in the spaces *between* cells, or *inside* the cell between the cell wall and the protoplast.

Water found outside of cells contains fewer solutes and so inevitably freezes first. This concentrates the solutes by removing liquid water which, in turn, draws more water out of the cell where it freezes. Consequently, water inside the cell is depleted, the solutes inside become more concentrated and the volume of the protoplast is substantially reduced. Sub-zero temperatures therefore have a similar effect on the protoplasm as desiccation (Larcher 2003). In sensitive species, this dehydration damages the cells' membranes by causing excessive contraction and can lead to the solutes reaching toxic concentrations. In these circumstances, even rehydration of the cell is not sufficient to overcome the irreversible damage to the protoplasm: cell death is inevitable.

Fortunately, the cell is protected in several ways in species that are tolerant of freezing: first, ice forms outside of the cell membrane rather than within it; and, secondly, specialised proteins, amino acids, sugars and sugar alcohols are used to protect the delicate cellular structures, allowing them to work again once the ice thaws and the cell is rehydrated (Körner 2012). The two factors together result in cells that can safely dehydrate and rehydrate. Ultimately, the greater the cold tolerance of a species, the greater its ability to tolerate freezing-induced dehydration.

Frost Injury

If the temperature drops below the critical hardiness threshold of the tree, or its flowers, fruits or leaves, then injury will occur regardless of the time of year. Indeed, cold injury may be so severe that it kills the entire tree if it is poorly adapted to the climate. This can happen in over-ambitious exotic plantings or as a consequence of extreme weather.

In temperate deciduous species, young leaves in spring are probably the most likely to be injured. Late frosts can often catch trees out, particularly if they do not have much of a winter chilling requirement and warm spring weather promotes early shoot development. After frost injury, young leaves often appear black (Figure 10.3) whilst more mature leaves tend to turn reddish to dark brown (Costello *et al.* 2003). Providing

Figure 10.3 Frost injury to young shoots of European ash *Fraxinus excelsior* has caused the damaged shoots to turn black. *Source:* Courtesy of Duncan Slater.

the stem and bark are not killed, new shoots can arise from axillary or epicormic buds (see Chapter 2). Early autumn frosts can also damage shoots that have not yet been able to cold-harden. For this reason, lammas growth, brought about by warm late-summer temperatures or late shoot growth, stimulated by incorrect fertiliser applications, are particularly vulnerable to cold injury. However, all shoots will have some level of susceptibility prior to cold acclimation.

Flower buds are more vulnerable to low temperatures than vegetative buds. Clearly, damage to the flowers can have serious consequences for fruit (and seed) production, which can be economically devastating for fruit growers. However, for amenity species the damage is often more of an aesthetic problem: the magnificence of an early flowering magnolia is often brought to an abrupt end by a late frost in the UK.

Low temperatures can also damage woody stems. *Winter sunscald* is a freezing injury on the southern sides of trunks that receive more sun than the shaded northerly sides. During the day, the sun can warm up the sunny side of the tree by at least 20 °C compared with the shaded side. After the sun disappears behind a cloud, or after sunset, the temperature drops so rapidly that the cambium tissues can be killed by rapid freezing. This causes lesions on the trunk, characterised by dead patches of bark (Kozlowski *et al.* 1991). In serious cases, frequent alternation of freezing and thawing causes the bark to separate from the xylem; this disrupts sugar movement in the phloem and exposes the surface of the xylem to infection by pests and diseases. Many northern birch *Betula* spp. have a white compound, betulin, in their bark that helps to reflect away the sun's energy and so reduces the occurrence of sunscald. To help achieve the same effect, it is a tradition in some regions to paint the lower trunk of some susceptible species white. Wrapping some form of insulating material, such as straw or hessian, around the trunk has also been used to help prevent great fluctuations in temperatures and so reduce the likelihood of injury (Figure 10.4).

Severe frosts can also cause vertical *frost cracks* (splitting), acting through a complex interaction of factors. Part of this is caused by the tendency for the sapwood to shrink in the cold more than the warmer underlying heartwood, causing tension. Dehydration of the outer sapwood as the water freezes also causes the outer part of the wood to contract (Kubler 1983). The result is a long vertical crack that closes up as the tree warms up. Many different trees are susceptible to frost cracks, including: *Abies, Acer, Alnus, Betula, Fraxinus, Ostyra, Platanus, Populus, Quercus* and *Ulmus* but most species will also develop frost cracks given the right conditions. Frost cracks are also more likely when bacterial wetwood of the heartwood is present (Sakai and Larcher 1987; see also Chapter 9). Callus tissue formed by the cambium at the margins of these cracks lead to characteristic *frost ribs* that protrude slightly from the surface of the stem (Figure 10.5). These frost cracks can reopen each winter for many years until, eventually, enough callus tissue builds up to keep it closed (often aided by a mild winter or two), but the raised rib persists as evidence of the original injury.

If freezing occurs during secondary growth, *frost rings* can occur inside the xylem as a consequence of injury to the cambium. Damage may be around an entire ring or confined to a portion of the stem, where collapsed cells, callus tissue and bent rays can be seen (Schweingruber 2007). Depending on the severity of the frost rings, this can disrupt water movement in the xylem and can locally reduce the wood's strength, making their presence undesirable in trees grown for their wood.

Figure 10.4 Trunks wrapped in hessian to protect them from excessive fluctuations in temperature. Ornamental *Prunus*, Southern Hokkaido, Japan.

Figure 10.5 Frost rib on sessile oak *Quercus petraea* in southern Scotland.

Soil and, if present, snow cover helps protect roots from exposure to very low temperatures. For this reason, roots tend to be less adapted to freezing which will halt physiological activity and cause lethal damage to many of the fine roots. In temperate species, freezing damage to fine roots occurs at −1 to −3 °C and to larger woody roots at −5 to −25 °C. Some boreal conifers may even extend this range to −35 °C (Sakai and Larcher 1987). Root damage is particularly common in containerised trees, as roots are not insulated by surrounding soil and so are exposed to cold temperatures from the side, as well as the top, of the soil. This effectively excludes above-ground containerised plant production from many cold climates.[1]

A further problem associated with frozen soils is that of *winter desiccation*. On sunny winter or early spring days, the leaf temperature may rise sufficiently to stimulate photosynthesis in evergreen trees. Obviously, this involves some transpiration and water loss from the leaves. If the soil is still frozen, water supply cannot keep pace with demand and the foliage gradually dries out, causing desiccation injury. The problem can be particularly acute on mountainsides where the high elevation keeps the soil frozen for much of the winter, resulting in red, dehydrated needles seen across whole mountainsides. Injury can be especially severe if soils have not recharged with water during early winter.

In addition to these physical injuries caused by freezing temperatures, a very important physiological consequence of winter freeze–thaw cycles is cavitation of xylem, resulting in loss of hydraulic conductivity (see Chapter 2).

High Temperatures

All trees require a certain amount of warmth if they are to grow. The rate of all major biological processes will increase with temperature up to some optimum value that is often well correlated with average summer-time temperatures of a species' native region. Summer heat may be essential for some species to thrive, even if they are capable of surviving in regions with cooler summers. In temperate climates with a hot summer (≥22 °C), typical of the eastern deciduous forest region of North America, a rapid transition from spring to summer over a few weeks, combined with higher summer temperatures, means that trees here experience much more heat than trees growing in an oceanic temperate climate as found in the UK and other parts of western Europe. Here, the spring is rather protracted (months) and the summer is defined as having more than 4 months with temperatures ≥10 °C (Peel *et al.* 2007). Such differences in summer temperatures can be important to the performance of a species originating from one region but planted in the other. For example, white oak *Quercus alba*, widespread in the eastern USA, enjoys summer heat and is cold tolerant enough for the UK, but does not do well in the cooler summers. Conversely, pedunculate oak *Quercus robur* is perfectly at home with the cooler UK summers but does not seem to perform well in the warmer summers found in the eastern USA. Climate matching of species to both winter and summer temperatures of the planting area is therefore essential when planting non-native species in amenity landscapes or commercial forests.

---

1 A system known as pot-in-pot container production may still be possible in some regions. This system grows trees in containers but places them into a socket pot within the ground, which helps buffer the root-zone against low temperatures.

Tolerance of high temperatures is important to a tree's performance throughout the year, and may become increasingly so as a changing climate increases the frequency and intensity of heatwaves (Teskey *et al.* 2015). Trees can adjust (acclimate) their photosynthesis to heat but damage to photosystem II,[2] caused by high temperatures, appears to be irreversible above ~40 °C (Yordanov 1992). This can be prevented by allowing transpiration to cool leaves but is obviously dependent upon water being available. Stems may also benefit from convective heat transfer as the sap moves through the tree. Nevertheless, for the whole tree, the threshold for substantial heat injury in temperate areas tends to be around 50 °C, but in the tropics and sub-tropical regions this can increase up to 60 °C (Larcher 2003). However, even within temperate areas, sensitive species will show visible damage between 40 and 45 °C, whereas heat-tolerant species will be able to withstand temperatures in excess of 50 °C. Although these figures sound high, vegetation and soils can reach temperatures well above the ambient air temperature as a result of high light (solar radiation), particularly in very still conditions or when evapotranspirational cooling is limited. Heat stress is therefore possible on days when the ambient air temperature is comparatively cool. Exposure to additional heat emitted from built structures, or released from equipment, vents or pipes, can exacerbate heat stress for some urban trees.

As with cold temperatures, the extent of the injury varies within a species as a result of genetic differences, the time of year, the duration of the heat exposure, the level of acclimation and the maturity of the tree. Injury to leaves is usually seen as blanching or necrosis (dead patches), but leaf mortality, leaf shedding and reduced leaf expansion have also been observed in young, recently emerged leaves (Filewood and Thomas 2014). Non-lethal high temperatures can cause a decline in net photosynthesis because of increased respiration rates, reduced performance of key photosynthetic enzymes and damage to cell membranes (Bita and Gerats 2013; also see Chapter 7). Growth is then reduced primarily because of the negative impact of heat on photosynthesis and an overall reduction in total leaf area (Table 10.1).

Coping with High Temperatures

When trees are exposed to temperatures at least 5 °C above their optimal growing conditions, a number of cellular and physiological responses occur to aid their tolerance of heat (Table 10.1). *Heat shock proteins* are produced to help stabilise membranes that otherwise become more fluid as temperatures increase. Other compounds, such as proline, glycine betaine, soluble sugars, abscisic acid, ethylene, hydrogen peroxide and salicylic acid, have also been implicated in improving heat tolerance in plants (Song *et al.* 2012; Bita and Gerats 2013).

Some trees produce *volatile organic compounds* (VOCs) in response to abiotic stress such as high temperature. For example, at least some species in the genera *Bauhinia, Eucalyptus, Liquidambar, Picea, Populus, Pterocarpus, Quercus* and *Salix* have been found to produce high levels of *isoprene* in response to short-term heat stress (Monson *et al.* 2013). Other species, such as pines *Pinus* spp., release *monoterpene* under stress.

---

2 Photosystem II is a vital protein complex in which excitation energy from photons (light) is passed to an electron in the light-dependent reactions of photosynthesis. If this protein complex is damaged, then photosynthesis is reduced.

Table 10.1 Known effects of high temperatures on major cellular, leaf and whole tree processes. An increase, decrease or no change in a process in response to high temperatures is indicated by +, – and 0, respectively. More than one symbol associated with a process indicates between- or within-species variation. 'Yes' indicates that acclimation in response to high temperatures has been reported in the literature; 'No' indicates that no acclimation has been reported; '?' indicates that acclimation may exist but evidence is limited. PSII is photosystem II; VOCs are volatile organic compounds; thylakoid membranes surround the green chloroplasts in plant cells; rubisco is a protein used in photosynthesis. Many of the cellular and leaf processes are described in Chapter 3.

	Process	Response to high temperatures	Acclimation
Cellular	PSII functioning	–	Yes
	Thylakoid membrane fluidity	+	Yes
	Rubisco activity	–	Yes
	Heat shock proteins produced	+	Yes
Leaf	Photosynthesis	–	Yes
	Dark respiration	+	Yes
	Photorespiration	+	Yes
	VOC emission	+/0	?
	Stomatal conductance	+/–	Yes
	Transpiration	+/–	No
Whole tree	Leaf area development	–	No
	Leaf shedding	+	?
	Early budburst	0/+	No
	Growth	–	Yes
	Mortality	+	Yes
	Fecundity	–	Yes

Source: Adapted from Teskey *et al.* (2015). Reproduced with permission of John Wiley and Sons.

These VOCs help protect photosynthetic apparatus and may have an important antioxidative role in trees exposed to heat stress (Sharkey *et al.* 2008). This response has also inspired the naming of landscapes: the Great Smoky Mountains and Blue Ridge Mountains in the Appalachians (USA) as well as the Blue Mountains in New South Wales (Australia) can attribute their names to the haze of VOCs emitted by the trees, particularly during hot summers.

It is often challenging to separate the effects of high temperatures from those of the drought (see Drought and Water Deficits) and high light (see Chapter 7) that typically accompany them. Extreme high temperatures associated with high levels of soil moisture are rare. Much more typically, high temperatures are associated with simultaneous soil water deficits. During dry periods, cloud cover is reduced which increases solar radiation and therefore the heating of land surfaces and the vapour pressure deficit (see Chapter 6). This tends to increase transpiration, further drying the soil and so providing a positive feedback loop that amplifies the effect of both high temperatures and drought. This combination of heat and drought has been demonstrated to be an important

Figure 10.6 *Vachellia* spp. (previously a member of the *Acacia* genus), seen here growing in Kenya, have very small leaves to help them prevent damaging leaf temperatures. When there are lots of browsers around, having small leaves also means that they can be well protected by defensive thorns.

driver of forest mortality, even in relatively humid temperate regions (Allen *et al.* 2010). Such stresses also predispose trees to pests and pathogens that accelerate tree mortality rates further (Anderegg *et al.* 2015).

As well as the cellular and physiological adaptions to heat already described, a number of trees from hot climates alter the way their leaves grow. Many of these adaptations involve the avoidance of the sun's energy that can rapidly heat leaves to dangerous temperatures. In savannah trees, such as *Vachellia* spp. and *Acacia* spp. (Figure 10.6), small leaves (microphylls) help to reduce the radiation absorbed as well as being able to lose heat more rapidly than species with larger leaves. Other hot climate specialists, such as many *Eucalyptus* spp., allow their leaves to hang vertically to reduce radiation load, which helps to prevent them overheating and creates the so-called *shadeless forests* (look back at Figure 3.16). Even species from more temperate regions have developed strategies to avoid their leaves overheating. Silver lime *Tilia tomentosa* inverts its leaves during hot spells to reveal the characteristic silvery underside of the leaf (made up from leaf hairs), which helps reflect light and therefore helps prevent injury from high leaf temperatures (Figure 10.7).

Drought and Water Deficits

Assuming that the tree is sufficiently hardy to cope with low temperatures, it is the availability of water that is most likely to limit tree development because it is important for every major physiological process as well as for growth itself (Pallardy 2008). Nutrient availability is important but water is ultimately the most common limiting

(a)

(b)

Figure 10.7 Silver limes *Tilia tomentosa* twist their leaves so that the silvery underside of the leaf reflects light to reduce the heat load on their crowns: (a) shows a small group of silver limes growing in southern Sweden; (b) inverted leaves displaying the silvery underside of the leaves characteristic of this species.

factor controlling growth. In natural environments, limitations to a tree's water supply are usually a consequence of *climatic drought*, where seasonal rainfall falls below potential evaporation, and soil moisture becomes seriously depleted. This occurs seasonally in Mediterranean and many sub-tropical climates, but can occur in all other forest types during extreme weather events.

Water shortage for amenity trees can be made worse by root loss during transplantation or construction activities, limitations to soil volumes, impermeable surfaces over rooting areas, and land drainage, because these disrupt the absorption or the availability of water (see Chapter 6). As none of these factors necessarily reflects a climatic drought, it is much more accurate to refer to this environmental stress as a *water deficit* rather than *drought*, as the latter refers to a meteorological event rather than a physiological condition. The term *water stress* has its own limitations because it gives no indication as to whether there is too much or too little water.

Basic tree water relations and the management of soil water availability is covered in Chapter 6. Here, we focus on the impacts of water deficits, adaptations to cope and the selection of trees based on their tolerance to water deficits.

Water Deficits and Tree Development

One of the most prominent effects of water deficit is to reduce tree height: trees growing on dry sites are shorter than those of the same species growing on moist sites. For example, coastal redwood *Sequoia sempervirens* can grow over 100 m on the moist alluvial soils of lowland valleys in northwest California, but only reach about 30 m on the shallow and relatively dry soil on adjacent uplands (Kozlowski *et al.* 1991). Similar trends can be seen across other landscapes that have large gradients in water availability. Vegetation in Mediterranean and savannah biomes tends to be shorter than in temperate areas.

As described in Chapter 6, plant cells are dependent upon being full of water (at full turgor or pressure) to maintain their shape and size. Under conditions where the root

Figure 10.8 Leaf wilting is one of the first clear signs of water deficit. Here, rowan *Sorbus aucuparia* (middle left) shows clear signs of wilting, whilst the associated sessile oak *Quercus petraea* and juniper *Juniperus communis* (middle right and foreground, respectively) are apparently coping quite well with the level of soil moisture on this site in southern Sweden. Interactions such as these affect the establishment of trees and which species can grow on a site.

system fails to acquire enough water to meet the needs of the crown, the tree experiences a decline in water potential as the water in the xylem is placed under increasing tension. This makes it harder to maintain turgor pressure, and a gradual loss of turgor occurs as cells dehydrate. In trees with relatively soft leaves, a reduction in turgor results in leaf wilting (Figure 10.8). Wilting tends to occur later in the drying cycle and is much less noticeable in the leathery leaves of broadleaved evergreens as they contain more supportive tissue.

Even before wilting is seen, the reduction in cell turgor will slow growth throughout the tree. Leaf expansion, extension of shoots and radial growth are reduced by water deficits, but the timing and extent of the deficit will affect any reduction in growth. Cell division is much less affected by water deficits than expansion of cells, so, if the period of deficit is short-lived (days), recovery of growth can occur as cells will simply delay their expansion. However, when the water deficit continues for longer periods (weeks), a permanent reduction in growth can often occur as the activity of meristems is suppressed. For both gymnosperms and angiosperms, there is a consistent link between radial growth and soil moisture availability (Schweingruber 2007). In temperate environments, 70–80% of the variation in radial growth has been associated with the availability of water (Zahner 1968) although this does vary among species. This is used in dendrochronology to reconstruct past climates and is valuable for comparing tree growth across precipitation gradients (Fonti *et al.* 2010).

In addition to these more obvious signs of water deficiency, a number of vital cellular processes are also affected. As cells dehydrate and reduce in volume, key processes (such as protein synthesis) become impaired, the activity of enzymes is hampered and every aspect of cell metabolism is inhibited. With continued dehydration, the cell turgor pressure reaches zero. In trees, the disruption caused by cellular dehydration to the point of zero turgor is so significant that the cell becomes permanently damaged. For this reason, the leaf water potential at zero turgor (Ψ_{P0}) is seen as a highly useful value as it defines the soil water potential threshold below which the tree will be unable to recover from wilting. Species differ in their turgor loss point, so it is used as a quantitative measure of drought tolerance as trees with a lower (more negative) Ψ_{P0} can maintain photosynthetic activity and growth at a lower soil water potential (i.e. drier soil) than trees with a higher (less negative) Ψ_{P0}. The potential for using leaf turgor loss point as a trait to aid tree selection is explored below.

The seasonal timing of drought or water deficit is also relevant to the overall effect on growth. In species with determinate growth, where the new shoot develops in a single flush in spring, a decline in soil moisture during summer is less likely to affect shoot growth. However, late summer water deficits can still affect bud formation, which in turn will affect shoot development in the next growth season. Indeterminate species, which are capable of shoot elongation throughout the growing season, respond much more to fluctuations in soil moisture during the growing season.

Water deficits, particularly during periods of leaf expansion, will reduce individual leaf size and result in a lower leaf area for the entire crown. Indeed, trees in drought-prone areas have evolved small leaves to reduce the water demand and so increase the likelihood of survival, even though this inevitably reduces the carbon gain of the whole crown.

A major physiological implication of water deficit is the reduction in photosynthesis. As soil moisture declines, various hydraulic and non-hydraulic mechanisms (i.e. chemical signalling from plant growth regulators, such as abscisic acid, ABA) act to close stomata in order to conserve water (Augé *et al.* 2000). Inevitably, this reduces photosynthesis but helps preserve the hydraulic integrity of the tree by significantly reducing water loss from the crown. Whilst all trees have control of their stomata, the relationship between stomatal water loss (i.e. stomatal conductance; g_s) and leaf water potential differs widely across species. For this reason, it is impossible to give a general value for the leaf water status that will cause stomatal closure.

In response to soil drying and a declining leaf water potential (Ψ_{leaf}), *isohydric* species will begin to close their stomata very promptly (at Ψ_{leaf} around -1 MPa) in order to keep leaves from wilting and reduce the likelihood of cavitation in the xylem (see Chapter 2). This is effective in conserving water but photosynthesis quickly reduces as it runs out of carbon dioxide, and so, particularly under high light conditions, photoinhibition can become a problem (see Chapter 7). *Anisohydric* species keep their stomata open for much longer as the leaf water status declines and some species may still have high levels of stomatal conductance at Ψ_{leaf} less than -3 MPa. These species can remain productive for longer through the drying cycle but run the risk of serious cavitation if the shoot water status continues to decline unchecked. Importantly, different species are not necessarily of one type or the other but form a continuum between these two opposing stomatal behaviours (Klein 2014).

An emerging trend from studies of the dieback of temperate trees under drought is that species with low wood density tend to be more isohydric, whilst species with higher wood density tend to be more anisohydric (Hoffmann *et al.* 2011). Comparisons of the leaf water potential at 50% of maximum stomatal conductance (Ψ_{gs50}) across species suggests that stomata are less sensitive to leaf water potential in ring-porous species than in diffuse-porous species, and that trees from more arid climates have a lower Ψ_{gs50} than those from wetter climates (Klein 2014) (Figure 10.9). Evaluation of a wider range of species from different climates would help confirm these findings.

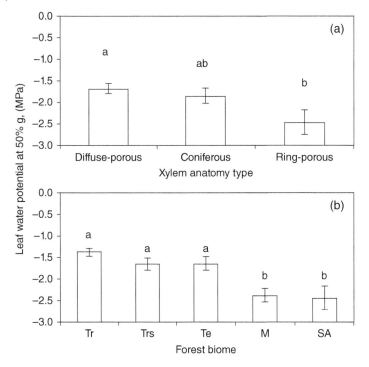

Figure 10.9 Variations in leaf water potential at 50% stomatal conductance ($\Psi g_s 50$) between trees with: (a) different xylem anatomy type; and (b) between forest biomes. Biomes are tropical (Tr), tropical seasonal (Trs), temperate (Te), Mediterranean (M) and semi-arid (SA). Letters indicate significant ($P < 0.05$) differences between xylem types or biomes. Error bars represent the standard error of the mean. *Source:* Klein (2014). Reproduced with permission of John Wiley and Sons.

Understanding these strategies can be useful when planting urban trees. If urban trees are intended to help cool the environment through evapotranspiration, then isohydric species will not help because they close their stomata after only modest declines in leaf water potential. These conditions are likely to occur during the same periods that cooling is most desirable. It would be much better to plant anisohydric species that will continue to transpire water over a much larger range of leaf water potentials. Admittedly, both types of tree, assuming that they are able to maintain a healthy crown, would still cast valuable shade. Conversely, if planting trees on a shrinkable clay soil, anisohydric species will continue to extract soil water and may be more likely to cause a volume change in the soil. In turn, this could lead to subsidence and damage to property. In this scenario, it would be better to have isohydric species that would shut their stomata earlier in the drying cycle, reducing water loss and, consequently, limiting shrinkage of the soil. However, it should be noted that other characteristics, such as crown leaf area and rooting depth, will also have a significant influence on the amount of water extracted from the soil. Unfortunately, the stomatal behaviour of amenity tree species is rarely known, let alone considered, when selecting trees for urban and semi-urban sites.

Where stomatal closure does not sufficiently control water loss, or soil drying continues unabated, transpirational water losses will exceed supply. This leads to increased tension within the water column, and eventually embolism of the xylem occurs with an associated loss of hydraulic conductivity (see Chapter 2 for more details). This leads initially to leaf wilting but extensive loss of hydraulic conductivity within trees is also accompanied by *crown dieback*. The highest shoots operate under higher tension than shoots lower down the crown, so higher twigs and branches experience more cavitation than lower branches. Subsequent loss of hydraulic conductivity in the periphery of the crown is the main cause of crown dieback. Crown injury can be made worse by drought, loss of roots from root decay fungi or low-quality rooting environments (e.g. highly compacted soil) that reduce the supply of water to the crown.

Vascular wilt diseases, such as Dutch elm disease (*Ophiostoma novo-ulmi*), also greatly reduce the conduction of water through stems as the tree reacts to the fungus and blocks vessels with tyloses and/or gels (see Chapter 9). This causes a rapid decline in conductivity in stems, increased tension within the water column and widespread embolism. The net result is extensive dieback within the crown and, in many cases, death of the entire tree. Again, while the cause is nothing to do with soil moisture, the effect on the tree mimics that of a water deficit.

Further detail on the effects of drought and water deficits in plants can be found in Larcher (2003), Schulze *et al.* (2005), Lambers *et al.* (2008) and Pallardy (2008), or specialist texts such as Aroca (2012).

Resistance of Water Deficits Using Avoidance and Tolerance Strategies

Plants have evolved a number of *traits* that allow them to resist water deficits either through avoidance or tolerance of the stress (Figure 10.10). These strategies are not necessarily mutually exclusive and many species will combine a range of adaptations and/or responses to remain competitive in water-limited environments.

As shown in Figure 10.10, a water deficit can be avoided either by maximising water acquisition or by reducing water use. Improvements in water acquisition come about by increasing absorptive area through root growth, increasing the hydraulic conductivity of the roots and developing deep root systems. The advantage provided by deeper roots is clear when comparing maximum rooting depth from different biomes (see Figure 4.14): trees from regions with seasonally dry periods tend have much deeper roots than those from regions where soil moisture is more plentiful (Canadell *et al.* 1996). A *dimorphic* root system, consisting of shallow horizontal roots to intercept sporadic rainfall and a taproot to draw on ground water, is also common in drier regions. This type of root morphology also helps facilitate hydraulic lift (see Chapter 6), which may be key to survival in dry environments. Clearly, the major value of deep roots is that they allow trees to somewhat decouple their water supply from that of rainfall and therefore maintain water supply for much longer periods throughout the year. Figure 10.11 shows that the deep-rooting habit of *Vachellia* spp. allows them to hold on to some green leaves long after other vegetation has died or become dormant.

Even within a temperate forest community, variation in rooting depth can have important implications for tree performance during a drought. For example, it has been found that deep-rooting Mahaleb cherry *Prunus mahaleb* was largely unaffected by an extreme summer drought in north-eastern Italy, while shallower rooted downy oak *Quercus pubescens* and hop hornbeam *Ostrya carpinifolia* suffered an average of 60% loss of hydraulic conductivity and extensive dieback (Nardini *et al.* 2016).

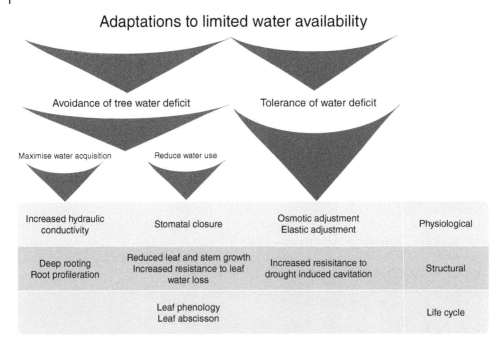

Adaptations to limited water availability

Avoidance of tree water deficit   Tolerance of water deficit

Maximise water acquisition   Reduce water use

Increased hydraulic conductivity	Stomatal closure	Osmotic adjustment Elastic adjustment	Physiological
Deep rooting Root profileration	Reduced leaf and stem growth Increased resistance to leaf water loss	Increased resisitance to drought induced cavitation	Structural
	Leaf phenology Leaf abscisson		Life cycle

Figure 10.10 An outline of important strategies for temperate tree resistance to water deficit.

Figure 10.11 The deep-rooting habit of *Vachellia* sp. (previously members of *Acacia*) in Samburu National Reserve, Kenya, means that these savannah specialists can keep some working leaves, even during prolonged periods of drought. Other vegetation has long since succumbed to the water deficits.

The reduction of water use also helps a tree avoid the development of deleterious water deficits. Seasonal water deficits can be coped with by closing the stomata. If the water deficit becomes more extreme, leaves can be prematurely lost to reduce total leaf area and conserve water, as seen in a number of drought-sensitive temperate trees, such as many birch *Betula,* willow *Salix* and poplar *Populus* spp. In more arid environments, water use can be reduced by having smaller leaves and by reducing the overall leaf area of the canopy. The tree's life cycle may be modified so that it does not have any leaves during the driest periods of the year; being *drought-deciduous* is common in areas with seasonal drought (Archibald 1995; Mooney and Miller 1985). Drought-deciduous species often have green stems so that they can continue some photosynthesis whilst minimising water loss.

Other leaf characteristics that can help reduce water loss include epicuticular waxes and leaf hairs (trichomes). Such *xeromorphic* features are important for the success of tree species in dry environments, such as those from the Brazillian savannah or cerrado (Bieras and Graças Sajo 2009). This is partly because they reduce leaf water loss by increasing reflectance and the boundary layer resistance of leaves (holding a moist layer of air against the leaf, reducing further evaporation), but also because they deter herbivores. However, even trees from relatively moist environments may use leaf trichomes for the same purposes.

Trees also tolerate water deficits by accumulating osmotically active compounds to reduce the osmotic potential (Ψ_π) of their cells, in order to attract and hold water inside the cells more strongly: a process known as *osmotic adjustment.* As a result, cells have a higher turgor potential when fully hydrated (Morgan 1984) so they can maintain turgor for longer during the drying cycle. Osmotic adjustment also helps trees to continue to take up water for longer during soil drying and extends the range of soil water potential that the trees can grow in without substantial injury. Whilst most trees will be able to store some water in their xylem parenchyma and use this to buffer short-term water deficits (a process known as *capacitance*), this strategy is of particular importance for some dryland specialists, such as baobabs *Adansonia* spp. (Figure 10.12). Elastic cell walls of the xylem parenchyma allow these trees to store enough water in their stems to survive in a leafless condition for up to 18 months without taking up water from the soil (Wickens 2008).

Tolerance to water deficit is also seen in the vulnerability to cavitation, which is hugely variable across species (see Chapter 2). In general terms, vulnerability to cavitation is closely associated with the conditions that the tree would typically experience in its natural habitat. Trees tend not to over-engineer their hydraulic system, so species from more xeric (dry) environments are less vulnerable to cavitation than those from mesic (moist) environments but, interestingly, the safety margin of species may be similar between these habitats (Choat *et al.* 2012). Microscopic adjustments in xylem anatomy, particularly the pit structure, govern the susceptibility to cavitation and the range of water potentials over which hydraulic conductivity is lost. So, even closely related species can respond quite differently to water deficits based on their xylem anatomy. For example, on the southern margin of its distribution, pedunculate oak *Quercus robur* is often associated with holm oak *Quercus ilex. Quercus robur* is more vulnerable to cavitation and has a narrower hydraulic safety margin than Q. *ilex.* This means that Q. *robur* is less competitive on drier sites than Q. *ilex* and is at greater risk of mortality during drought. Given the fact that predictions of climate change suggest an increase in

Figure 10.12 *Adansonia grandidieri*, a baobab, growing near Morondava on the west coast of Madagascar. These have swollen trunks that are capable of storing large volumes of water to help the tree survive the long periods of drought.

frequency and intensity of drought, it is likely that *Q. robur* will become increasingly less competitive on many sites that it currently shares with *Q. ilex* (Urli *et al.* 2015). Similar effects will be seen elsewhere in the world as precipitation patterns are altered and global temperatures rise.

Drought Tolerance for Difficult Urban Sites

In recently planted trees, root loss during transplanting reduces the tree's ability to supply water to its crown, even when there is plenty of available water in the surrounding soil (i.e. soil water potential more than −0.05 MPa). This often leads to transpirational demand for water exceeding the supply from the roots. If water deficits are allowed to develop, tree decline results from a combination of loss of hydraulic conductance, nutrient deficiency and reduced photosynthesis. Even when root systems have recovered and are growing well, water deficiencies can result from impermeable surfaces preventing water from penetrating into the soil, restricted soil volumes, various climatic factors (such as high temperatures or low rainfall) or a combination of these. As a result, one of the primary reasons of premature mortality in both newly planted and established amenity trees is water deficit. Not surprisingly then, higher survival rates have been observed for drought-tolerant trees (Roman *et al.* 2014).

The prominence of drought as an ecological factor led to the development of drought tolerance scales for species under defined environmental conditions. For example,

Table 10.2 Scale used by Niinemets and Valladares (2006) to rank 806 temperate tree species according to their drought tolerance. Trees were allocated a ranking based on their ability to survive on a site, with <50% foliage damage and dieback. P: PET is the ratio of precipitation to potential evapotranspiration.

Scale ranking	Annual precipitation (mm)	Distribution of precipitation (coefficient of variation)	P : PET ratio	Soil water potential (MPa)	Duration of dry period
1 Very intolerant	>600	Minimal	>3.0	> −0.3	A few days
2 Intolerant	500–600	<10%	1.5 : 3	−0.3 to −0.8	A few weeks
3 Moderately tolerant	400–500	10–15%	0.8–1.5	−0.8 to −1.5	Up to a month
4 Tolerant	300–400	20–25%	0.5 : 0.8	−1.5 to −3	2–3 months
5 Very tolerant	<300	>25%	<0.5	< −3	More than 3 months

Niinemets and Valladares (2006) developed a five-point scale for the drought tolerance of temperate trees, shown in Table 10.2. Trees were allocated a rank based on their ability to survive with less than 50% foliage damage. This type of scale is very valuable when seeking to understand how trees fit into natural ecosystems and can be equally instructive when selecting trees for planting in natural environments. However, because species rankings are based on the performance of trees on a particular site, they are unable to discriminate between species that have avoidance strategies and those that have tolerance strategies. As a result, trees may fairly be ranked as drought-tolerant, even when they avoid the development of water deficits by deep rooting to groundwater sources or closing stomata early in the drying cycle. In fact, when you compare the drought tolerance ranking of Niinemets and Valladares (2006) with a physiological drought tolerance trait, such as the water potential at 50% loss of hydraulic conductivity, then it is clear that many trees that perform well on drought-prone sites are not tolerant of low water potential (Figure 10.13).

The fact that there is such variation in the physiological tolerance of trees to water deficits has important implications for the use of ecological drought tolerance scales when selecting trees for many urban planting sites. In streets, courtyards, car parks and many other urban sites, limited soil depth means that deep rooting cannot be used to reduce water deficits. Many routine practices, such as root pruning, transplanting and containerisation will further diminish the value of many avoidance strategies. Therefore, for many amenity trees, a physiological tolerance to low water availability is much more effective than morphological adaptations (such as deep rooting) as it helps the tree, even in restricted soil volumes and after root loss.

A useful measure of physiological drought tolerance is the leaf turgor loss point (Ψ_{P0}) because this relates to the minimum water potential that a tree can experience without permanent injury. Plants that have a low (more negative) Ψ_{P0} tend to maintain leaf gas exchange, hydraulic conductance and growth at lower soil water potentials, so have a competitive advantage when soil moisture is depleted (Mitchell *et al.* 2008; Blackman *et al.* 2010).

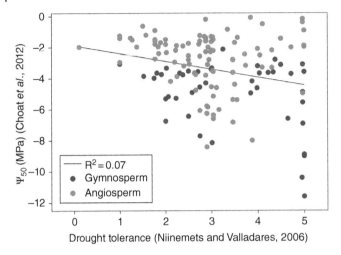

Figure 10.13 Relationship between drought tolerance (see Table 10.2) and water potential at which 50% loss of hydraulic conductivity occurs (Ψ_{50}). This shows that although there is a significant (P = 0.003) relationship between drought tolerance and Ψ_{50}, there is so much variation (R^2 0.07) that it is difficult to predict one trait from the other. Some species ranked as very tolerant to drought (5) are very susceptible to cavitation induced by water deficits (high, i.e. less negative Ψ_{50}). *Source:* Adapted from Niinemets and Valladares (2006) and Choat *et al.* (2012).

By evaluating a range of plant characteristics that contribute towards drought tolerance, Bartlett *et al.* (2012a) demonstrated that the osmotic potential at full turgor ($\Psi_{\pi100}$) is a key variable driving Ψ_{P0} and can therefore be used to predict Ψ_{P0} in plants. A major advantage of assessing $\Psi_{\pi100}$ is that it can be rapidly determined using techniques such as vapour pressure osmometery (Bartlett *et al.* 2012b). Consequently, it is now possible to screen traditional and novel tree species efficiently for a drought tolerance trait that is highly relevant for the urban environment (Sjöman *et al.* 2015).

By assessing the $\Psi_{\pi100}$ and subsequently predicting the Ψ_{P0}, Sjöman *et al.* (2015) evaluated the Ψ_{P0} of 27 different maple *Acer* species and cultivars. As might be expected, species that have evolved to grow in the shady, humid forest understories are much more sensitive to water deficits than those that naturally grow in much drier environments (Figure 10.14). In summer, the mean Ψ_{P0} for all the *Acer* spp. evaluated was −3 MPa but it varied by 2.7 MPa, from −1.6 MPa in mountain maple *Acer spicatum* to −4.3 MPa in Montpellier maple *Acer monspessulanum* (Figure 10.15). Those species that had the lowest turgor loss point all demonstrated marked osmotic adjustment between spring and summer, confirming the importance of this drought-tolerance strategy in maples.

This type of drought-tolerance data can be of great help in selecting trees for urban sites where water deficits are likely to occur. For example, Figure 10.16 shows urban planting beds with highly restricted soil volumes. Sites such as these will never be able to supply adequate water for tree development unless permanent supplementary irrigation is used. Trees that are unable to tolerate water deficits would be unsuitable for these sites, so drought-tolerant species should be preferentially selected as they will be much more likely to be able to perform well. However, selecting trees that are tolerant to any environmental stress does not mean that fundamental tree maintenance can be neglected.

Figure 10.14 Species of maple *Acer* differ widely in their habitat preference. (a) Understorey species, such as mountain maple *Acer spicatum* have been shown to have a high (less negative) leaf turgor loss point whilst maples (b), such as bigtooth maple *Acer grandidentatum*, from relatively dry environments produce much lower (more negative) leaf turgor loss values during summer. *Source:* (b) Courtesy of Henrik Sjöman.

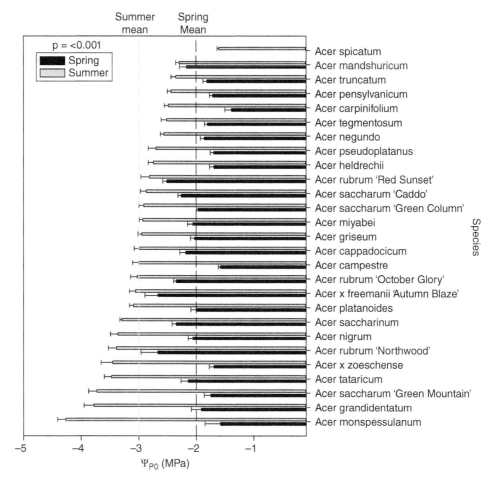

Figure 10.15 Predicted turgor loss of leaves in fully expanded spring and summer leaves based on the assessment of osmotic potential at full turgor of 27 maple *Acer* species and cultivars. Dashed lines represent the mean for all trees in spring and in summer. Bars show standard error. *Source:* Sjöman *et al.* (2015). Reproduced with permission of Elsevier.

Figure 10.16 Many urban sites, such as these two planting beds in Itacha, New York, USA, provide highly restricted rooting environments for trees. This inevitably leads to water deficits becoming a serious constraint for tree development. When selecting trees for this type of site, choosing a species with a low (more negative) leaf turgor loss point would be valuable as these species are capable of tolerating dry conditions.

Parks and garden locations without such severe rooting constraints can be planted with a much broader range of species. These sites will be less vulnerable to rapidly declining soil moisture so species with a higher Ψ_{P0} may be able to establish perfectly well. Therefore, the evaluation of key drought-tolerance traits, such as Ψ_{P0}, is of inherent interest to those involved in specifying or producing trees for the urban environment (Sjöman *et al.* 2015).

Flooding and Waterlogging Tolerance

In natural environments, *flooding* (water above the soil level) and *waterlogging* (only the soil flooded) have a profound influence on the tree species that grow adjacent to streams and rivers in *riparian* habitats,[3] as well as in forested *wetlands* or other poorly drained areas. Trees that persist in these environments have to cope with long periods of inundation, sediment deposition, shear stress caused by fast-moving water and physical abrasion from debris moving downstream. In wetlands, flooding may be present for much of the year, whereas in strongly seasonal river systems or higher up the river where gradients are steeper, flooding is often characterised by relatively short periods of inundation followed by a rapid decline in water levels and periods of water deficit.

---

3 The term *riparian* comes from the latin word *riparius* meaning 'belonging to the river bank' and refers to habitats on the shores of streams, rivers, ponds, lakes and some wetlands (Naiman *et al.* 1998).

The nature of floodwater can also vary substantially. Moving, turbulent water is more oxygenated that still water; nutrient contents can also differ widely. In addition, the duration and timing of the flood in relation to tree growth processes can have a profound influence on the tree's ability to survive inundation. Therefore, the nature of this ecological disturbance can be quite different along the course of a river as it depends a great deal on the site topography and the proximity of the tree to the water channel.

The main effect of flooding is caused not so much by the abundance of water itself but by the sudden loss of oxygen as water replaces the air in soil pores. Oxygen moves through water 10 000 times more slowly than in air, so plant tissues below the water find it extremely difficult to acquire oxygen. The sudden reduction in gas diffusion also causes the accumulation of carbon dioxide and the plant stress hormone ethylene. Various soil properties are also changed as a result of oxygen depletion, either as a result of chemical reactions that produce toxins or because of its impact on microbial communities, altering the availability of nutrients. Stagnant water tends to be more harmful because there is little mixing of oxygen into the water as happens in flowing water, so oxygen runs out even more quickly, producing *hypoxic* (low oxygen) or *anoxic* (no oxygen) conditions. In most trees, a shortage of oxygen in the soil reduces the development and growth of roots. For this reason, sites with high water tables are characterised by shallow spreading root systems with few deeper roots.

Flooding Injury

Although the first signs of flooding are seen in trees above ground, these are caused by the sudden change in the aeration around the roots. Paradoxically, initial visual symptoms of flooding injury are similar to those of water deficits as they include leaf wilting and shedding (Crawford 1982). Clearly, there is not an issue with the availability of water per se but the sudden reduction in oxygen availability disrupts the ability of the roots to take up water (Kreuzwieser and Rennenberg 2014). If flooding is sustained for more than a few days, roots begin to die in flood-sensitive species, leading to even less water uptake. This leads to water limitation in the shoots, reducing photosynthesis which, in turn, leads to a reduction in shoot growth and, if flooding continues, the shedding of leaves. Over prolonged periods, shoot dieback also develops in flood-sensitive species (Kozlowski 1997).

Oxygen deficiency in the rooting environment also causes an energy crisis in the roots as aerobic respiration is halted (Bailey-Serres and Voesenek 2008). Sugars can no longer be broken down to release their energy and key processes, such as carbohydrate transport and nutrient acquisition, become much more difficult. During the growing season, the effect is greater as demand for energy is higher and warmer temperatures increase respiration rates. Energy can be produced anaerobically via fermentation pathways and glycolysis (the partial breaking down of sugar without oxygen), but these are much less efficient and so can rapidly deplete carbohydrates stored in the roots. Switching to anaerobic metabolism also creates other problems such as causing the cell contents to become increasingly acid as lactic acid accumulates (just as it does in our muscles if oxygen cannot be delivered quickly enough), and the formation of toxins, such as ethanol and acetaldehyde, produced by fermentation (Lambers *et al.* 2008). The build-up of acetaldehyde presents a problem after reoxygenation, when it provides a source of reactive oxygen species (see Chapter 7) which can lead to serious membrane damage (Blokhina *et al.* 2003). This *post-anoxic* stress can be a major cause of injury during re-aeration after a period of flooding. Stress hormones, such as abscisic acid and ethylene, produced in the root system under oxygen deficiency also alter normal development processes.

Flooding and Soils

In most trees, the lack of oxygen around the roots leads to reduced growth and initiation of roots. However, it also leads to a rapid decline in mycorrhizal fungi and a wide range of other services provided by soil microorganisms. The metabolic challenges and loss of beneficial soil organisms also predisposes the tree to root pathogens. Specialist root pathogens, such as many *Phytophthora* spp., are well adapted to low oxygen environments and are able to increase their activity in wet conditions. *Phytophthora* spores are also highly mobile in flooded soils so the risk of infection increases. Diseased roots further exacerbate the extent of injury to the tree and will often lead to a downward spiral in root health. The speed and severity of such a decline will depend upon many factors but, clearly, from a practical management perspective, improving site drainage can reduce the duration of flooding and its associated problems.

Flooded soils therefore present a number of immediate challenges to root vitality and function. As well as causing serious challenges to aerobic respiration, they often lead to major disruption in water and nutrient acquisition, and can also increase the susceptibility of trees to pathogens. Where these effects result in extensive root mortality, the tree will also be more vulnerable to post-flooding injury from windthrow and water deficits.

Variation in Tolerance to Flooding

Trees vary a great deal in their tolerance to flooding during the growing season. The most sensitive species may be injured by just a few hours of waterlogging, while the most tolerant can survive flooding for more than a year. Even within the same species there can be genetic differences in tolerance, and mature trees tend to be more resistant to injury than very young or very old trees (Kozlowski 1984). In general, angiosperm trees are more tolerant to flooding than gymnosperms – think of willow *Salix* and poplar *Populus* spp. (Kozlowski and Pallardy 1997) – although the boreal species tamarack *Larix laricina* and the wetland specialists pond cypress *Taxodium ascendens,* swamp cypress *T. distichum* (Figure 10.17), Montezuma bald cypress *T. mucronatum* and white cedar *Chamaecyparis thyoides* are notable exceptions to this rule.

Except for the most flood-sensitive species, trees tend to be able to tolerate a few weeks of saturated soils during dormancy. However, flooding during the growing season poses a greater risk as demand for oxygen is much higher. Trees that thrive in habitats that regularly flood or experience prolonged flooding need to possess adaptive traits. Unsurprisingly, these are primarily associated with the avoidance or tolerance of low oxygen.

Structural Adaptations to Flooding

Trees that have to survive prolonged periods of flooding often have *aerenchyma* in their roots and lower stems. This modified tissue takes the form of extensive air spaces

Figure 10.17 Swamp cypress *Taxodium distichum* is one of the few wetland conifer species. It can survive flooding for over a year if necessary, in part because of the presence of pneumatophores that protrude above the water line and help ventilate the roots. *Source:* Courtesy of Henrik Sjöman.

between the cells within the parenchyma that allow oxygen to diffuse from the well-aerated tissues higher up the stem into the roots, as well as allowing the diffusion of toxins and carbon dioxide out from the roots. This allows riparian species belonging to genera such as alder *Alnus*, ash *Fraxinus*, poplars *Populus* and willows *Salix* to avoid serious oxygen deficiency in their roots. These structures also enable the release of oxygen into a thin layer of soil surrounding each flooded root (the rhizosphere – the process is referred to as *rhizospheric oxidation*) to aid nutrient absorption, and help to keep the toxins that develop in anoxic soils away from the root.

The ventilation of gases to and from the lower portion of the stem is also enhanced by the production of *hypertrophied* (enlarged) *lenticels.* These work best when associated with aerenchyma, but they often develop in species that seem not to produce aerenchyma as they can increase oxygen in the roots in their own right. Species from the genera *Abies, Acer, Alnus, Araucaria, Betula, Eucalyptus, Fraxinus, Larix, Melaleuca, Picea, Pinus, Platanus, Populus, Pyrus, Quercus, Salix, Taxus* and *Ulmus* have hypertrophied lenticels (Kozlowski 1997; Glenz *et al.* 2006; Iwanga and Yamamoto 2008).

Perhaps the most common adaptation associated with flood tolerance is the ability to produce *adventitious roots* on submerged parts of the original root system and stem. These roots, arising from parenchyma cells, help compensate for the loss of deeper roots. Some species have developed particularly impressive adventitious roots, known as *pneumatophores*, which protrude like snorkels above the water line to ensure that gases can be exchanged proficiently between submerged roots and the atmosphere (Martin and Francke 2015). Good examples of pneumatophores are found in mangrove

(a) (b)

Figure 10.18 Mangrove species, such as (a) grey mangrove *Avicennia marina* subsp. *australasica*, found in Kaipara Harbour, New Zealand, have developed (b) pneumatophores that are covered with lenticels to help provide ventilation to the root system. Whilst these modified roots may become submerged at high tide, they significantly extend the period of root aeration during the day.

species, such as grey mangroves *Avicennia marina* (Figure 10.18), to limit the impact of tidal inundation, and in inland wetland specialists such as swamp cypress *Taxodium distichum* (Figure 10.17).

Adventitious roots also help trees cope with sediment deposition caused by flooding. Many alluvial soils are built up by sediment washed from upstream. This buries the lower portion of the tree stem and the roots, often leaving them in hypoxic conditions. However, in many floodplain species, adventitious roots from the base of the tree can generate a new root system closer to the surface of the soil where oxygen and nutrient supplies are more plentiful (Figure 10.19). Long-lived coastal redwoods *Sequoia sempervirens* have been known to cope with single deposits of 76 cm and the ground level being raised up to 9 m over their lifetime (Olson *et al.* 1990). Without the ability to develop an adventitious root system, these trees would never be able to survive in such a habitat. As a result, mature coastal redwoods can end up with a number of tiers to their root systems that reflect the frequency of substantial deposition events during their lifetime. These multi-layered systems provide excellent anchorage that help secure these gargantuan trees into the deep alluvial soils of California.

For trees to survive in very active river channels, they also need to cope with abrasion from debris and drag imposed by the flow of the water. Many riparian trees (e.g. alder *Alnus*, poplar *Populus* and willow *Salix* species) that grown close to the active channel have flexible stems to help them reduce the loading on their stems. It is also noticeable that many riparian trees have narrow leaves that help reduce the drag exerted by fast-flowing rivers (Rood *et al.* 2010). In relatively dry regions where the rivers are either seasonal or even dry for several years, the actual riverbed is an attractive place to establish for some trees because water can flow for long periods below the surface. However, these species must be able to cope will the inevitable inundation of fast flowing, debris-laden water. Tea-trees *Melaleuca* spp. in Australia have come up with a series of adaptations for this perilous habitat: they are able to recline, almost to the point of being prostrate, to reduce drag on their stems; they possess a flexible crown that is able to fold with the flow of water; and they have thick, spongy bark to absorb the impact of debris (Naiman *et al.* 2005).

Figure 10.19 The stump and roots of a 250-year-old daimyo oak *Quercus dentata* that had been partly buried by ash from a volcanic eruption in northern Japan. New adventitious roots had grown below the new surface, shown in the cut-away section. The stump is preserved in the Museum of the Tomakomai Citizens Park, Tomakomai, Hokkaido, Japan.

Physiological Adaptations to Flooding

The physical adaptations of aerenchyma, hypertrophied lenticels and adventitious roots have evolved to avoid serious oxygen deficiency within the roots and lower stem. However, some trees are also able to adjust physiologically to low oxygen levels. Many flood-tolerant species conserve energy during hypoxic or anoxic conditions by slowing energy-intensive processes, such as nutrient uptake and growth. Energy efficiency by itself is unlikely to help survival for long periods, so some trees produce energy via fermentation and glycolysis. Glycolysis is only about 10% as efficient as aerobic respiration (Bailey-Serres and Voesenek 2008) but can be enough to keep the tree functioning. The downside is that this inefficiency requires a greater supply of carbohydrates to yield the same amount of energy, so an adequate supply of carbohydrates is crucial for survival. Consequently, many flood-tolerant species have a great capacity to store and convert starch, fatty acid and lipid reserves to usable energy (Kreuzwieser and Rennenberg 2014).

Conversely, flood-sensitive species tend to deplete soluble sugars after a few days of waterlogging which inevitably increases the energy crisis in the roots. This is made worse because the movement of carbohydrates from the shoot system to the roots via the phloem is likely to be severely impaired in flood-sensitive species, whereas it can continue to be effective in more tolerant species (Ferner *et al.* 2012). An ability to sustain positive carbon relations, particularly in roots, is therefore likely to have an important role in surviving flooding. Even closely related species can vary in flood-tolerance: field elm *Ulmus minor* seedlings show poorer survival than European white elm *U. laevis* when flooded, largely because of their reduced ability to maintain a positive carbon balance over a number of weeks (Li *et al.* 2015).

Some trees, such as species of alder *Alnus,* birch *Betula* and poplar *Populus,* also contain chlorophyll in their stems. These tissues are able to photosynthesise using light that penetrates the bark (Pfanz *et al.* 2002). In species such as alder *Alnus glutinosa,* stem photosynthesis is so effective that this internally produced oxygen is likely to improve root aeration (Armstrong and Armstrong 2005). The multi-stemmed habit of *Alnus* species may increase the bark surface area close to the surface of the water and help root aeration via stem photosynthesis (Crawford 2008). This may, at least partly, explain why *Alnus* species growing close to water tend to have a multi-stemmed form, while those in less flood-prone areas are single stemmed.

Riparian Trees Adapted to Urban Environments

It would be easy to think that trees associated with abundant water may not make ideal candidates for the water-deficient urban environment. However, many riparian trees are actually very useful in urban sites, because their adaptations are less to do with coping with excess water and more to do with managing oxygen deficiency. Many urban soils can be hypoxic as a result of compaction and/or impermeable surfaces, so trees that possess adaptations to flooding often establish well in marginal urban sites. They have strategies to conserve energy, maintain physiological activity under low oxygen conditions, have a greater capacity for root regeneration and may even be able to cope with changes in soil level. Those species at home alongside strongly seasonal water courses, caused by annual snow melts or seasonal precipitation, make particularly good candidates for urban trees as they often have traits that can manage periods of low oxygen as well as periods of water deficit. Species such as Italian alder *Alnus cordata,* grey alder *A. incana,* river birch *Betula nigra,* yellow catalpa *Catalpa ovata,* sweetgum *Liquidambar styraciflua,* American plane *Platanus occidentalis* and oriental plane *P. orientalis* can be found in these highly dynamic environments.

The ability of riparian trees to produce new roots rapidly also helps the root system to recover after transplanting. For this reason, floodplain species, such as red maple *Acer rubrum,* silver maple *A. saccharinum,* European ash *Fraxinus excelsior,* green ash *F. pennsylvanica,* honey locust *Gleditsia triacanthos,* bird cherry *Prunus padus,* pin oak *Quercus palustris,* willow oak *Q. phellos,* alders *Alnus* spp. and poplars *Populus* spp. have proven easy to transplant into urban environments. Acknowledging the relevance of these ecological adaptions to specific habitats can aid species selection and help explain why trees from habitats without such periodic disturbances may be more challenging to establish.

Salt Tolerance

Saline soils (Box 10.1) arise naturally where there is an accumulation of salts derived from natural weathering of parental rocks. Sodium chloride (NaCl) is the most important of these but chlorides of calcium, magnesium and, to a lesser extent, sulfates and carbonates also occur. These salts can accumulate in depressions with very poor drainage (e.g. salt flats), and can be brought into the rooting zone of plants by high water tables (e.g. salt marshes) and by evaporation of water leading to the salts accumulating

> **Box 10.1 The Definition of a Saline Soil**
>
> Soils are classified as saline when a saturated paste of the soil has an electrical conductivity measurement of 0.4 Siemens per metre (S m^{-1}) or more. This corresponds to 40 mmol NaCl (the concentration of sodium chloride) and generates an osmotic potential of -0.14 MPa. For comparison, seawater contains around 500 mmol NaCl with a conductivity of 5.4 S m^{-1}, generating an osmotic potential of approximately -2.5 MPa. Water in a saline soil therefore starts at about 8% the concentration of seawater. Water used for irrigation should not have an electrical conductivity above 0.2 S m^{-1} (= -0.07 MPa), just 4% of the concentration of seawater (i.e. you would need to add 24 buckets of fresh water to make one bucket of seawater the maximum salinity for irrigation).

at the soil surface in the process of *salinisation*. Sea salts deposited in coastal regions by wind, rain, spray or tidal inundation also cause saline soils.

Humans enhance soil salinisation by using saline irrigation water. Dry regions tend to experience this problem more acutely for a number of reasons: demand for irrigation water is high whilst quality irrigation water is often scarce; low levels of rainfall mean that salts are not naturally leached through the soil profile. Cool, humid regions typically have fewer problems with saline soils as higher rainfall helps to flush salts through soils. However, during winter, significant amounts of salt are often used on roads and paths to reduce the risk of ice. This leads to adjacent soils becoming contaminated with high levels of salts that can cause significant stress to trees and other vegetation. It is also possible that salt stress is imposed on plants by importing saline soil into a planting site during landscape development. Under these circumstances, diagnosis of salt stress is less obvious because there is no obvious source of contamination. Regardless of their origin, where high levels of salts occur in soils, plants that lack strategies to cope will be adversely affected.

Dehydration and Toxicity Injuries in Saline Soils

Salts in the soil solution can cause a physiological water deficit by lowering the osmotic potential of the soil and therefore soil water potential. This has an immediate effect on the tree as it makes it harder for roots to extract water from the soil, even if the soil appears quite moist. Unsurprisingly, the effects on the tree are very similar to those caused by water deficiencies in dry soils: growth will be reduced, photosynthesis limited and various metabolic processes compromised (Kozlowski 1997). Trees can overcome this stress by osmotic adjustment of their cells which, at least partly, explains observable differences in salt tolerance. Generally, trees will be able to continue to grow whilst the osmotic potential of their cells is below that of the soil solution.

When sodium chloride (NaCl), the main component of seawater, dissolves in water, the atoms of sodium and chlorine come apart and each has an electrical charge – sodium has a positive charge (Na$^+$) and chlorine is negative (Cl$^-$) – and are referred to as *ions*. Other compounds or salts can similarly produce ions in water. Sodium and chlorine ions (and other ions such as boron) can accumulate inside plant tissues and, once above a species-specific threshold, can cause *ion toxicity*. It often takes a few weeks of exposure to saline conditions but once excessive levels of Na$^+$ or Cl$^-$ accumulate, they interfere with normal cell functions and signs of damage become visible. Leaf necrosis (death of patches of cells) and chlorosis (loss of the green colour) are common symptoms, particularly in the older leaves where ions have had longer to accumulate (Costello *et al.* 2003). In more severe cases of ion toxicity, defoliation occurs as the tree sheds badly injured leaves. The pattern of defoliation often helps to indicate the origin of the salt. If the whole rooting environment is saline, symptoms will occur across the entire crown. However, if the salt is mainly from spray from salted roads, the lower crown closest to the road may be noticeably more affected. Similarly, at coastal sites, the side of the crown closest to the salt-laden prevailing wind will be most affected. Leaf tolerance to salt-spray varies substantially between species, as those species with thicker, more robust leaves, often with leaf hairs, tend to be more tolerant. Coastal trees, such as the pohutukawa *Metrosideros excelsa* from New Zealand, with oblong leathery leaves covered with dense white hairs on their underside are typical of salt-resistant trees.

Salt-tolerant trees are able to avoid ion toxicity by being highly selective in the type of ion they allow into their roots, so keeping the inside of the plant salt-free. Although such selectivity has a metabolic cost, preventing the accumulation of potentially toxic ions in sensitive plant tissues helps to ensure survival. Some species are also able to achieve the same goal by sequestering (locking up) the ions in their root parenchyma and so avoiding their transport up to the leaves. Individual leaves can avoid ions in the same way by sequestering the ions within the cell vacuoles. However, this causes problems because water will move into the vacuole by osmosis, drying out the rest of the cell. The cell balances this by accumulating non-toxic compounds, such as sugars (*compatible solutes*), inside the cytosol – the liquid part of the cell (Munns and Testa 2008). As well as excluding salts and developing a degree of salt tolerance, many mangrove species, such as the grey mangrove *Avicennia marina* (Figure 10.18), are also able to avoid the toxic build-up of salts within their leaves by secreting salt out of the plant through specialist glands. Similar glands are also found on tamarisk *Tamarix* spp. which are well known for tolerating saline conditions. Many trees growing on challenging coastal sites would be unable to survive the highly saline conditions without good control over the uptake of salts, some tolerance to salts within cells and specialist adaptations for expelling salt through their leaves.

Managing Saline Soils in Amenity Tree Planting

If excess salt is suspected to be the cause of a decline in tree condition (e.g. alongside roads or paths that are salted during winter), a soil analysis can be used to confirm it. If confirmed, use of relatively salt-tolerant trees may be vital for the success of a planting scheme. In contrast to common agricultural species, much of what is known about the salt tolerance of amenity trees comes not from carefully designed research, but from practitioner experience. Consequently, tolerance scales are somewhat arbitrary, but many standard texts on tree selection contain some guidance on the perceived salt tolerance of different species: useful lists are published in Flint (1997), Costello *et al.* (2003), Trowbridge and Bassuk (2004) and Dirr (2011), as well as in literature produced by many tree nurseries. By far the best approach to managing saline sites or new planting sites that are vulnerable to salt input is to plant species know to have a reasonable tolerance to salt.

In addition to careful selection of species, good site drainage is essential to reduce the salinity by leaching the salt away from the root-zone. This may be achieved naturally in areas of high rainfall, but often requires irrigation with significant volumes of non-saline water before tree growth starts in spring. On sites with poor drainage, salt will not be effectively flushed through the rooting zone, and there is a risk of waterlogging, so irrigation may not be practical. Extensive irrigation may also lead to the leaching away of valuable nutrients: nutritional amendments would therefore be prudent after flushing. Clearly, in regions of water shortage it would also be ethically inappropriate to use large quantities of good-quality water to flush salts away. However, flushing of soils could be considered an option for high-value amenity trees in regions that are not prone to water shortages.

If significant contamination by de-icing salt occurs on a site that is unsuitable for soil flushing, adding gypsum (calcium sulfate dihydrate; $CaSO_4.2H_2O$) can ameliorate the soil (Roberts *et al.* 2006). This can be applied to the soil around trees but is likely to be more effective if the gypsum is integrated into the soil using air cultivation techniques (see Decompaction in Chapter 4).

Storage of salt, even in salt bins, should be avoided around tree root-zones. Many instances of salt damage to trees have occurred as a consequence of inappropriate salt storage. Care should also be taken when using de-icing salt around trees in hard landscapes if subsequent damage is to be avoided. This should include training those responsible for salt spreading on the impact of salt on trees.

References

Allen, C.D., Macalady, A.K., Chenchouni, H., Bachelet, D., McDowell, N., Vennetier, *et al.* (2010) A global overview of drought and heat-induced tree mortality reveals emerging climate change risks for forests. *Forest Ecology and Management*, 259: 660–684.

Anderegg, W.R., Hicke, J.A., Fisher, R.A., Allen, C.D., Aukema, J., Bentz, B., *et al.* (2015) Tree mortality from drought, insects, and their interactions in a changing climate. *New Phytologist*, 208: 674–683.

Archibald, O.W. (1995) *Ecology of World Vegetation*. Chapman and Hall, London, UK.

Armstrong, W. and Armstrong, J. (2005) Stem photosynthesis not pressurized ventilation is responsible for light-enhanced oxygen supply to submerged roots of alder (*Alnus glutinosa*). *Annals of Botany*, 96: 591–612.

Aroca, R. (2012) *Plant Responses to Drought Stress: From Morphological to Molecular Features*. Springer, Berlin, Germany.

Augé, R.M., Green, C.D., Stodola, A.J.W., Saxton, A.M., Olinick, J.B. and Evans, R.M. (2000) Correlations of stomatal conductance with hydraulic and chemical factors in several deciduous tree species in a natural habitat. *New Phytologist*, 145: 483–500.

Bailey-Serres, J. and Voesenek, L.A.C.J. (2008) Flooding stress: Acclimations and genetic diversity. *Annual Review of Plant Biology*, 59: 313–339.

Barnes, B.B., Zak, D.R., Denton, S.R. and Spurr, S.H. (1998) *Forest Ecology*, 4th edition. Wiley, New York, USA.

Bartlett, M.K., Scoffoni, C., Ardy, R., Zhang, Y., Sun, S., Cao, K., *et al.* (2012a) Rapid determination of comparative drought tolerance traits: using an osmometer to predict turgor loss point. *Methods in Ecology and Evolution*, 3: 880–888.

Bartlett, M.K., Scoffoni, C. and Sack, L. (2012b) The determinants of leaf turgor loss point and prediction of drought tolerance of species and biomes: A global meta-analysis. *Ecology Letters*, 15: 393–405.

Bieras, A.C. and Graças Sajo, M. (2009) Leaf structure of the cerrado (Brazilian savanna) woody plants. *Trees*, 23: 451–471.

Bita, C. and Gerats, T. (2013) Plant tolerance to high temperature in a changing environment: Scientific fundamentals and production of heat stress-tolerant crops. *Frontiers in Plant Science*, 4: 273: 1–18.

Blackman, C.J., Brodribb, T.J. and Jordan, G.J. (2010) Leaf hydraulic vulnerability is related to conduit dimensions and drought resistance across a diverse range of woody angiosperms. *New Phytologist*, 188: 1113–1123.

Blokhina, O. Virolainen, E. and Fagerstedt, K.V. (2003) Antioxidants, oxidative damage and oxygen deprivation stress: A review. *Annals of Botany*, 91: 179–194.

Brady, M. (2008) Hardiness zone maps of the northern hemisphere. *The Plantsman*, September: 170–176.

Canadell, J., Jackson, R.B., Ehleringer, J.R., Mooney, H.A., Sala, O.E. and Schulze, E.D. (1996) Maximum rooting depth of vegetation types at the global scale. *Oecologia*, 108: 583–595.

Choat, B., Jansen, S., Brodribb, T.J., Cochard, H., Delzon, S., Bhaskar, R., *et al.* (2012) Global convergence in the vulnerability of forests to drought. *Nature*, 491: 752–756.

Costello, L.R., Perry, E.J., Matheny, N.P., Henry, J.M. and Geisel, P.M. (2003) *Abiotic Disorders of Landscape Plants: A Diagnostic Guide.* University of California, Agricultural and Natural Resources Publication 3420, CA, USA.

Crawford, R.M.M. (1982) Physiological responses to flooding. In: Lange, O.L., Nobel, P.S., Osmond, C.B. and Zieger, H. (eds) *Physiological Plant Physiology II; Water Relations and Carbon Assimilation.* Springer, Berlin, Germany, pp. 453–477.

Crawford, R.M.M. (2008) *Plants at the Margin; Ecological Limits and Climate Change.* Cambridge University Press, Cambridge, UK.

Dee, D.P., Uppala, S.M., Simmons, A.J., Berrisford, P., Poli, P., Kobayashi, S., *et al.* (2011) The ERA-Interim reanalysis: Configuration and performance of the data assimilation system. *Quarterly Journal of the Royal Meteorological Society*, 137: 553–597.

Dirr, M.A. (2011) *Dirr's Encyclopedia of Trees and Shrubs.* Timber Press, Portland, OR, USA.

Ferner, E., Rennenberg, H. and Kreuzwieser, J. (2012) Effect of flooding on C metabolism of flood-tolerant (*Quercus robur*) and non-tolerant (*Fagus sylvatica*) tree species. *Tree Physiology*, 32: 135–145.

Filewod, B. and Thomas, S.C. (2014) Impacts of a spring heat wave on canopy processes in a northern hardwood forest. *Global Change Biology*, 20: 360–371.

Flint, H.L. (1997) *Landscape Plants for Eastern North America, Exclusive of Florida and the Immediate Gulf Coast*, 2nd edition. Wiley, New York, USA.

Fonti, P., von Arx, G., Garcia-Gonzalez, I., Eilmann, B., Sass-Klaassen, U., Gartner, H., *et al.* (2010) Studying global change through investigation of the plastic responses of xylem anatomy in tree rings. *New Phytologist*, 185: 42–53.

Fujikawa, S., Kasuga, J., Takata, N. and Arakawa, K. (2009) Factors related to change of deep supercooling capability in xylem parenchyma cells of trees. In: Gusta, L.V., Wisniewski, M.E. and Tanino, K.K. (eds) *Plant Cold Hardiness; From the Laboratory to the Field.* CABI, Wallingford, UK, pp. 29–42.

Glenz, C., Schlaepfer, R., Iorgulescu, I. and Kienast, F. (2006) Flooding tolerance of Central European tree and shrub species. *Forest Ecology and Management*, 235: 1–13.

Hoffmann, W.A., Marchin, R.M., Abit, P. and Lau, O.L. (2011) Hydraulic failure and tree dieback are associated with high wood density in a temperate forest under extreme drought. *Global Change Biology*, 17: 2731–2742.

Iwanga, F. and Yamamoto, F. (2008) Flooding adaptions of wetland trees. In: *Sakio, H.* and Tamura, T. (eds) *Ecology of Riparian Forests in Japan*. Springer, Berlin, Germany, pp. 237–247.

Kasuga, J., Hashidoko, Y., Nishioka, A., Yoshiba, M., Arakawa, K. and Fujikawa, S. (2008) Deep supercooling xylem parenchyma cells of katsura tree (*Cercidiphyllum japonicum*) contain flavonol glycosides exhibiting high anti-ice nucleation activity. *Plant Cell Environment*, 31: 1335–1348.

Keddy, P.A. (2007) *Plants and Vegetation; Origins, Processes, Consequences*. Cambridge University Press, Cambridge, UK.

Klein, T. (2014) The variability of stomatal sensitivity to leaf water potential across tree species indicates a continuum between isohydric and anisohydric behaviours. *Functional Ecology*, 28: 1313–1320.

Körner, C. (2012) *Alpine Treelines: Functional Ecology of the Global Elevation Tree Limits*. Springer, Berlin, Germany.

Kozlowski, T.T. (1984) Responses of woody plants to flooding. In: Kozlowski, T.T. (ed.) *Flooding and Plant Growth*. Academic Press, New York, USA, pp. 129–164.

Kozlowski, T.T. (1997) Responses of woody plants to flooding and salinity. *Tree Physiology Monograph*, 1: 1–29.

Kozlowski, T.T., Kramer, P.J. and Pallardy, S.G. (1991) *The Physiological Ecology of Woody Plants*. Academic Press, San Diego, CA, USA.

Kozlowski, T.T. and Pallardy, S.G. (1997) *Growth Control in Woody Plants*. Academic Press, San Diego, CA, USA.

Kozlowski, T.T. and Pallardy, S.G. (2002) Acclimation and adaptive responses of woody plants to environmental stresses. *The Botanical Review*, 68: 270–334.

Kreuzwieser, J. and Rennenberg, H. (2014) Molecular and physiological responses of trees to waterlogging stress. *Plant Cell Environment*, 37: 2245–2259.

Kubler, H. (1983) Mechanism of frost crack formation in trees: A review and synthesis. *Forest Science*, 29: 559–568.

Lambers, H., Stuart Chaplin, F.S.III and Pons, T.L. (2008) *Plant Physiological Ecology*, 2nd edition. Springer, Berlin, Germany.

Larcher, W. (2003) *Physiological Plant Physiology*, 4th edition. Springer, Berlin, Germany.

Levitt, J. (1980) *Responses of Plants to Environmental Stresses. Volume* II, 2nd edition. Academic Press, San Diego, CA, USA.

Li, M., López, R., Venturas, M., Pita, P., Gordaliza, G.G., Gil, L., *et al.* (2015) Greater resistance to flooding of seedlings of *Ulmus laevis* than *Ulmus minor* is related to the maintenance of a more positive carbon balance. *Trees*, 29: 835–848.

Martin, C.E. and Francke, S.K. (2015) Root aeration function of baldcypress knees (*Taxodium distichum*). *International Journal of Plant Science*, 176: 170–173.

Mitchell, P.J., Veneklaas, E.J., Lambers, H. and Burgess, S.S.O. (2008) Leaf water relations during summer water deficit: Differential responses in turgor maintenance and variation in leaf structure among different plant communities in south-western Australia. *Plant Cell and Environment*, 31: 1791–1802.

Monson, R.K., Jones, R.T., Rosenstiel, T.N. and Schnitzler, J.P. (2013) Why only some plants emit isoprene. *Plant, Cell and Environment*, 36: 503–516.

Mooney, H.A. and Miller, P.C. (1985) Chaparral. In: Chabot, B.F. and Mooney, H.A. (eds) *Physiological Ecology of North American Plant Communities*. Chapman and Hall, New York, USA, pp. 213–231.

Morgan, J.M. (1984) Osmoregulation and water stress in higher plants. *Annual Review of Plant Physiology*, 35: 299–319.

Munns, R. and Testa, M. (2008) Mechanisms of salt tolerance. *Annual Review of Plant Biology*, 59: 651–681.

Naiman, R.J., Décamps, H. and McClain, M.E. (2005) *Riparia: Ecology, Conservation, and Management of Streamside Communities*. Elsevier, Academic Press, Amsterdam, The Netherlands.

Naiman, R.J., Fetherston, K.L., McKay, S.J. and Chen, J. (1998) Riparian forests. In: Naiman, R.J., Bilby, R.R. and Kantor, S. (eds) *River Ecology and Management: Lessons from the Pacific Coastal Ecoregion*. Springer, New York, USA, pp. 289–323.

Nardini, A., Casolo, V., Dal Borgo, A., Savi, T., Stenni, B., Bertoncin, P., *et al.* (2016) Rooting depth, water relations and non-structural carbohydrate dynamics in three woody angiosperms differentially affected by an extreme summer drought. *Plant, Cell and Environment*, 39: 618–627.

Niinemets, Ü. and Valladares, F. (2006) Tolerance to shade, drought, and waterlogging of temperate northern hemisphere trees and shrubs. *Ecological Monographs*, 76: 521–547.

Olson, D.F. Jr., Roy, D.F. and Walters, G.A. (1990) *Sequoia sempervirens* (D. Don) Endl. In: Burns, R.M. and Barbara, H. (eds) *Silvics of North America: 1. Conifers*. Agriculture Handbook 654, United States Department of Agriculture, Forest Service, Washington, DC, USA, pp. 541–551.

Pallardy, S.G. (2008) *Physiology of Woody Plants*, 3rd edition. Academic Press, San Diego, CA, USA.

Peel, M.C., Finlayson, B.L. and McMahon, T.A. (2007) Updated world map of the Köppen-Geiger climate classification. *Hydrology and Earth System Sciences Discussions*, 4: 439–473.

Pfanz, H., Aschan, G., Langenfeld-Heyser, R., Wittmann, C. and Loose, M. (2002) Ecology and ecophysiology of tree stems: corticular and wood photosynthesis. *Naturwissenschaften*, 89: 147–162.

Roberts, J., Jackson, N. and Smith, M. (2006) *Tree Roots in the Built Environment*. Research for Amenity Trees No. 8, The Stationary Office, London, UK.

Roman, L.A., Battles, J.J. and McBride, J.R. (2014) Determinants of establishment survival for residential trees in Sacramento County, CA. *Landscape and Urban Planning*, 129: 22–31.

Rood, S.B., Nielsen, J.L., Shenton, L., Gill, K.M. and Letts, M.G. (2010) Effects of flooding on leaf development, transpiration, and photosynthesis in narrowleaf cottonwood, a willow-like poplar. *Photosynthesis Research*, 104: 31–39.

Sakai, A. and Larcher, W. (1987) *Frost Survival of Plants: Responses and Adaptation to Freezing Stress*. Springer, Berlin, Germany.

Schulze, E.-D., Beck, E. and Müller-Hohenstein, K. (2005) *Plant Ecology*. Springer, Berlin, Germany.

Schweingruber, F.H. (2007) *Wood Structure and Environment*. Springer, Berlin, Germany.

Sharkey, T.D., Wiberley, A.E. and Donohue, A.R. (2008) Isoprene emission from plants: why and how. *Annals of Botany*, 101: 5–18.

Sjöman, H., Hirons, A.D. and Bassuk, N.L. (2015) Urban forest resilience through tree selection: Variation in drought tolerance in *Acer*. *Urban Forestry and Urban Greening*, 14: 858–865.

Song, L., Jiang, Y., Zhao, H. and Hou, M. (2012) Acquired thermotolerance in plants. *Plant Cell, Tissue and Organ Culture*, 111; 265–276.

Teskey, R., Wertin, T., Bauweraerts, I., Ameye, M., McGuire, M.A. and Steppe, K. (2015) Responses of tree species to heat waves and extreme heat events. *Plant, Cell and Environment*, 38: 1699–1712.

Thomashow, M.F. (1999) Plant cold acclimation: Freezing tolerance genes and regulatory mechanisms. *Annual Review of Plant Biology*, 50: 571–599.

Trowbridge, P.J. and Bassuk, N.L. (2004) *Trees in the Urban Landscape: Site Assessment, Design and Installation*. Wiley, Hoboken, NJ, USA.

Urli, M., Lamy, J.B., Sin, F., Burlett, R., Delzon, S. and Porté, A.J. (2015) The high vulnerability of *Quercus robur* to drought at its southern margin paves the way for *Quercus ilex*. *Plant Ecology*, 216: 177–187.

Wickens, G.E. (2008) *The Baobabs: Pachycauls of Africa, Madagascar and Australia*. Springer, Berlin, Germany.

Yordanov, I. (1992) Response of photosynthetic apparatus to temperature stress and molecular mechanisms of its adaptations. *Photosynthetica*, 26: 517–531.

Zahner, R. (1968) Water deficits and growth of trees. In: Kozlowski, T.T. (ed.) *Water Deficits and Plant Growth*. Academic Press, New York, USA, pp. 191–254.

Index

Applied Tree Biology, First Edition. Andrew D. Hirons and Peter A. Thomas.
© 2018 John Wiley & Sons Ltd. Published 2018 by John Wiley & Sons Ltd.